Volcanic Reservoirs in Petroleum Exploration

Volcanic Reservoirs in Petroleum Exploration

Zou Caineng,
Zhang Guangya,
Zhu Rukai,
Yuan Xuanjun,
Zhao Xia,
Hou Lianhua,
Wen Baihong,
Wu Xiaozhi

Research Institute of Petroleum exploration & Development,
PetroChina,
No. 20 Xueyuan Road, Haidian District, Beijing 100083
People's Republic of China

ELSEVIER

AMSTERDAM • BOSTON • HEIDELBERG • LONDON • NEW YORK • OXFORD
PARIS • SAN DIEGO • SAN FRANCISCO • SINGAPORE • SYDNEY • TOKYO

Elsevier
225 Wyman Street, Waltham, MA 02451, USA
525 B Street, Suite 1900, San Diego, CA 92101-4495, USA

First edition **2013**

Library of Congress Cataloging-in-Publication Data
Zou, Caineng, 1963-
    Volcanic reservoirs / Zou Caineng, Zhang Guangya, Zhu Rukai, Yuan Xuanjun, Zhao Xia, Hou Lianhua, Wen Baihong, Wu Xiaozhi. — First edition.
        pages cm
    Includes bibliographical references and index.
    ISBN 978-0-12-397163-0 (alk. paper)
1. Hydrocarbon reservoirs. 2. Volcanic soils. 3. Petroleum–Geology–China. I. Title.
    TN870.57.Z68 2013
    622′.338–dc23

                                                                    2012047025

British Library Cataloguing in Publication Data
A catalogue record for this book is available from the British Library

For information on all Elsevier publications
visit our web site at store.elsevier.com

ISBN: 978-0-12-397163-0

Contents

Preface

The volcanic reservoir is a type of unconventional reservoir with volcanic rock as the reservoir bed. Overseas oil and gas exploration in volcanic rock has taken place over the past 120 years and yielded several discoveries in many countries. However, most of the discoveries occurred by chance, and they are not regarded as primary exploration targets. Volcanic reservoirs mainly occur in Mesozoic- and Cenozoic-age rocks. They are formed in passive continental margins, with primary lithologies of intermediate to basic basalt and andesite. Fractures in these reservoirs play an important role in improving the reservoir properties. Typically, volcanic reservoirs are of small scale, with occasional exceptions of large oil and gas fields.

In the past 50 years, volcanic reservoir exploration in China has gained importance. A series of late generation-early storage type oil and gas fields with Permian source rocks was first discovered in the 1950s in the Carboniferous volcanic rock at the northwest margin of Junggar Basin. Then, some volcanic reservoirs were discovered successively in 11 basins, including Bohai Bay Basin. Volcanic reservoirs in eastern Bohai Bay Basin belong to the self-generation-self-storage type reservoirs. Exploration of volcanic reservoirs went from fortuitous or local to large-scale exploration in the twenty-first century and gave birth to some important discoveries in the deep Songliao Basin, Ludong-Wucaiwan region in the Junggar Basin, Santanghu Basin, and others. Volcanic reservoirs have been explored as a key petroleum domain and have become an important percentage of the reserves' increase. Two volcanic hydrocarbon provinces in Songliao Basin and northern Xinjiang, respectively, have significant reserves scale. There are Mesozoic, Cenozoic, and Paleozoic volcanic reservoirs. In terms of the formation environment, they are classified as inland rift type, rift type after collision and orogeny, and island arc type. Besides self-generation-self-storage reservoirs, there are also late generation-early storage type reservoirs. Supporting exploration techniques for volcanic rock distribution and reservoir prediction have been developed.

There are great differences between volcanic rocks and sedimentary rocks in formation conditions and hydrocarbon accumulation rules. First, volcanic rock itself cannot generate hydrocarbon, so the key to hydrocarbon accumulation is the effective matching between reservoir rock and source rock in time and space. Second, volcanic rock developed in endogenic geologic processes is distinct from sedimentary rock in lithofacies, diagenesis, genesis, distribution rules, and prediction of effective reservoirs. Third, hydrocarbon traps are developed irregularly in volcanic rock, showing complex reservoir types and distribution rules. Fourth, it is necessary to integrate different methods in volcanic reservoir characterization and prediction, including gravity magnetic, electrical, and seismic methods.

In recent years, along with the progress of volcanic reservoir exploration in Songliao, Junggar, Santanghu, and Bohai Bay basins, this kind of unconventional reservoir has become one of the major targets in onshore exploration in China. In view of this development, the research teams of the Research Institute of Petroleum Exploration and Development (RIPED) that typically specialize in lithologic-stratigraphic reservoirs have shifted focus and conducted extensive investigations into the development and distribution of volcanic rocks; types, genesis, and distribution of volcanic reservoirs; hydrocarbon accumulation conditions; and reservoir types and distribution rules. Unique techniques and methods for volcanic reservoir evaluation have also been developed. These achievements have played an important role in volcanic exploration and indicate the direction for future volcanic exploration. Based on the present situation of volcanic reservoir exploration and investigation at home and abroad, as well as some detailed case studies in major basins, this book systematically describes the geologic features of volcanic reservoirs in China as well as future exploration and relevant techniques.

This book is composed of six chapters. Chapter 1 focuses on overseas volcanic reservoir exploration history, features of discovered oil and gas fields and some case studies, review of domestic reservoir exploration history, and major research findings and features of volcanic reservoir exploration.

Chapter 2 discusses the generation period and distribution of volcanic rocks in China, especially in hydrocarbon-bearing basins, based on the analyses of global volcanic rock distribution and tectonic setting. In this chapter, we show that volcanic rocks in hydrocarbon-bearing basins mainly occur in four basin groups: north China to northeast China, northern Xinjiang, Tarim, and Sichuan to Tibet. The volcanic rocks in eastern China mainly developed in inland continental rifts in the Mesozoic and Cenozoic eras, while those in central and western China mainly developed in postorogenic continental facies, transitional facies, and marine rift environments in the late Paleozoic era. Volcanic rocks of the island arc type developed at basin margins.

Chapter 3 discusses the development features, reservoir types, reservoir characteristics, and main controlling factors of favorable reservoir development of volcanic rocks in hydrocarbon-bearing basins. It is believed that domestic volcanic rocks in depositional basins are mainly developed along faults in centered and compound eruptions, which gave birth to layered volcanoes with developed explosive facies and eruptive-effusion facies. The volcanic massif is small when isolated, but groups of volcanic massifs distribute extensively. Volcanic rocks under three kinds of geologic processes would form four kinds of reservoirs: lava reservoir, volcaniclastic reservoir, dissolved reservoir, and fractured reservoir. Volcanic rocks in eastern China are mainly intermediate to acidic in composition, with volcaniclastics of primary explosive facies and lava of eruptive-effusion facies being the most favorable reservoir belts. In western China, volcanic rocks are mainly intermediate to basic in composition. Other lithologies may form good dissolved reservoirs after later weathering and leaching. Formation compaction has little influence on volcanic reservoirs. Therefore, there is no clear relationship between reservoir properties and burial depth.

Chapter 4 summarizes the assemblages of hydrocarbon accumulation in volcanic rock, as well as reservoir types and oil and gas distribution rules, and points out that the assemblage of reservoir rock and effective source rock are the key to forming reservoirs. The hydrocarbon-generating center controls the distribution of oil and gas, and the inner source and proximal assemblages would be the most favorable. The distal assemblages need faults or unconformity surfaces for communication. In fault depressions in eastern China, proximal assemblages are dominant. Reservoirs of explosive facies are developed at structural highs along fractures, forming structural-lithologic reservoirs. Eruptive-effusion facies are widely distributed at the slope, and fractured facies are favorable for forming lithologic reservoirs. In western China, proximal and distal assemblages exist. Reservoir beds that have been weathered and leached along unconformity surfaces form large stratigraphic reservoirs. Lithologic reservoirs also exist inside volcanoes. In addition, gas with high CO_2 content in the

Songliao Basin mainly has an inorganic mantle-sourced origin and distributes near deep and major fault belts that were active in later periods.

Chapter 5 points out future targets of volcanic reservoir exploration in China and puts forward some suggestions. The first is to construct two major oil and gas provinces in the deep Songliao Basin and the Carboniferous system of Junggar Basin. The second is to strengthen volcanic reservoir exploration in the Santanghu and Bohai Bay basins to build a hundred-million-ton-level oil and gas field. The third is to actively explore the Carboniferous-Permian strata in Tuha Basin, other peripheral basins in northern Xinjiang, the Permian strata in Sichuan and Tarim basins, as well as Ordos Basin to achieve new breakthroughs.

Chapter 6 summarizes relevant techniques in volcanic region prediction, play prediction, reservoir prediction, and fluid prediction.

The book was written by Zou Caineng, Zhang Guangya, Zhu Rukai, Yuan Xuanjun, Zhao Xia, Hou Lianhua, Wen Baihong, Wu Xiaozhi, and others. Zou Caineng and Zhang Guangya were responsible for the final compilation and editing.

The compilation of this book received great support and instruction from Zhao Zhengzhang, Vice President of PetroChina, and Wang Daofu, Chief Geologist of PetroChina. Also, Zhao Wenzhi and Du Jinhu, Vice Presidents of PetroChina Exploration & Production Company; Feng Zhiqiang, Vice President of PetroChina Daqing Oilfield Company; Zhao Zhikui, Chief Geologist of PetroChina Jilin Oilfield Company; Kuang Lichun, Vice President of PetroChina Xinjiang Oilfield Company; and Liang Shijun, Chief Geologist of PetroChina Tuha Oilfield Company have offered much instruction and assistance. Academicians Dai Jinxing and Jia Chengzao, as well as Professors Gao Ruiqi, Zhao Huakun, and Gu Jiayu have rendered instruction and support. Academician Liu Jiaqi took time out of his busy schedule to compose the foreword for the book. Zhang Yan, Hu Suyun, Tao Shizhen, Jia Jinhua, Wang Lan, Gao Xiaohui, Yang Chun, Fang Jie, Wei Yanzhao, Zhang Qingchun, Zhang Guosheng, Liu Xiao, Yang Hui, Mao Zhiguo, Wei Yuanjiang, Liu Lei, Feng Youliang, Li Wei, Zhao Yimin, Fang Xiang, and Luo Beiwei have aided in the compilation. The authors are very grateful to them all. For any oversights or omissions in this book, comments and corrections would be greatly appreciated.

The authors

Foreword

Igneous rock (including volcanic rock), forming in a high-temperature environment, is usually considered an unlikely candidate for hydrocarbon exploration because oil and gas are conventionally known to accumulate in sedimentary rock. Although volcanic reservoirs were discovered 100 years ago in various parts of the world, they have received little attention because of the lack of significant discoveries. This has resulted in volcanic reservoirs constituting a very small percentage of all reservoir types.

Oil and gas exploration in China has made great progress in the past half century. However, as far as volcanic reservoir exploration is concerned, no significance had been attached to it until the 2000 discovery of an oil and gas field in the deep Mesozoic and Cenozoic volcanics of Songliao Basin and in the Paleozoic volcanics of Junggar and Santanghu basins in Xinjiang Province. These discoveries broke down the traditional theory that volcanics were not favorable for hydrocarbon accumulation. After this, volcanics were regarded as major targets for onshore hydrocarbon exploration in China and represents another advance in Chinese oil and gas exploration after the theory of the continental origin of oil.

Volcanic reservoirs, considering volcanic rock as a reservoir type or being closely related to volcanism, are more complex than conventional sedimentary reservoirs in their unique hydrocarbon generation, migration, accumulation, and preservation. Their hydrocarbon accumulation may be independent of sedimentary rock, and as such they complement sedimentary reservoirs, which expands the province of oil and gas exploration.

Volcanic rocks are widely distributed in China and diverse reservoir types exist. In terms of source-reservoir assemblage, there are the late generation-early storage type and the self-generation-self-storage type; in terms of genesis, there are the inland rift type, rift type after collision and orogeny, and island arc type. Therefore, the immediate task is to establish theories and methods for volcanic reservoir exploration to guide related discovery and exploration.

Professor Zou Caineng and his research team have been engaged in oil and gas exploration and development for years. They have accumulated rich experience and research findings on volcanic reservoirs from the Songliao Basin to Tianshan Mountain, involving field surveys and laboratory studies. The book *Volcanic Reservoirs in Petroleum Exploration* is the outcome of their years of efforts. Starting from a review of volcanic reservoir exploration and research history at home and abroad, they expound the genesis and distribution of volcanic rocks, volcanic reservoir types and characteristics, hydrocarbon accumulation mechanisms and distribution rules of volcanic reservoirs, and evaluation methods and techniques for volcanic reservoirs. By thorough analyses on volcanic reservoirs in major basins in China, they have unveiled the geologic background, tectonic setting, lithology, lithofacies features, reservoir types, and main controlling factors for favorable reservoir development and summarized the hydrocarbon accumulation assemblages and hydrocarbon distribution rules in volcanic rock, and the technology of volcanic reservoir characterization and exploration, which indicates the direction and key domains for volcanic reservoir exploration. This book presents an update on hydrocarbon accumulation mechanisms and exploration technologies of volcanic reservoirs, as well as high-level theoretical and application values. I hope this book will be of great instructive significance to volcanic reservoir exploration in China and abroad, and I look forward to its publication.

CAS Academician: Liu Jiaqi
August 2010

Chapter 1

Exploration History and Features of Volcanic Reservoirs

Volcanic reservoirs, regarded as unconventional reservoirs with volcanic rock as the reservoir bed, have been widely discovered in many hydrocarbon-bearing basins at home and abroad for more than 100 years. Overseas volcanic reservoirs have experienced a long history of exploration with some discoveries of large oil and gas fields, but most of which were discovered by chance or in very localized areas. Volcanic reservoirs have not garnered much interest on the part of oil and gas explorers. However, in China, volcanic reservoirs have received more attention recently and have become one of the major targets for exploration. Hydrocarbon exploration has made breakthroughs in volcanic rocks in the deep Songliao Basin and the Carboniferous-Permian Formations in the Junggar and Santanghu Basins. Reserves have increased on a large scale, demonstrating the potential of oil and gas exploration in volcanic rocks.

1.1 OVERSEAS VOLCANIC RESERVOIR EXPLORATION

1.1.1 Exploration History and Features

Volcanic rocks are widely distributed in many hydrocarbon-bearing basins at home and abroad and could function as important reservoir rocks. The first reported discovery overseas (1887) was in the San Joaquin Basin in California, United States, which has since experienced 120 years of exploration and development. There are altogether more than 300 cases of discoveries of oil and gas reservoirs or oil and gas shows in volcanic rock (Figure 1.1) at present, among which 169 reservoirs have proven reserves.

1.1.1.1 Overseas Exploration for Volcanic Reservoirs and Division of Research Period

Most reservoirs in volcanic rocks were the accidental by-products of traditional shallow oil reservoirs, and were considered to be of no commercial value, and therefore were ignored without appraisal or study.

Researchers eventually became aware that hydrocarbon accumulation in volcanic rocks is not an abnormal phenomenon and conducted oil and gas exploration intentionally in volcanic rocks in some local areas. La Paz oil field, discovered in Venezuela in 1953 with the highest oil production in a single well (up to 1828 $m^3 d^{-1}$), was the first successful case in the world of exploration aimed at volcanic rock. From this point on, the study of volcanic oil reservoirs entered a new phase.

After the discoveries of oil fields in volcanic rock, volcanic reservoir exploration was conducted extensively around the world, giving birth to many discoveries of oil and gas reservoirs (fields) in the United States, Mexico, Venezuela, Argentina, former Soviet Union, Japan, Indonesia, and Vietnam. Well-known cases include the tuff oil reservoir in Samgori (Georgia), oil reservoir in eruptive rock in Muradkhanli (Azerbaijan), andesitic oil reservoir in Jatibarang (Indonesia), rhyolitic reservoir in Yoshii to east Kashiwazaki (Japan), and granitic reservoir in White Tiger oil field in shallow seas (south Vietnam).

1.1.1.2 Features of Overseas Research

Overseas volcanic reservoirs have undergone a long history of exploration with many discoveries of oil and gas reservoirs, but most were discovered by chance or in localized areas. Little attention has been paid to this kind of reservoir; therefore, volcanic reservoirs have not been extensively explored as a whole and their contribution to the global total of reserves only accounts for 1% or so. Research remains at a low level. Relevant papers and studies specializing in volcanic reservoirs are very few at home and abroad, resulting in a lack of understanding.

Neogene, Paleogene, and Cretaceous volcanic rocks are relatively rich in oil and gas, while little occurs in rocks Jurassic and older. Volcanic rocks that have formed reservoirs are likely to occur in strata from hundreds of meters to 2000 m deep, and seldom deeper than 3000 m.

Discovered volcanic reservoirs mainly occur in a circular distribution around the Pacific Rim, from the United States and Mexico in North America, to Cuba, Venezuela, Brazil, and Argentina in South America, and then to China, Japan, and Indonesia in Asia. They also have been discovered from Central Asia to Eastern Europe, including Georgia, Azerbaijan, Ukraine, Russia, Romania, and Hungary. Some countries in North Africa and Central Africa, e.g., Egypt,

Volcanic Reservoirs in Petroleum Exploration. http://dx.doi.org/10.1016/B978-0-12-397163-0.00001-4

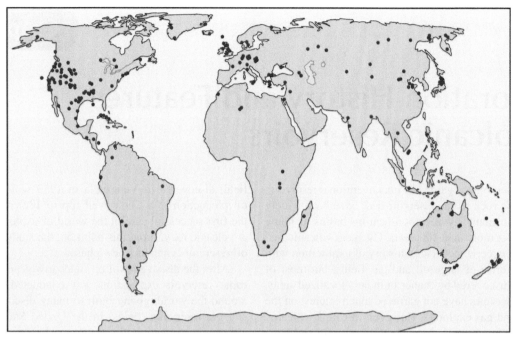

FIGURE 1.1 Global oil and gas distribution in volcanic rock, among which there are 169 reservoirs, 65 cases with oil and gas shows, and 102 cases with oil seepage.

Libya, Morocco, and Angola, have reported discoveries of volcanic reservoirs as well.

In view of the structural setting of hydrocarbon-bearing basins, volcanic reservoirs mainly distribute at passive continental margins, e.g., those discovered in North America, South America, and Africa, and can also develop in inland rift basins.

With respect to rock type, volcanic reservoir rocks mainly include intermediate to basic rock, such as basalt

and andesite. Primary and secondary pores serve as reservoir space. These develop from fractures and dissolved pores of various geneses and are crucial to the improvement of reservoir properties.

In general, oil and gas reservoirs have been reported to be limited in volcanic rock, but large, high-yield oil and gas fields do exist. Table 1.1 lists 11 overseas volcanic oil and gas fields with recoverable reserves all higher than 2000×10^4 tons oil equivalent, among which Jatibarang

TABLE 1.1 Reserves of Overseas Large Volcanic Oil and Gas Fields

Country	Oil and gas field	Basin	Fluid	Gas ($10^8 m^3$)	Oil (10^4 tons)	Lithology
Australia	Scott Reef	Browse	Oil and gas	3877	1795	Effusive basalt
Indonesia	Jatibarang	NW Java	Oil and gas	764	16,400	Basalt, tuff
Namibia	Kudu	Orange	Gas	849		Basalt
Brazil	Urucu area	Solimoes	Oil and gas	330	1685	Diabase
Congo	Lake Kivu	?	Gas	498		?
United States	Richland	Monroe Uplift	Gas	399		Tuff
Algeria	Ben Khalala	Triassic/Oued Mya	Oil		>3400	Basalt
Algeria	Haoud Berkaoui	Triassic/Oued Mya	Oil		>3400	Basalt
Russia	Yaraktin	Markovo-Angara Arch	Oil		2877	Basalt, diabase
Georgia	Samgori	?	Oil		>2260	Tuff
Italy	Ragusa	Ibleo	Oil		2192	Gabbro

TABLE 1.2 Production of Overseas Volcanic Oil and Gas Fields

Country	Oil and gas field	Basin	Fluid	Production Oil (t d^{-1})	Gas (10^4 m^3 d^{-1})	Lithology
Cuba	Cristales	North Cuba	Oil	3425		Basaltic tuff
Brazil	Igarape Cuia	Amazonas	Oil	68-3425		Diabase
Vietnam	15-2-RD 1X	Cuu Long	Oil	1370		Altered granite
Argentina	YPF Palmar Largo	Noroeste	Oil and gas	550	3.4	Porous basalt
Georgia	Samgori	?	Oil	411		Tuff
United States	West Rozel	North Basin	Oil	296		Basalt, agglomerate
Venezuela	Totumo	Maracaibo	Oil	288		Volcanics
Argentina	Vega Grande	Neuquen	Oil and gas	224	1.1	Fractured andesite
New Zealand	Kora	Taranaki	Oil	160		Andesitic tuff
Japan	Yoshii-Kashiwazaki	Niigata	Gas		49.5	Rhyolite
Brazil	Barra Bonita	Parana	Gas		19.98	Effusive basalt, diabase
Australia	Scotia	Bowen-Surat	Gas		17.8	Cataclastic andesite

oil and gas field in the northwest Java Basin, Indonesia, is the largest oil field with recoverable oil reserves of 1.64×10^4 tons. Scott Reef oil and gas field in Browse Basin, Australia, is the largest gas field with recoverable gas reserves of 3877×10^8 m^3. Table 1.2 lists the production of 12 overseas volcanic oil and gas fields, among which daily oil production of Cristales oil field in the North Cuba Basin is the highest, up to 3425 tons, and daily gas production of the Yoshii-Kashiwazaki gas field in Niigata Basin, Japan, is the highest, up to 49.5×10^4 m^3.

1.1.2 Typical Volcanic Reservoirs

1.1.2.1 Scott Reef Oil and Gas Field, Australia

Browse Basin lies in the northwestern sea area of Australia with seawater depth between 80 and 300 m and a basin area of 14×10^4 km^2. Several oil and gas fields have been discovered in the basin, among which Scott Reef volcanic oil and gas field has the largest reserves (Table 1.2).

Browse Basin has a Paleozoic to Cenozoic deposition thickness greater than 15,000 m, with the development of several stages of a Mesozoic depositional basin. Most of the basin lies in continental shelf. There were six periods of structural evolution: (1) inward extension of the craton into a half-graben basin in the later Carboniferous to the early Permian, (2) heat subsidence in the later Permian to the Triassic, (3) tectonic reversal in the later Triassic to the early Jurassic, (4) extension in the early and middle Jurassic, (5) heat subsidence in the later Jurassic to the Cenozoic, and (6) reversal in the middle and later Miocene.

Source rocks mainly developed in the lower Permian, Jurassic, and lower Cretaceous, which were predelta onshore-plain, fine-grained sediment of coal-bearing mudstone. Reservoir rock includes river delta sandstone and Campanian-Maastrichtian marine sandstone of the middle and lower Jurassic, among which there are large amounts of volcanic interbeds in the Jurassic with good reservoir properties. Burial depth of the reservoir rock can reach 4000-5000 m. The depth of the basin margin is 3000-3500 m. In Scott Reef field, the burial depth of the reservoir rock lies between 3934 and 4695 m. Reservoir trap types primarily include fault, anticline, buried hill, and stratigraphic overlap.

1.1.2.2 Jatibarang Oil and Gas Field, Indonesia

Jatibarang oil field, northwest of Java, Indonesia, was discovered through well JTB-44 drilled in 1969. with zone of interest at 2011 m. The field entered the development stage in 1973. Daily oil production of a single well is from 250 to 3000 bbl and burial depth is from 2000 to 2300 m.

The Upper Cretaceous Jatibarang Formation pay zone is 1200 m thick and composed of andesite and dacite with interbeds of clay, glutinite in the lower part, and andesitic volcaniclastic rocks and altered volcanics in the upper part. Volcanic rock develops in an onshore fluvial environment. The reservoir is fractured volcanic rock, including lava (andesite/basalt), tuff, volcanic breccias, and agglomerate. Fractures, intergranular pores, and intercrystalline pores form the reservoir space. The reservoir rock is characterized by high heterogeneity and well-developed fractures with porosity varying between 16% and 25% and permeability of 10×10^{-3} μm^2. The burial depth of the reservoirs lies between 2700 and 4000 m. There are structural reservoirs and lithologic-stratigraphic reservoirs in this field.

1.1.2.3 Kudu Gas Field, Namibia

Kudu gas field in Namibia, discovered in 1974, lies in the western sea area of Namibia at water depths of approximately 170 m. Basalt of effusive facies distributes extensively at the African and South American continental margins with lateral extension from 60 to 100 km. This extension inclines seaward and forms the well-known seaward-inclined reflection sequence (SDRS), distributing within the range of sea bed depths of 200-4500 m. Kudu gas field has at least 849×10^8 m^3 natural gas reserves in the SDRS. The source rock, which developed in the Permian system, consists of lacustrine mudstone of the prerift stage, and marine or continental mudstone of the synrift stage. Reservoir rocks with burial depths of 4400 m are sandstones interbedded with plateau basalts.

1.1.2.4 Muradkhanli Oil Field, Azerbaijan

The oil field in the east-central Kula Basin, Azerbaijan, was a major discovery in the early 1970s. Crude oil mainly accumulated in eruptive rock at the top of a buried hill.

At the early stage of the later Cretaceous, a volcanic eruption in the Muradkhanli uplift resulted in the accumulation of trachybasalt and andesite with a maximum thickness of 1950 m. The volcanic eruption, accompanied by transgression, gave rise to alternating layers of volcanic rock and sedimentary rock. These layers were then eroded, developing a weathering crust of 50-100 m at the top of the uplifted buried hill, and then subsequent deposition from the Paleocene to Quaternary. As a result, eruptive rock was overlapped by younger sediments. The Muradkhanli buried hill is 20 km long, 15 km wide, and 1000-1600 m in relief. The axis is cut by vertical faults, and the overlying strata demonstrate syngenetic structures that inherited the ancient landform of the buried hill.

The reservoir rock mainly consists of eruptive rocks, such as andesite, basalt, and porphyrite. Porosity varies from 0.6% to 20% with an average of 13%. The porosity of microfractures accounts for 0.44% of the total volume. Large pores 2 cm long and 1.5 cm wide and many small 1-mm pores, as well as secondary leached pores of 0.05 mm $\times$ 0.97 mm, have been seen in samples. There are large fractures and microfractures in the upper part of eruptive rocks, which may result in mud loss during drilling and also high yield in oil recovery.

The field consists of five oil reservoirs, one of which is an eruptive reservoir above the erosion surface. The reservoir is situated at the top of an uplift and northwest periclines and belongs to a massive, stratified oil reservoir. The porosity of the eruptive rock varies from 10% to 16%. Matrix permeability is close to zero. Oil production from a single well is controlled by fractures; high yield is possible in the case of well-developed fractures. Each well has unique production profiles because of uneven fracture development. The fracture system in the eruptive oil reservoir is well connected because of the fractures in the weathering crust, which may lead to high yield in oil production.

1.1.2.5 Yoshii-East Kashiwazaki Gas Field, Japan

This gas field, lying northeast of Kashiwazaki City, Japan, is a long, narrow, anticlinal, trapped green tuff gas field that stretches from West Mountain to the central oil and gas province in Niigata Basin. The northwest high belongs to the east Kashiwazaki gas field of Teikoku Oil Co., and the southeast high belongs to the Yoshii gas field of Japan Petroleum Exploration Co., Ltd. The anticline is 16 km long and 3 km wide, with a gas-bearing area of 27.8 km^2, original recoverable gas reserves of 118×10^8 m^3, oil reserves of 225×10^4 tons, and well depth of 2310-2720 m.

Green tuff developed in the middle Miocene Epoch, i.e., the Qigu Age, with effective reservoir thickness of 5-57 m, porosity of 7-32%, and permeability of 5×10^{-3} to 150×10^{-3} μm^2. The source rock is mudstone from the Qigu Formation with an organic carbon content of 1-1.5% and Type I kerogen. The Qigu Formation was buried at depths shallower than 2000 m in the early Xishan Age, at temperatures of approximately 100 °C. Crude oil generated in this period migrated to and accumulated in the volcanic rocks of the anticlinal trap. Subsequently, the Qigu Formation continued to subside. When the underground temperature rose above 130 °C, crude oil in the original oil reservoirs began to be pyrolyzed to generate gas and condensed oil with a gas/oil ratio from 4000 to 5000.

Originally, tuff has poor porosity and permeability with low productivity. The ultimate capacity of the green tuff gas zone relates to a secondary pore and fracture system. The overall gas reservoir has a strong water drive with high gas well pressure and small pressure drop.

1.2 VOLCANIC RESERVOIR EXPLORATION IN CHINA

1.2.1 Exploration History

Domestic volcanic reservoirs were discovered in the Junggar Basin for the first time in 1957, which was followed by successive discoveries in 11 basins over the next 50 years, including Bohai Bay, Songliao, Erlian, and Santanghu Basins, among others. Volcanic reservoir exploration in

China has experienced three stages: accidental discoveries, local exploration, and overall exploration. The progress in each exploration stage is closely related to the increasing knowledge of volcanic reservoirs and technical advancements in exploration.

1.2.1.1 Stage of Accidental Discoveries

From the 1950s to 1980, volcanic reservoirs had primarily been discovered by chance at the northwestern margin of Junggar Basin and in the Liaohe and Jiyang depressions, Bohai Bay Basin.

During the preliminary prospecting and adjustment of oilfield development in Karamay (from 1957 to 1973), there were favorable oil and gas shows in the Carboniferous volcanic rock system or lower Permian series in the Zhongguai uplift, Hongshanzui oil field, and Blocks 1, 5, 6, and 7 in the Karamay oil field. In 1957, well 222 in Block 9, Karamay oil field, yielded a commercial oil stream from the Carboniferous system for the first time, with daily production of $7.25\ m^3$. However, it was disregarded, and the volcanic rock was only considered to be the basement of the depositional basin.

During the major development of Karamay oil field (from 1973 to 1980), well Gu3 in Block 9 at the upper wall of the Ke-Bai fault belt yielded a high volume, of commercial oil stream from Carboniferous tuff and andesite with daily production of $177\ m^3$, which made it clear that Carboniferous volcanic rocks may act as effective reservoir rocks. Limited by the shortage of useful seismic data because basement reflections could not be extracted from single-point analog signals, basement volcanic rocks had not been characterized. In addition, affected by the misunderstanding of the day, which regarded volcanic rock only as basin basement, Carboniferous volcanic rock had not been considered an effective reservoir, and the exploration did not proceed.

The period from the early 1960s to the early 1980s witnessed major discoveries of middle to large oil and gas fields in east China. The hydrocarbon exploration targets in this period were clastic rocks of the Paleogene to Neogene systems and buried hill reservoirs of pre-Paleogene. The discoveries of oil layers or reservoirs in volcanic rock had only been contingent events during conventional exploration. From 1970 to 1972, volcanic oil layers had been discovered in six structures in the east of Liaohe depression, including the Yulou, Huangjindai, and Dapingfang structures. There were 24 wells with commercial oil streams or oil shows. From 1972 to 1980, an SC13 diabase oil reservoir, Lin41 diabase oil reservoir, and Xia8 oil reservoir in a volcanic cone draping structure were drilled successively in the Jiyang depression, as well as the Shijiutuo 428 volcanic oil reservoir in the sea area of Jiyang depression. Magmatic activity mechanisms have a certain relationship with complicated fractured anticlines in rift basins; therefore, it is inevitable to find oil reservoirs or oil and gas shows in volcanic rocks frequently in this period. On the other hand, volcanic reservoirs had not been taken seriously in the 1970s. The proven oil-bearing area was only $2.7\ km^2$ with proven reserves of 172×10^4 tons to the end of the 1970s in volcanic oil reservoirs in Bohai Bay Basin.

1.2.1.2 Stage of Local Exploration

From 1980 to 2002, because of our increased understanding of volcanic rocks and technical progress, exploration focused on volcanic rocks was conducted in several basins, including Bohai Bay and Junggar Basins.

In the 1980s, volcanic rock exploration centered in the Jiyang, Huanghua, and Jizhong depressions and Subei Basin. Twelve volcanic oil reservoirs were discovered, including the Bin 338 reservoir (1982), Fenghuadian reservoir (1986), Yibei dike rock reservoir (1986), and Heqiao diabase reservoir (1989). Volcanic reservoir exploration reached a peak around 1986. Cumulative proven reserves reported over 10 years time amounted to 1781×10^4 tons with an areal extent of $6.5\ km^2$. In general, the exploration of volcanic rock was regarded as successful during this time. However, this experience of exploration and exploitation uncovered the complexities and peculiarities of volcanic reservoirs. During the extensive drilling of the Xinbin 348 volcanic oil reservoir in the Jiyang depression in 1989, five exploratory wells, including well Bin349, were dry holes. Volcanic reservoir exploration then fell into a standstill in the early 1990s. The activities concerning volcanic reservoirs mainly focused on favorable reservoirs and enrichment rules, which laid the foundation for exploration in the middle to late 1990s.

In 1994, well Da 72 in the Jiyang depression was the first to discover a volcanic oil reservoir. Then, in 1995, the Zaobei volcanic oil reservoir was reported to contain proven reserves of 1050×10^4 tons, and wells Shang741, Luo151, and Ou 26 yielded high daily oil streams of more than 150 tons in 1997. Volcanic reservoir exploration again reached a new peak. Proven reserves in volcanic rock continued to rise, and probable and possible reserves increased continuously. The integration of different disciplines in exploration was integral at this stage, and formation microimaging (FMI), nuclear magnetic resonance, seismic reservoir prediction, and numerical reservoir simulation were applied extensively in volcanic reservoir exploration.

Along with the introduction and popularization of digitized seismic and fracturing techniques, detailed seismic surveys were conducted, focusing on major fault belts around Junggar Basin, which deepened the knowledge of the fault belts. The large thrusting fault belt at the northwest margin was thought to be the most favorable region in Junggar Basin for hydrocarbon accumulation, and oil and gas would have accumulated in different kinds of traps.

Aiming at the Carboniferous volcanic basement at the upper walls of the thrusting fault belt and the brim of volcanic rock at the lower wall of the faults, large-scale exploration was conducted intentionally and resulted in great progress in volcanic reservoir exploration. The achievements included (1) the discovery of an uncompartmentalized oil reservoir in the Carboniferous basalt in Block 1, Karamay oil field, in 1984, with large scale, proven oil reserves; (2) an exploration breakthrough in the Carboniferous volcanic rock at Hong-Che fault belt and in the Permian volcanic rock at Wu-Xia fault belt, which increased the standby reserves in the basin.

Along with the popularization of digitized seismic techniques in the desert, regional 3D seismic prospecting, fracturing techniques, and *in situ* immediate well logging evaluation after 1993, exploration was aimed at hydrocarbon enrichment systems in desert areas and the hinterland of basins on a large scale. This further deepened our understanding of basin structures, tectonic frameworks, oil- and gas-bearing systems, hydrocarbon accumulation in different series of strata at different stages, and oil and gas distribution. Within the effective depth of prospecting, extensive exploration was aimed at Carboniferous and Permian volcanic rocks, and a great breakthrough took place in several second-order structural belts, such as the Carboniferous andesite and rhyolite oil reservoir in the Shixi uplift, Luliang upheaval in 1994, the gas field discovered in the Carboniferous andesite and welded volcaniclastic rock in Wucaiwan sag in 1998, and other discoveries in basement volcanic rock in Ke-Bai, Hong-Che and Wu-Xia fault belts at the northwestern margin, Zhangbei faulted fold zone, and Jimusa'er sag.

During the drilling of well A2 in A'nan sag, Erlian Basin, 1981, eruptive rock of the A'ershan Formation, Bayanhua Group, lower Cretaceous series yielded a commercial daily oil stream of 27 tons, resulting in the discovery of Abei oil field. Abei oil field is located east of the Manite depression, Erlian Basin. Abei is one of the four oil fields in A'ershan, which is a massive basalt oil reservoir with an oil-bearing area of 15 km^2, oil layer thickness of 33.8 m, and a certainty of proven reserves. Other findings include the Minzhuang volcanic oil reservoir in Subei Basin, and well Zhougong1 in Sichuan, which has a high gas yield of 20×10^4 m^3.

In this stage, deep natural gas exploration in Songliao Basin mainly centered on glutinite instead of volcanic rock, whereas there were still discoveries of hydrocarbon gas reservoirs in eastern Wangjiatun and a CO_2 gas reservoir in the Fangshen9 well block in the Xujiaweizi area, Songliao Basin, which laid the foundation for volcanic rock exploration on a large scale after 2000.

1.2.1.3 Stage of Overall Exploration

Volcanic reservoir exploration has been conducted on a large scale in Bohai Bay, Songliao, Junggar, and Santanghu Basins. The progress in volcanic exploration and great

increase in reserves made volcanic rock a major target for hydrocarbon exploration in China.

Volcanic rock exploration in the middle section of the eastern sag in Liaohe depression, Bohai Bay Basin, is a representative example of overall exploration in China. The proven and probable reserves have reached more than 10 million tons, and the oil- and gas-bearing areas of Oulituozi and Huangshatuo oil fields have merged.

The middle member of the eastern sag has its southern border at Rehetai oil field and northern border at Zhujiafangzi. The Rehetai, Oulituozi, Huangshatuo, and Tiejianglu structures occur in succession from south to north, with a total area near 800 km^2. This member has shallow subsidence with rapid facies changes and a small-scale depositional system. Adjacent to the Rehetai-Tiejianglu and Yujiafangzi source rock, the member has favorable oil source conditions. The reservoir rock mainly includes clastic and volcanic rock types. Affected by regional dextral strike-slip tectonic activities since the Oligocene, structural fractures are developed in the volcanic reservoir rock. Therefore, the volcanic reservoir rock in the member is regarded to be favorable for hydrocarbon accumulation.

In 2000, a breakthrough was made during exploration of the trachyte of Member 3, Shahejie Formation, via two drilled wells in the Huangshatuo area, which gave birth to the Huangshatuo oil field. Then, five additional wells were drilled in the Oulituozi area from 2003 to 2004 for further exploration, three of which yielded high commercial oil and gas flow.

Based on oil testing in previously drilled wells, research into volcanic lithologies, lithofacies distribution, fracture growth, structural interpretation, and volcanic reservoir prediction was strengthened in 2005. At the junction of the Huangshatuo and Oulituozi, three wells were drilled successfully, which enlarged the oil- and gas-bearing areas in Oulituozi volcanic rock. The oil- and gas-bearing areas of the Oulituozi and Huangshatuo oil fields have been merged to form a volcanic reservoir block with reserves of 10 million tons.

Discoveries in exploration aiming at volcanic reservoirs before 2000 included the volcanic gas reservoir in eastern Wangjiatun in 1996, and volcanic CO_2 gas reservoir in well block Fangshen 9 in 1997.

From 1996 to 2000, proven reserves of deep natural gas were reported for the first time in Changde, Shengping, Buhai, Gudian, Xiaochengzi, Xiaohelong, Gujiazi, and Qinjiatun (oil) gas fields, with discoveries of middle scale of fields of the Changde and Shengping gas fields and increasing the original gas in place. Deep gas exploration in Songliao Basin then began.

Deep volcanic reservoir exploration in Songliao Basin followed after the success of well Xushen1 in 2002, with a daily gas yield of 54×10^4 m^3 in volcanic rocks of the Yingcheng Formation. Since 2004, exploration in

Xujiaweizi has increased rapidly, with an increase in proven gas reserves of hundreds of billion cubic meters to the present day. In 2005, venture well Changshen1 was drilled in the Changling fault depression using the exploration experience of Xujiaweizi for reference, which succeeded in a high gas yield in volcanic rocks of the Yingcheng Formation and made a breakthrough in deep exploration in the south. The two natural gas provinces have natural gas reserves of 100 billion cubic meters to the present day. In general, the deep Songliao Basin has the potential of $5000 \times 10^8 \text{ m}^3$ oil and gas in place (OGIP), which is of great strategic significance to the relatively stable oil and gas production in the basin long term.

1.2.1.3.1 Well Xushen1

The drilling of well Xushen1 made a breakthrough in the deep Xujiaweizi fault depression, resulting in gas reserves of hundreds of billion cubic meters.

The Xujiaweizi fault depression has an exploration area of 5350 km^2 and includes the Anda subdepression, Xuzhong structural belt, as well as the Xuxi, Xudong, and Zhaozhou subdepressions.

Exploration activities in the Xujiaweizi fault depression before 1998 had focused on the evaluation of the fault depression, and drilling was conducted primarily at the paleohigh and the structural trap at the depression margin. Through systematic investigation during the eighth and ninth Five-Year-Plan periods, the resource potential of Xujiaweizi was understood more clearly, and exploration targets have turned to the volcanic lithologic trap and glutinite stratigraphic-lithologic trap inside the fault depression.

In 1998, the Xingcheng nose-like structure, an uplifted structure of Shengping uplift extending southward, was discovered via 2D seismic data interpretation. In 2001, well Xushen1 was drilled to the Huoshiling Formation based on the interpretation of newly acquired 3D seismic data. This well yielded a high commercial gas stream of $50 \times 10^4 \text{ m}^3 \text{ d}^{-1}$ in volcanic rocks of the Yingcheng Formation and a commercial gas stream in conglomerate. Two subsequent wells, i.e., Xushen2 and Xushen6, were drilled south of the Xingcheng structural belt (the middle volcanic rock belt) and yielded a high commercial gas stream in volcanic and conglomerate reservoirs of the Yingcheng Formation.

After that, preliminary prospecting has been strengthened and Xujiaweizi has yielded gas reserves on a scale of hundreds of billion cubic meters to the present day.

1.2.1.3.2 Well Changshen1

The drilling of well Changshen1 made a breakthrough in the deep Changling fault depression, initiating the exploration of deep strata in the southern Songliao Basin. Two natural gas provinces with reserves of 100 billion cubic meters have resulted at present.

The Changling composite fault depression is located in the central fault depression belt of southern Songliao Basin with an area of 7044 km^2. Before 2005, most deep exploratory wells were drilled into strata below Member 1 of the Quantou Formation, distributed in Fulongquan, Shuangtuozi, and Dalaoyefu areas at the slope belt, east of the fault depression, with discoveries of several small-scale volcanic or glutinite gas reservoirs.

Volcanic reservoir exploration in the Changling fault depression was strengthened in 2005 with reference to the exploration experience of Xujiaweizi. Venture well Changshen1 was drilled at the Ha'erjin structure and yielded a gas stream of $47 \times 10^4 \text{ m}^3 \text{ d}^{-1}$ in the Yingcheng Formation, making a breakthrough in natural gas exploration. The Ha'erjin structure is located at the central uplift of the Changling fault depression, with good superposition of structural trap and volcanic rock. Volcanic rock types include tuff and rhyolite, as well as a small amount of volcanic breccia and andesite. The gas reservoir in well Changshen1 is mainly controlled by the structure and is a structural gas pool with bottom water.

After drilling the Changshen1 well, there were 29 exploratory wells drilled deeply in the Changling sag, among which 10 wells yielded hydrocarbon gas, 6 wells yielded CO_2 gas, and 6 wells had a low-yield gas stream. The proven hydrocarbon gas reserves exceeded 100 billion cubic meters, along with CO_2 gas reserves of tens of billion cubic meters. In addition, five wells yielded a commercial gas stream and could reach the reserves level of $1000 \times 10^8 \text{ m}^3$.

Overall, exploration aimed at the Carboniferous and Permian volcanic rock has been conducted in the deep Ludong-Wucaiwan since 2005 and made a breakthrough in natural gas exploration. Oil and gas exploration also made great progress in the northwestern margin.

In 2005, well Dixi10 was drilled in the Ludong-Wucaiwan area, aiming at the structural trap of Carboniferous volcanic rock and yielding a high commercial gas stream in rhyolite. The Carboniferous rock mainly consists of intermediate and acidic eruptive rock, including dacite, welded rhyolite, and a small amount of tuff. Thus, reservoir exploration had made great breakthroughs in the Carboniferous self-generation-self-storage reservoirs in northern Xinjiang.

After 2005, with the upper Carboniferous large stratigraphic gas reservoirs as primary exploration targets and the volcanic rock belt around well Dixi5 as a major interest, the gas reservoirs in blocks of well Dixi14, well Dixi18, and well Dixi17, as well as the gas reservoir in well Dixi10, were discovered in succession, making an overall breakthrough in the exploration of Carboniferous volcanic lithologic-stratigraphic reservoirs. Kelameili has been confirmed to be a major gas field, with reserves of 100 billion cubic meters through reservoir evaluation. Another major onshore gas province emerges in China.

The extensive exploration directed at undeveloped areas in the basin and peripheral basins since 2008 gave birth to some new discoveries in Wucaiwan, Beisantai, Zhundong, Dibei, and Junggar Basins, such as commercial gas streams in well Cai55 in Wucaiwan, internal lithologic oil reservoir in well Bei32, and a commercial gas stream or favorable gas show in well Quan6 in Dibei and well Dajing1 in Zhundong.

Besides further exploration of Carboniferous self-generation-self-storage reservoirs, the exploration of late generation-early storage reservoirs also resulted in new discoveries and reserves increases. It is recognized that the Carboniferous reservoir at the northwestern margin of Junggar Basin is controlled by a large stratigraphic unconformity, reservoirs are distributed as belts-apparent and different belts are related in the whole area. Exploration contributed to the reserves increase at the upper wall of the fault belt, and the discoveries of commercial oil flows in wells Che 91, Che 912, Chefeng 3, and Chefeng 6 in the Chepaizi area. At the lower wall of the Wu-Xia fault belt, a welded tuff oil reservoir was discovered in the Fengcheng Formation of the lower Permian via drilling of well Xia 72, with proven oil reserves of more than 10 million tons, which would be a great breakthrough in deep volcanic exploration in the northwestern margin of Junggar Basin. This discovery will have significant influence on the direction of hydrocarbon exploration in this area.

Tuha oil field focused more on the exploration of a lower assemblage in Santanghu Basin in 2006. The Malang sag in Santanghu Basin was considered a major target. The preliminary prospecting well Ma 17 was drilled at the No. 2 Niudong structure and indicated favorable oil and gas shows in the Jurassic and Carboniferous systems. There was high oil and gas yield in the Kalagang Formation in the formation testing that followed well completion, indicating the success of exploration in the lower assemblage.

The Carboniferous strata in the Malang sag consist of alternating layers of volcanic rock, volcaniclastic rock, and clastic rock. Well Tangcan 3 discovered favorable oil source rocks in the Carboniferous system, as well as oil and gas shows in the upper Carboniferous series. Oil and source rock correlation indicates that the hydrocarbons originated in the Carboniferous system and the lower assemblage of the basin provides the conditions of hydrocarbon accumulation. The Carboniferous strata in the Malang sag have shallow burial depths, with faulted fold zones distributed in belts and zones, and would be the major exploration interest of the lower assemblage.

Preliminary prospecting wells were deployed to explore the oil-bearing scale of the lower assemblage in Niudong block. The drilling results showed a stable distribution of oil layers in the Kalagang Formation and also brought forward more evidence of the large reserves scale in the Kalagang Formation. The reservoir space in the Kalagang Formation is mainly fractures and pores.

There are five faulted fold structural belts developed in the lower assemblage of the Malang sag, descending westward to the main body of the sag. The oil reservoir in the Kalagang Formation is a structural-stratigraphic reservoir related to volcanic rock weathering and leaching. The effective reservoir mainly distributes around the denudation area of the overlying Lucaogou Formation, which would be the most favorable to explore of the lower assemblage in Santanghu Basin.

Through volcanic reservoir exploration in Santanghu in recent years, oil reserves in the volcanic reservoir of Block Ma 17 have been confirmed. Exploration activities were extended to the lower assemblage at the peripheral area of Niudong. Drilled venture well Fang1 encountered the Carboniferous oil source rock, indicating good potential of the lower assemblage.

1.2.2 Achievements and Features

1.2.2.1 Achievements

After 50 years of exploration, tens of volcanic reservoirs have been discovered in 11 hydrocarbon-bearing basins, including Songliao, Bohai Bay, Junggar, and Santanghu Basins. PetroChina Company Limited (PetroChina) has proven oil reserves of 6×10^8 tons and proven natural gas reserves of $4700 \times 10^8 \, m^3$ in volcanic reservoirs to the present day.

There were three peaks in PetroChina's reserves increase in volcanic reservoirs (Figure 1.2). The first was from 1984 to 1986, with proven reserves of more than 100 million tons reported in the northwestern margin of Junggar Basin. The second was from 1994 to 1997, with proven reserves approximating 100 million tons reported in the Hongshanzui, Chepaizi, and Shixi at the northwestern margin of Junggar Basin, and Rehetai and Zaoyuan oil and gas fields of Bohai Bay Basin. The third is 2005 to the present day, with proven oil and gas reserves reported in the northwestern margin of Junggar Basin, deep Songliao Basin, Kelameili at Junggar Basin, and Niudong at Santanghu Basin. These three peaks indicate that volcanic reservoir exploration has entered the stage of major discoveries and rapid reserves increase.

Since 2002, the overall deployment and exploration of volcanic reservoirs have been strengthened, and a series of major discoveries have been made. Two volcanic reservoir provinces, the deep Songliao Basin in east China and the Carboniferous volcanics in northern Xinjiang in west China, have provided a reliable scale of reserves.

In 2002, the drilling of well Xushen1 discovered the Xushen gas field in the deep fault depression of Songliao Basin, which is the largest deep volcanic gas field in the world. Proven natural gas reserves were predicted to be

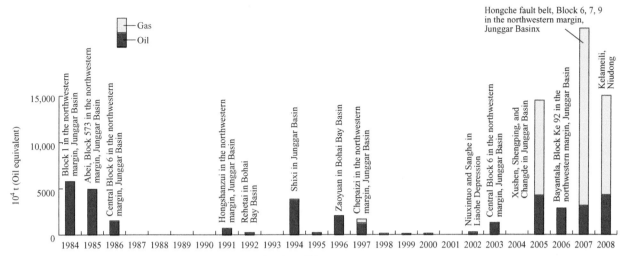

FIGURE 1.2 Reserves increase histogram of volcanic reservoirs, PetroChina.

$1018 \times 10^8 \, m^3$ until 2006. In 2005, the drilling of well Changshen1 discovered the Changshen gas field, followed by the discovery of Yingtai gas field. The natural gas reserves in deep Songliao Basin have amounted to several hundred billion cubic meters to the present day.

The northern Xinjiang has focused more on volcanic reservoir exploration in recent years with the achievements of the Kelameili gas field in Junggar Basin and volcanic oil field in the northwestern margin. Proven oil reserves in the northwestern margin were reported to be 1.43×10^8 tons, and proven gas reserves in Kelameili gas field were $1000 \times 10^8 \, m^3$. The Carboniferous volcanic reservoir exploration in Santanghu Basin has achieved reserves of more than 100 million tons, indicating the resource potential of the Carboniferous-Permian volcanic reservoir in northern Xinjiang.

1.2.2.2 Exploration Features

From the 1980s to 1990s, some volcanic reservoirs were discovered in succession in Junggar, Bohai Bay, and Subei Basins, such as the Karamay basalt reservoir in the northwestern margin, Junggar Basin; the Abei andesite reservoir in Erlian Basin, Inner Mongolia; the Mesozoic andesite reservoir in Fenghuadian of Huanghua depression; the basalt reservoir in Member 3 of the Shahejie Formation in Zaobei, Bohai Bay Basin; and the Shang741 diabase reservoir in the Jiyang depression.

Since 2000, China focused more on volcanic reservoir exploration and discovered oil and gas reservoirs in succession in deep Songliao Basin and the Carboniferous of Junggar and Santanghu Basins, causing volcanic reservoirs to become a major exploration target in China. In particular, the Xushen gas field in the Xujiaweizi fault depression and Changshen gas field in the Changling fault depression in deep Songliao Basin, Kelameili gas field in Junggar Basin,

and Niudong gas field in Santanghu Basin have made the deep Songliao Basin and the Carboniferous Formation in northern Xinjiang become two major natural gas provinces with volcanic reservoirs in the majority in China.

The following conclusions can be made as a result of analyzing the discovered volcanic reservoirs in China:

1. **Basin type:** volcanic reservoirs may develop in inland continental rift basins, such as Bohai Bay and Songliao basins, and also in continental, transitional, and marine spreading rifts after collision and orogeny, as well as in remnant-ocean basins, such as Ludong-Wucaiwan and the northwestern margin in Junggar and Santanghu Basins. Volcanic reservoirs mainly come into being in rift basins, indicating that the proximal assemblage of volcanic rock with lacustrine, transitional, and marine source rock would be crucial to the formation of volcanic reservoirs.
2. **Geologic age:** volcanic reservoirs in eastern China primarily developed in the Mesozoic and Cenozoic, and those in western China primarily developed in the late Paleozoic.
3. **Volcanic rock type:** rocks in the east are mainly intermediate and acidic volcanic rock and those in the west are mainly intermediate and basic volcanic rock. However, all types of volcanic rocks may act as reservoir beds to accommodate hydrocarbon.
4. **Reservoir type and scale:** reservoirs in the east are mainly lithologic reservoirs that could merge to form lithologic oil and gas fields distributed over a large area, such as the Xushen and Changshen gas fields. Those in the west are mainly stratigraphic reservoirs that could form uncompartmentalized oil and gas fields on a large scale, such as the Kelameili gas field, Northwestern margin oil field, and Niudong oil field.

Volcanic reservoir exploration has benefited from the development of exploratory techniques in recent years.

Since the 10th Five-Year-Plan period, a suite of techniques have been applied to the exploration of volcanic reservoirs, such as the high-precision gravity-magnetic-electric technique and 3D seismic technique for volcanic rock distribution prediction and reservoir prediction, target description and evaluation, logging technology for lithology identification, reservoir and hydrocarbon evaluation, underbalanced drilling, and large-scale hydraulic fracturing techniques. These all provide technical support for volcanic reservoir exploration.

1.2.2.3 Future Development

There are six development trends in volcanic reservoir exploration in China. The exploration activities would extend (1) from Bohai Bay Basin in the east to deep Songliao Basin and Junggar and Santanghu Basin in the west, (2) from the Mesozoic and Cenozoic strata in the east to the upper Paleozoic strata in the west, (3) from shallow and middle layers to medium-deep layers or even deep layers, (4) from structural highs to the slopes and sags, (5) from single lithologies and lithofacies to various lithologies and lithofacies and from near-crater facies to far-crater facies, and (6) from structural reservoirs to lithologic and stratigraphic reservoirs.

Volcanic rocks are distributed widely in China, with a total area of $215.7 \times 10^4\,km^2$ and predictive favorable exploration area of $39 \times 10^4\,km^2$, indicating great potential of volcanic reservoir exploration. According to a preliminary estimate based on the exploration achievements at present, the total oil resources in volcanic rock would be $19\text{-}26 \times 10^8$ tons, total natural gas resources at $4.2 \times 10^{12}\,m^3$, and total oil equivalent at 52×10^8 to 59×10^8 tons. The ratio of proven oil reserves to total oil resources is estimated to be 19-25%, ratio of proven gas reserves to total gas resources is 2%, and ratio of proven reserves to total resources is 6-7%. Therefore, there are abundant remaining resources with a great potential for volcanic reservoir exploration. At present, volcanic reservoirs are one of the important replacement reservoirs in China.

Chapter 2

Formation and Distribution of Volcanic Rock

Global oil and gas fields of volcanic rock are primarily distributed in the Phanerozoic volcanic rock zone, which is intimately connected with the most active tectonic provinces in the world. The oil and gas fields in the Bohai Bay Basin, China, are associated with the Cenozoic volcanic massif; the gas fields in the Sacramento Basin, United States, intersect with or are superimposed on the modern crater. Both show that these oil and gas accumulations are intimately connected with volcanism. Volcanism can supply a great variety of oil and gas sources for volcanic rock plays, and sedimentary rock and volcanic rock can form a reservoir-caprock-trap assemblage, which is favorable for the formation of oil and gas fields.

2.1 GLOBAL DISTRIBUTION OF VOLCANIC ROCKS

2.1.1 Overview of Present Distribution of Volcanic Rocks in the World

Volcanic phenomena appear to be individually isolated and diversified. However, since the plate structure theory was established, scholars built a global volcanic model that can organically link the isolated and diversified volcanic individuals. Most volcanics are distributed at plate boundaries and form the four worldwide volcanic rock distribution areas (Figure 2.1), i.e., east Africa-Red Sea continental rift zone, west Pacific intraoceanic island arc zone, Cordilleran continental margin volcanic arc zone, and North Atlantic midoceanic ridge zone.

2.1.1.1 East Africa-Red Sea Continental Rift Zone

As the largest rift zone on the continent, the east Africa-Red Sea continental rift zone is divided into two branches. The east branch starts from the estuary of Shire River in the south, passes through Malawi, middle part of the east African plateau and middle of Ethiopia from south to north, and dies out at the north end of the Red Sea, with a total length of about 5800 km. It also connects with the Jordan Valley northward. The west branch starts from the northwest end of Lake Malawi, passes through Lake

Tanganyika, Lake Kivu, Lake Edward, Lake Albert, and dies out in the Albertan Nile Valley, with a total length of about 1700 km. The rift zone is generally 1000-2000 m deep and 30-300 km wide and forms a series of long, narrow, deep-set valleys and lakes. For instance, the surface of Lake Assal located in the Great Rift Valley at the eastern side of the Ethiopian Plateau is 150 m below sea level, and it is the lowest point on African land.

The east Africa-Red Sea continental rift is a Mesozoic rift, where volcanism has been frequent and more active since the late Cenozoic. Statistical data show that there are more than 30 active volcanoes in Africa, and most of them occur in the vicinity of the rift faulting. Hot springs are widely developed in modern active volcanic areas, with apparent volcanic exhalation activities, and all these are evidence of present-day volcanism.

There are two types of volcanic eruption in the east Africa-Red Sea continental rift zone: one is the fissured eruption and the other is the centered eruption. The fissured eruption mainly occurred at both sides of Ethiopian rift system and formed a basaltic lava highland with a thickness of about 4000 m, which accounts for two-thirds of the Ethiopian territory. The fissured eruption also formed a 1000-m thick lava plateau in northwest Kenya, but it is older than that of the Ethiopian lava plateau, being formed 140,000-230,000 years before present (bp). Later, phonolite was formed, and a 300-km long phonolite lava plateau was formed 110,000-130,000 years bp. The centered eruption mostly occurs at the fringe of the rift zone where there are still numerous active volcanoes, such as the Ethiopian Eitel volcano. However, some volcanic eruptions only formed an explosion crater, crater bowl, or crater lake. For example, the diameter of Ngorongoro crater bowl is 19 km, and it has an area of about 304 km^2.

2.1.1.2 West Pacific Intraoceanic Island Arc Zone

The west Pacific island arc zone mainly refers to the Kuril islands, Japan islands, Ryukyu, China Taiwan island, the Philippines, and Indonesia islands. Volcanism is frequent in this zone, where there are more than 400 active volcanoes. Based on historical records, 80% of the volcanoes

Volcanic Reservoirs in Petroleum Exploration. http://dx.doi.org/10.1016/B978-0-12-397163-0.00002-6

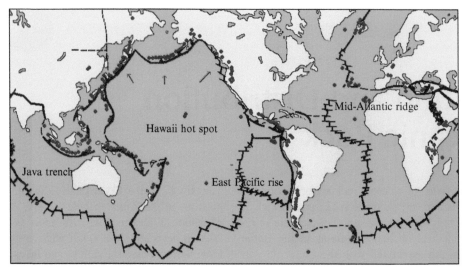

FIGURE 2.1 Global distribution of volcanic rocks.

active in the present day occur in North America, Kamchatka, Japan, Philippines, and Indonesia.

The west Pacific volcanic system is situated at the active continental margin of a trench-volcanic arc system. The volcanics are mainly the calc-alkalic rock erupted by intermediate magma, and the modal volcanic rock type is andesite. The andesite occurs in an arc parallel to the trench within 150-300 km away from the trench axis and forms the so-called "andesite line." Another characteristic of the west Pacific volcanic system is that the rock has a general horizontal zonality in the direction from trench to land, and with increasing distance from the trench, tholeiitic, calc-alkalic, and alkalic series rocks occur in turn. The intraoceanic island arc far from the continent is a series of chain islands chiefly composed of tholeiite. Its geologic age is generally younger than Paleogene, and it is directly located on the 10- to 20-km-thick oceanic crust because of the absence of continental crust basement. The volcanic arc that comprises large islands like the Philippines and Java is located on the continental crust. Its rock is predominantly andesite, the activity time is late Mesozoic, and the crustal thickness is 25-40 km. This also shows that the island arc evolution has an internal relationship with the growth of continent crust. The volcanic eruption is mostly a centered eruption with an intensive volcanic explosion.

2.1.1.3 Cordilleran Continental Margin Volcanic Arc Zone

The Cordilleran continental margin volcanic arc mainly refers to the Cordilleran mountain range zone in South America, and it, together with the west Pacific volcanic arc zone, composes the circum-Pacific volcanic circle, which has a total length of more than 40,000 km. There are 30 active volcanoes in the south section of the Andes,

the Cordillera mountain system, and 16 active volcanoes in the north section. The Llullaillaco volcano in the middle with an altitude of 6723 m is the highest active volcano in the world. The characteristics of volcanic rock in the Cordilleran continental margin volcanic arc are similar to those in the west Pacific intraoceanic arc; that is, dominated by intermediate andesite, the volcanic rock has apparent horizontal zonality in the direction from trench to land. The main eruption mode is centered eruption.

2.1.1.4 North Atlantic Midoceanic Ridge Zone

The midoceanic ridge is also called an oceanic rift and it has a "W" shape. From the Arctic province, the ridge goes from Iceland to the south Atlantic, dividing the Atlantic crust into equal parts and paralleling the shoreline of both coasts. Then it bypasses the south end of Africa southward, turns to the northeast, and connects with the mid-Indian ridge. The mid-Indian ridge extends northward to the north end of the African continent and connects with the east African rift valley, and bypasses Australia southward, then goes eastward, and connects with the south end of the mid-Pacific ridge. The mid-Pacific ridge leans to the east of the Pacific, extends northward, and enters the Arctic region. The whole "W" shape is essentially a global oceanic rift with a total length of more than 80,000 km. The midoceanic ridge is primarily uplifted oceanic ridge, which is 2-3 km higher than the oceanic plain at both sides. There are many 20- to 30-km wide and 1- to 2-km deep grabens in the middle of the oceanic ridge; therefore, it is also called an oceanic rift. The volcanoes in the ocean mainly occur at the oceanic rift zone, that is, the midoceanic ridge volcanic zone. Based on the study of the chronology of submarine rock, the oceanic rift was formed earlier, but the tensional fracture enlargement and violent activity occurred from the

Mesozoic to the Cenozoic, and has been more active since the Quaternary.

The North Atlantic midoceanic ridge zone is the region where volcanism is primarily concentrated and is mainly composed of basalt. Substantive volcanoes occur in this region, some being very active at present, even continuously erupting for up to 4 years and forming new volcanic islands. Only scattered volcanoes occur beyond the midoceanic ridge, and they occur as volcanic islands. For instance, the Hawaiian islands formed from the eruption of oceanic volcanoes in the Pacific and are composed of basalt, basically the same lithology as that of the volcanic rock in the oceanic rift zone.

2.1.2   Volcanic Rock Distribution in Geologic History

The peak time of global volcanic activity was the Mesozoic and Cenozoic. The Cenozoic volcanic rock zone occurs in the latitudinal structural belt crossing the Eurasian continent, that is, starting from the Pyrenean islands in the west, through the Alps, and ending up in the Himalayas in the east, with a total length of more than 100,000 km. The structural belt is a latitudinally trending fold-uplift zone resulting from south-north-trending compression and was mainly formed in the Quaternary. Volcanoes occur sporadically in this zone. In the west section, affected by the south-north-trending compressional forces at the time of forming the latitudinal tectonic uplift zone, the longitudinal tensional fracture and rift zone were also formed, such as the east African rift system extending from the south to the north at the south side of this zone. Due to rift faulting, inland seas such as the Mediterranean, Red Sea, and Gulf of Aden were formed in the transition area of the longitudinal and latitudinal structural belts. The volcanic activity in this region also has its own features, and numerous well-known volcanoes such as the Italian Vesuvius volcano, Etna volcano, Vulcano volcano, and Stromboli volcano were formed. Some islands in the Aegean sea are also volcanic islands, featuring intensive activity, violent explosion, and typical characteristics, and the volcanic eruption type is classified by the characteristics of these volcanoes. The lithology belongs to calc-alkalic series, mainly composed of andesite and basalt. The volcanic activity is weak in the middle section of the latitudinal structural belt. However, in the east section, the volcanic activity is intensified again at the northern foot of the Himalayas, and some volcanic groups like Makhacuo, Kardax, Yurba Tso, Wulanla lake, Hohxil, and Tengchong occur at the margin of the uplift and massif, which includes a total of more than 100 volcanoes. Among these, the Kardax and Hohxil volcanoes in China erupted in the 1950s and 1970s, respectively, and their lithology is andesite and alkali basalt.

Volcanic distribution in China is related to the margins of two plates in genesis: first, influenced by the westward subduction of the Pacific plate, substantive volcanoes were formed in eastern China; second, affected by the collision of the Indian plate, volcanoes were formed in the Qinghai-Tibet plateau and its peripheral areas. Therefore, Cenozoic volcanic activity was frequent in China and Cenozoic volcanic rocks are widely distributed. The middle stage of the Miocene and the Quaternary was two peak times of Cenozoic volcanic activity in China. At that time, violent volcanic activity existed in both eastern China and the Qinghai-Tibet area. In the Holocene, the volcanic activity weakened, but it still goes on in many regions.

2.2   TECTONIC SETTING FOR THE FORMATION OF VOLCANIC ROCK

Different types of volcanic rock have remarkable geographic distribution rules. In the 1960s, with the establishment of plate structures, the relationship between the variation and distribution rules of volcanic rock lithologic series associations and the global tectonic setting was studied more closely by geoscientists. For the moment, it is generally accepted that the magma series is composed of tholeiitic, calc-alkali, and alkalic components, and these three kinds of magma series and the paragenesis of volcanic rock have completely different distribution characteristics. Ringwood (1969) put forward an opinion of classifying the magma as per the plate structure setting, believing that the generation of magma was related to the plate structure. Great progress has been made on the study of the relationship between the formation and distribution of volcanic rocks and plate structure since the 1980s, and it is believed through systemic summarization that the volcanic rock was mainly formed in tectonic settings such as intraplate, plate margin, and island arc (Figure 2.2).

2.2.1   Rift Zone

A rift zone is one of the most important tectonically active zones on the earth's surface. It is a depressed land form developed roughly parallel to the faulting and belongs to a large extensional structure with deep influence and long extension (Ma, 1982). Because of the upwelling of hot mantle or magma, the lithosphere thins out in the course of extension. Arching action occurs first, with large domal structures or triplex structures and numerous faulted blocks being formed, and is simultaneously accompanied by the eruption of a large area of continental facies alkalic and subalkalic basalt. When the extension action is strong enough to affect the domal uplift region, the lower part of the lithosphere thins because of the diapir of asthenosphere matter and the downward caving of lithosphere, and the upper crust surface subsides, forming a continental rift (also called an

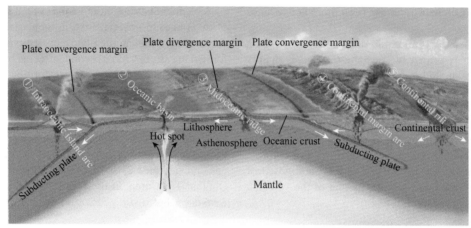

FIGURE 2.2   Sketch map of plate structure setting.

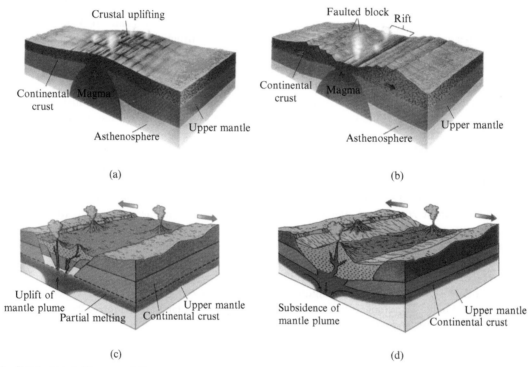

FIGURE 2.3   Sketch map of rift zone evolution.

intracontinental rift). The basement and both sides of the continental rift basin are all continental crust; furthermore, the crustal thickness of the basement is thinner than that of the landmass at both sides (Figure 2.3). The rifts such as east Africa, Rhine, and Baikal are at this stage at present.

Another type of rift is an intercontinental rift resulting from further evolution of a continental rift, and they have a close genetic link. On the basis of the continental rift, the upper part of the lithosphere extends and thins, and the asthenosphere matter penetrates and spills along the rift axis, and thus forms new oceanic crust. The continental lithosphere splits completely and is separated toward both sides, and the seawater submerges the whole valley floor

and forms either a narrow oceanic basin with a spreading ridge or an intercontinental rift basin. Both sides of the intercontinental rift are continental crust; however, the basin basement is not continental crust but transitional crust and oceanic crust, such as the Red Sea, Gulf of Aden, Gulf of California, and others.

There are a great variety of volcanic rocks in the rift zone, and the continental rift is characterized by an alkalic rock association or bimodal volcanic complex. Generally, strongly alkaline and alkali series predominate at the initial stage of rifting, while the alkalinity and tholeiitic series are the products of rift stage. During the expansion of rift, the early extrusive rock occurs on the platform at the edge of the rift, while the late extrusive rock occurs in the rift axis zone and forms a symmetrical zonal distribution. For instance, in the south section of the Kenya rift, the central type volcano composed of strongly basic nephelinite-calcium-rich phonolite-carbonatite occurs on the platform at the edge of the rift, which is formed mainly at the pre-rifting and arching stages. The moderately alkaline olivine basalt-mugearite-trachite-pantellerite-calcium-poor phonolite series occurs in the extensive effusive lava at the bottom of the rift and in the low shield volcano, which is the major magma type at the rift stage, and a bimodal association occurs (Wilson, 1989).

Notable tholeiitic magmation or even low-potassic tholeiitic series magmation similar to the midoceanic ridge basalt (MORB) occurs in the mean axis of the intercontinental rift, which shows that with the splitting of continental plates, the magmatic alkalinity steps down until the oceanic rift is formed and becomes the tectonic setting of midoceanic ridge or quasi-midoceanic ridge, and similar magmation and igneous rock associations occur.

2.2.2 Continental Margin Arc and Intraoceanic Island Arc

Based on the developmental position of the island arc, the tectonic setting of magmation, and the magmatogenic difference, the island arc is subdivided into intraoceanic island arc and continental margin arc.

The magmation of a subduction zone mainly occurs within the range of the magmatic arc, about 150-300 km away from the trench axis, and spreads in an arc shape parallel to the trench. The major rock series includes island arc tholeiite series, calc-alkaline series, and island arc alkali series (or shoshonite series). The island arc volcanic rock characterized by high K_2O and Al_2O_3 and low TiO_2 is different from the volcanic rock formed in other environments. The island arc tholeiitic series volcanic rock is mainly composed of tholeiite, andesite, and minor dacite; the calc-alkaline series volcanic rock mainly includes andesite, dacite, high-alumina basalt, and ryolite. Comparing with the island arc tholeiitic series, ferric enrichment is seldom seen, SiO_2 content is higher (59%), large ionic lithophile (LIL) elements is enriched, light rare-earth elements are enriched slightly, $^{87}Sr/^{86}Sr$ is slightly high, and with the increase of SiO_2 content, the K_2O content increases relatively rapidly. The island arc alkali series volcanic rock is represented by the shoshonite association, which is a representative rock association of mature island arcs.

The island arc volcanic rock is characterized by explosive facies, and the pyroclastic volume can account for 80% of the volcanics; however, this percentage is much lower in the midoceanic ridge and the oceanic island. In addition, the sandstone and mudstone composed of volcanic debris, intrusive rock, and metamorphic debris are often interbedded with volcanic rocks, and this interbedded series is also one of the important marks to identify the island arc volcanic rock series. The magmatic rock in subduction zones often has apparent horizontal zonality in the direction from the trench to the continent. With increasing distance away from the trench axis, the tholeiitic series, calc-alkaline series, and alkali series occur in turn. With increasing distance away from the trench axis and increasing depth of the subduction zone, the volcanic rock component varies regularly. This phenomenon is called the component polarity, which can be used to indicate the inclining direction of the subduction zone. With the occurrence, development, and evolution of subduction, the island arc will also experience an immature-semimature-mature evolution process.

2.2.3 Midoceanic Ridge Structural Belt

The midoceanic ridge is the most important divergent plate margin. It is the place where peak volumes of igneous rock in the oceanic province are produced, and is also the place where oceanic crust is created. The oceanic bottom is formed continuously at the midoceanic ridge and then migrates toward both sides.

The midoceanic ridge is characterized by the production of tholeiite and the absence of andesite. This tholeiite is usually called ridge tholeiite. Magma is produced at shallow depths along the midoceanic ridge and earthquake activity zone, forming alkali-poor tholeiitic magma (Jokat et al., 1992). The absence of andesite in the midoceanic ridge can be attributed to the absence of subduction. Therefore, moisture is not drawn into the upper mantle, or because the earth's crust is thin, some expanding fissures along the midoceanic ridge allow the moisture to dissipate. As a result, the hydraulic pressure is too low to generate andesitic magma.

The ridge tholeiite has the following characteristics: (1) the phenocrysts are olivine or (and) plagioclase and the matrix minerals are olivine, plagioclase, klinaugite, and iron minerals, which often contain the vitreous and crystallographic mineral crystallite, (2) low potassium (K_2O less than 0.4%), high titanium (TiO_2 is 0.7-2.3%), low P_2O_5 content (<0.25%), and (FeO + Fe_2O_3)/MgO are 0.7-2.2, and (3) distinguished from the continental and island arc tholeiite by relatively high Al_2O_3 and Cr contents, high LIL elements such as Rb, Cs, Sr, Ba, Zr, U, Th, and a deficit or flat of light rare-earth elements distribution.

The major rocks in the midoceanic ridge are gabbro and peridotite. There are two kinds of gabbros: one is the cumulate formed by crystallization differentiation at an early stage and the other is the gabbro formed by residual melting after violent differentiation. There are also two kinds of peridotites: one is that which remains after partial melting of pyrolite and the other is the product of crystallization differentiation. The ridge tholeiite is formed by 20-30% partial melting of some depleted mantle peridotites within 30 km depth, and the gabbro and peridotite are related to the fractional crystallization of the original melt.

2.2.4 Craton

Volcanic rock is not extensively developed in the continental craton region. The volcanic rock discovered at present mostly occurs in the form of small intrusive complexes such as dikes, sills, volcanic necks, pipes, or (on few occasions) small volcanic regions, and the compositional variation is more complicated than the oceanic crust region. It is generally agreed that the continental craton igneous rock is related to a certain intraplate tensional tectonic setting, and in the region where there are no apparent tectonic traces, the magmation is usually related to a hot spot or an ascending hot plume of the mantle. The igneous rock in the continental plate generally includes an association of four rock types: kimberlite, alkalic rock (high K rock series), plateau flood basalt, and carbonatite.

Kimberlite mostly concentrates in Siberia and South Africa. Because of rapid emplacement, kimberlite usually does not undergo a crystallization differentiation and cannot form a rock series. Kimberlite and carbonatite mostly concentrate in the vicinity of rifts or in the rift system, but some are seen at broken continental margins such as the east African rift valley, Baikal rift, Aldanian dome, Rhine graben, and Brazilian coast. The high alkalic rock series in the western part of the east African rift valley is the region where the high K lava most commonly occurs. The rock includes ugandite and katungite, and carbonatite lava and tuff are also seen locally. Large centered eruptions started from the early Miocene in the Kenya domal zone where the east African rift valley passes, and the product erupted includes nephelinite, dark nephelinite lava, volcaniclastic rock and minor basanite, basalt, phonolite, and trachite. The plateau basalt or torrential basalt on the continent belongs to the tholeiite rock series, and some contain a great deal of ryolite interbeds.

2.2.5 Oceanic Basin

The magmatic exhalation within the range of oceanic basin is exhibited mainly in the volcanic island and oceanic volcano system and has two fundamental occurrences: volcanic island chain and isolated volcanic island. The Hawaiian islands and the Marshall-Gilbert islands are typical volcanic island chains. The chain of volcanoes results from the mantle plume, and the oceanic island and volcanic island chain are formed after the magma generated in or above the mantle plume has erupted.

The volcanic rock of oceanic islands is mainly the product of tholeiitic magma and consists of a great deal of tholeiite and minor alkalic rocks. Tholeiite accounts for 85% of the volcanic rock of the Hawaiian islands, and the other rocks are intermediate rock, acidite, alkali basalt, and alkalic basite. The volcanic rock in Iceland also belongs to the tholeiite rock series, but is differentiated more than the rocks in the Hawaiian islands, and includes ryolite, volcaniclastic rock, and alkali basalt. The content of K_2O, TiO_2, and P_2O_5 is high in the oceanic island basalts; moreover, the content of incompatible elements with macroion radii, including the light rare-earth elements, is higher than that in the midoceanic ridge basalts. Generally, the chemical composition of the initial basaltic magma formed by the mantle plume depends on three factors: the physical makeup and mineral association in the mantle source zone, the level and mechanism of partial melting of the source materials, and the depth of magmatic differentiation.

2.2.6 Tectonic Setting of Passive Continental Margin and Collision Zone

The passive continental margin extends parallel to the continental contour and includes the quite extensive transitional belt between the oceans and the continents composed of continental shelf, continental slope, and continental rise. The passive continental margin is the product of continental rifting and subsequent sea floor expansion and continental drift. It originates at the initial phase of intercontinental rifting and is in an extensional state all times. The periphery of the Pacific is surrounded by a chain of volcanoes, and its rocks are the calc-alkalic series characterized by a basalt-andesite-ryolite association. On the contrary, there is no volcanic activity in the periphery of the Atlantic. The plate structure theory posits that the continental margin with volcanic activity is the active margin related to plate subduction, and the margin without volcanic activity is the passive margin with no plate subduction activity. Therefore, magmation is not developed at the passive continental margin. Apart from retaining the igneous rock association formed at the phases of early continental rift and intercontinental rift, some magmation related to the mantle hot spot or plume can also occur and forms the igneous rock association similar to that in the plate (continental plate and oceanic plate). This is dominated by the mafic and ultramafic intrusive activity and dike activity, accompanied by some central or fissured volcanic eruption action.

After the depositional basin underwent intracontinental rift-intercontinental rift-oceanic extension and oceanic rift-marginal rift-oceanic basin-subduction, the ocean was subducted and the arc trench system was formed, creating a suture zone and remnant ocean basin and forming a complete Wilson cycle. The further development of subduction results in the collision between island arcs and continents or between continents, forming a suture zone or collision orogen. The type and association of volcanic rock in the collision belt is complicated. The magmatism related to the continent-continent collision can be divided into four phases, and each phase includes a characteristic source zone.

1. Precollision phase: The magma sources from the volcanic arc before collision, which still belongs to the arc volcano activity type.
2. Syncollision phase: In this phase, the crust thickens and results in the emplacement of granite. In the period of obduction of the earth's crust, the volatile constituent discharged from the sedimentary wedge of fluid infiltrates the floating high-temperature thrust sheet, resulting in anatexis, and enrichment of Rb, F, and B, while elements such as rare earths, Zr, and Hf remain in the melted remnant and experience a loss. The developing level of syncollision granite is controlled by the thickening level of the earth's crust in the collision period.
3. From late stage of collision to post-collision phase: A calc-alkalic rock suite similar to the volcanic arc magmatism is formed. Although it results from the rich fluid underthrusted on the mantle wedge of the oceanic lithosphere, it may be contaminated and changed by the molten mass of the lower crust.
4. Post-collision phase: Characterized by alkalic rock, the alkali magmatism possesses intraplate geochemical properties and possibly develops in the incised mantle region that was not affected by subduction. Actually, magmatism can occur at any stage of collision events in the post-collision phase.

2.2.7 Marginal Sea Basin

A marginal sea basin, also called a back-arc basin, is a series of small sea basins located between a continent and island arc or among island arcs (Figure 2.4). It is the product of secondary sea floor expanding in the back-arc region and has the slightly thick (<20 km) oceanic crust as its basement.

The petrographic, geochemical, and evolutionary characteristics of tholeiitic rock in the marginal sea basin are all similar to those in the oceanic ridge, and the variation in geochemical characteristics depends on the following factors: partial melting level, homogeneous level of mantle source zone, and high crystallization differentiation level in the magmatic chamber. In addition, the fluid phase derived in the subducting plate may also be another important factor. The geochemical component of trace elements not only has the characteristics of MORB but also displays certain characteristics of island arc basalt. This shows that the component affected by subduction is involved in the formation of the basaltic magma in the expansion center of the marginal sea basin and thus characterizes this type of basalt as an island arc tholeiite. Usually, at the initial stage of extension of the marginal sea basin, the fluid phase derived in the course of subduction has the most important impact on the formation of magma. However, with the extension of the basin, its impact weakens gradually.

The plate motion is directly constrained by the deep action, and the volcanic eruption or intrusion is the performance of upper mantle and deep crust convection on the earth's surface or in the shallow crust. Therefore, different igneous rock associations or series and different rock geochemical characteristics are formed in different tectonic settings. It is important to study the distribution and characteristics of volcanic rock by taking the tectonic setting into consideration.

2.3 FORMATION AND DISTRIBUTION OF VOLCANIC ROCK IN DEPOSITIONAL BASINS OF CHINA

China's mainland is located where the Siberian, circum-Pacific, and Tethyan structural domains have connected

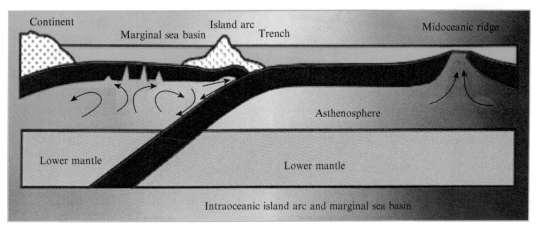

FIGURE 2.4 Sketch map of formation of marginal sea basin.

and interacted long term. The mainland is constrained by the tectonic evolution of the original Tethys, Paleotethys, and neo-Tethys and is superimposed and reconstructed by the Pacific plate subduction plus frequent magmation, all of which have resulted in widespread volcanic rocks. The type of volcanic rock is closely related to different kinds of plate motion. Geologists have conducted extensive petrographic studies to better understand the properties and characteristics of tectonic movement in China's geologic history. In recent years, with the discovery of volcanic reservoirs in basins such as Junggar and Bohai Bay, additional focus has been placed on the volcanic rock in the basin, and progress is continually being made.

Volcanic rocks are developed to different degrees in the hydrocarbon-bearing basins of China (Figure 2.5). China has mainly experienced three magmation peaks: the Paleozoic Hercynian, Mesozoic Yanshan, and Cenozoic Himalayan. Influenced by the magmation and reconstruction or superimposition of late-stage tectonic movement, it mainly formed three types of volcanic rock basin group: the Paleozoic

volcanic rock basin group, Mesozoic volcanic rock basin group, and Cenozoic volcanic rock basin group. The Permo-Carboniferous volcanic rock predominates in the west, and the Jura-Cretaceous and Paleogene volcanic rock predominates in the east. And a total volcanic rock distribution area is $39 \times 10^4 \, \text{km}^2$. The volcanic rocks in all hydrocarbon-bearing basins of China are developed under different regional tectonic settings, which are interdependent but also have their own characteristics (Figure 2.6). For example, the Paleozoic volcanic rock in the west is mainly intermediate and basic, while the Mesozoic and Cenozoic volcanic rock in the east is mainly intermediate and acidic, and the volcanic rocks in the hydrocarbon-bearing basins are all formed in the intracontinental rift environment related to extension action.

2.3.1 Paleozoic Volcanic Rock Basin Group

The Hercynian was affected by the collision of the Siberian and Indian plates and is a relatively strong magmation stage

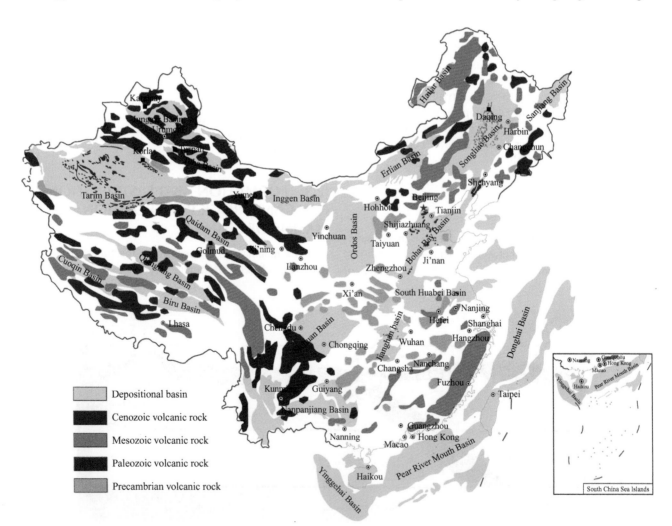

FIGURE 2.5 Distribution of volcanic rock in China's hydrocarbon-bearing basins.

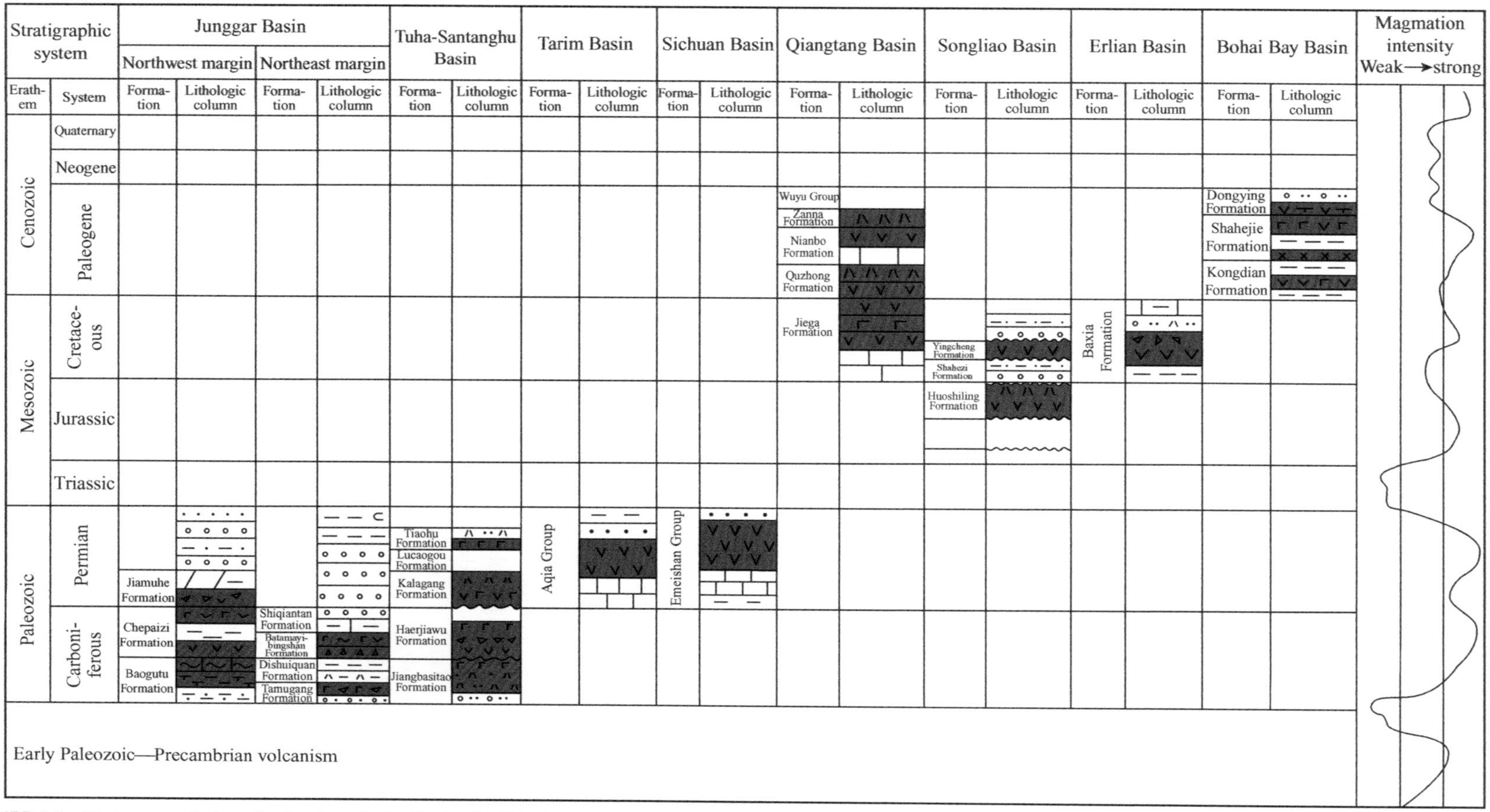

FIGURE 2.6 Horizon correlation of volcanic rock distribution in China's hydrocarbon-bearing basins.

in the geologic history of China. Volcanic rock primarily developed during the Permo-Carboniferous stage, and these rocks are mainly distributed in the basins of western China. Taking the Tianshan-Xingmeng trough and the Huabei-Tarim as the boundaries, the basins of western China are divided into three groups: northern Xinjiang, Tarim, and Chuanzang, with a total distribution area of $19 \times 10^4 \, \text{km}^2$. There are various kinds of volcanic lithologies in the western basins, where basic volcanic rock, intermediate and basic volcanic rock, and intermediate and acidic volcanic rock are developed. These form an approximately east-west trend mainly along the faulting, and this trend is related to the intraplate rift that occurred after the collision of the Siberian-Tarim plates (Figure 2.7).

2.3.1.1 Northern Xinjiang Basin Group

The Paleozoic northern Xinjiang basin group, comprising the Junggar, Santanghu, and Tuha basins, occurs in the Tianshan-Xingmeng structural belt as a part of the Siberian south margin and north margin of Huabei-Tarim continent (Figure 2.8). These basins, with similar tectonic settings, experienced some tectonic variations from the collision between the Siberian and Tarim plates to the dying out of the Pal Mongolian ocean and occur in the extension rift environment after the dying out of the Kelameili ocean (Figures 2.9 and 2.10). The volcanic rock in the northern Xinjiang Basin group spreads mainly along the faulting. Its lithology is complicated, with intermediate and basic volcanic rock in the majority, but also includes intermediate and acidic volcanic rock, volcaniclastic rock, and breccia lava. The distribution area is $9 \times 10^4 \, \text{km}^2$.

The volcanic rock in the Junggar Basin is found mainly along the Kewu, San'gequan, and Kelameili faults in the three foreland basins at the northwest, northeast, and southeast margins, respectively (Figure 2.11). The volcanic rock in the northeast mainly occurs in the Carboniferous Kalagang Formation and Batamayineishan Formation, and the volcanic rock in the northwest mainly occurs in the Carboniferous Baogutu Formation and Permian Jiamuhe Formation.

The volcanic rock is mainly composed of intermediate, basic, and acidic lava and volcaniclastic rock, among which the lava category includes ryolite, andesite, basaltic andesite, and basalt, and the volcaniclastic rock category includes agglomerate, breccia, welded tuff, breccia tuff, sedimentary tuff, and tuffaceous breccia. The lithofacies is a set of epimetamorphic volcaniclastic rock formations and local magmatic intrusive rock formations deposited in the intermediate stage of the Hercynian, and clastic rock and volcanic formations deposited in the transitional facies and continental facies.

Oceanic eruptive types dominated before the Carboniferous and then transited to continental eruption widely in the late Carboniferous-early Permian. The lithology of the upper Carboniferous is volcanic rock intercalated with sedimentary rock, and the volcanic activity was violent—vesicles in deep volcanic rocks are notably absent in celadon (dust color as a whole), while vesicles can be seen in shallow volcanic rock, which becomes light to brown in color. These volcanics have the characteristic of transitioning from oceanic eruption to continental eruption. The Carboniferous volcanic horizon in the periphery of the basin also presents the same characteristics, which shows that the Carboniferous volcanic rock was formed in an eruptive environment converting from continental margin to intracontinent. The volcanic rock is of continental facies predominantly in the west, subaquatic eruption facies in the southeast, and both continental facies and subaquatic eruption facies in the east (north). The Permian volcanic rock is mainly seen in the northwest and northeast of the basin, more rarely seen than the Carboniferous volcanic rock during drilling, and is all aquatic continental volcanic rock. Among which, tuff is mainly continental volcaniclastic rock, while in the southeast of the Junggar Basin, only a small amount of continental volcanic rocks were discovered.

The volcanic rock in the Junggar Basin is mainly formed in a rift environment. The tectonic setting discrimination diagram of trace elements of basic volcanic rock in the basin (Figure 2.12) shows that the volcanic rock has the characteristics of intracontinental rift. In the standardized trace and rare-earth element ratio spider diagram of primitive mantle (Figure 2.13), it appears as a slight enrichment of LIL elements, the overall characteristics being similar to the trace element distribution of P-MORB, and the Nb and Ta negative anomaly of volcanic rock is not obvious in the area. Therefore, it should not be related to the fluid metasomatism of the subduction zone. Geologic and geochemical proofs have shown that the volcanic rock in the Junggar Basin is mainly formed in an intracontinental rift environment related to the collision, resulting in the dying out of the Junggar oceanic basin.

The Permo-Carboniferous is the important period for the formation of a great volume of volcanic rock in the Santanghu-Tuha Basin. With the characteristics of extensive distribution, and thick and multiple lithologies, the volcanic rock is mainly bedded basalt, then basaltic andesite, basaltic trachite, trachite, ryolite, and rarely subvolcanic breccia and volcanic tuff.

Petrologic and geochemical studies on the volcanic rock in the basin show that its formation environment is still related to the intracontinental rift. The basalt in the Malang sag and Tiaohu sag in the middle of Santanghu Basin is mostly celadon-dark green and taupe-brown, tholeiitic or aplitic texture, and with a few vesicular amygdaloidal structures. The minerals mainly consist of plagioclase, clinoaugite, and few olivine, and the plagioclase content is high. The andesite is mostly gray, celadon, and dark red-dark purple,

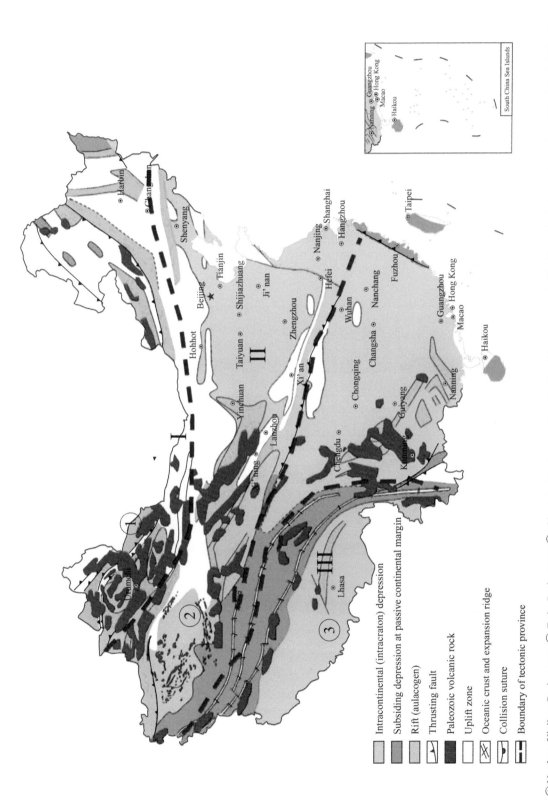

FIGURE 2.7 Distribution of Paleozoic volcanic rock basin groups in China.

① Northern Xinjiang Basin group; ② Tarim Basin group; ③ Chuanzang Basin group. I —Tianshan-Xingmeng trough; II —Tarim-Huabei tectonic province; III —Qinghai-Tibet Thetys

Intracontinental (intracraton) depression

Subsiding depression at passive continental margin

Rift (aulacogen)

Thrusting fault

Paleozoic volcanic rock

Uplift zone

Oceanic crust and expansion ridge

Collision suture

Boundary of tectonic province

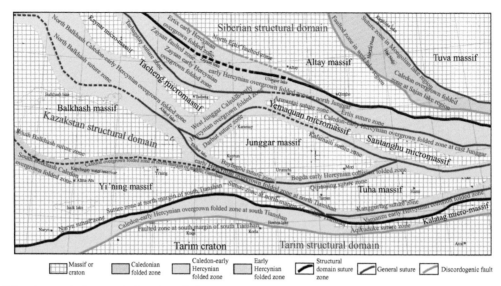

FIGURE 2.8 Paleozoic tectonic framework of northern Xinjiang and adjacent area.

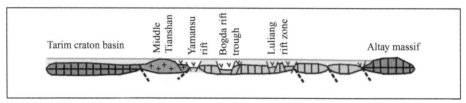

(2) Extension and collapse stage after collision phase of late Carboniferous

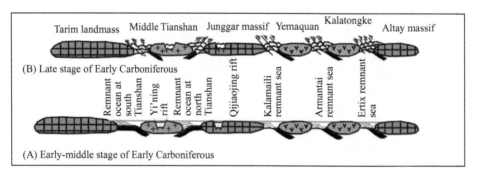

(1) Phase of Early Carboniferous collision intermission extension-remnant ocean closure and continent-continent collision

FIGURE 2.9 Evolution of the Carboniferous plate in northern Xinjiang.

with porphyritic texture, vesicular amygdaloidal structure, and fractures developed. The volcanic rock sample points all fall into the intraplate basalt region in the geochemical trace element discrimination diagram (Figure 2.14), and the trace element ratio spider diagram also exhibits the characteristics of intraplate volcanic rock, which show that the Permian volcanic rock in the Santanghu-Tuha Basin is formed in a post-orogenic extension environment, belonging to a post-colligion intracontinent rift environment, and not related to an island arc.

2.3.1.2 Tarim Basin Group

There are four volcanic rock zones in the Tarim Basin: (1) Kuluktag volcanic rock zone, (2) Tabei intermediate and acidic volcanic rock zone, (3) Tazhong basic volcanic rock zone, and (4) Ta'nan intermediate and acidic intrusive rock zone. These zones include the two Sinian-Cambrian and Permian horizons, with the Permian being the main period for the formation of volcanic rock in the basin. In the early- to mid-Permian, the Paleotethys ocean started to under-thrust and died out, and the magmatism in the Tarim Basin

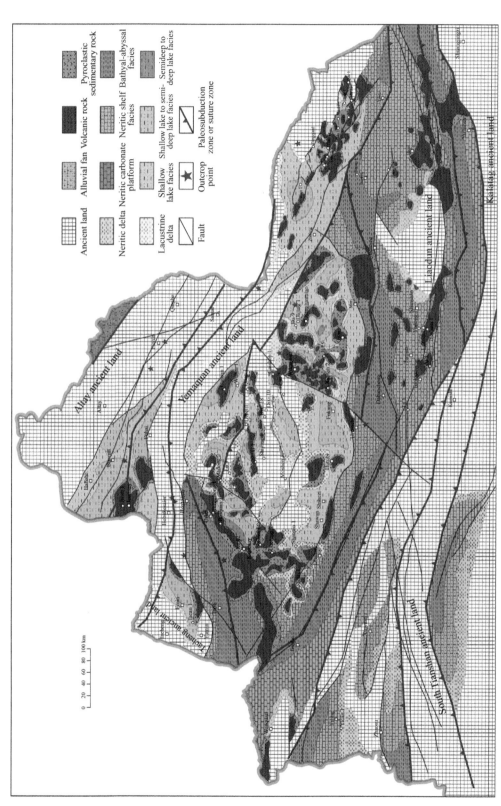

FIGURE 2.10 Late Carboniferous lithofacies paleogeography in northern Xinjiang.

Legend:

- Ancient land
- Neritic delta
- Lacustrine delta
- Alluvial fan
- Neritic carbonate platform
- Shallow lake facies
- Volcanic rock
- Pyroclastic sedimentary rock
- Neritic shelf facies
- Shallow lake to semi-deep lake facies
- Bathyal-abyssal facies
- Semideep to deep lake facies
- Fault
- Paleosubduction zone or suture zone
- Outcrop point

0 20 40 60 80 100 km

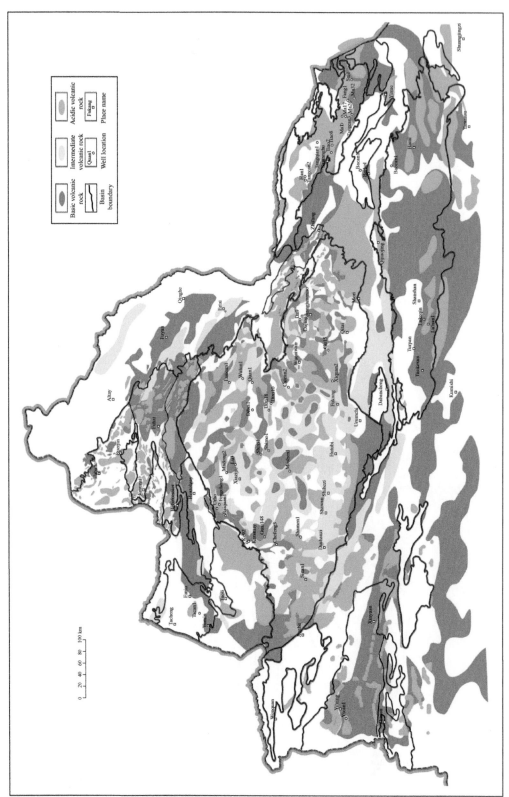

FIGURE 2.11 Prediction of volcanic rock distribution in Junggar Basin.

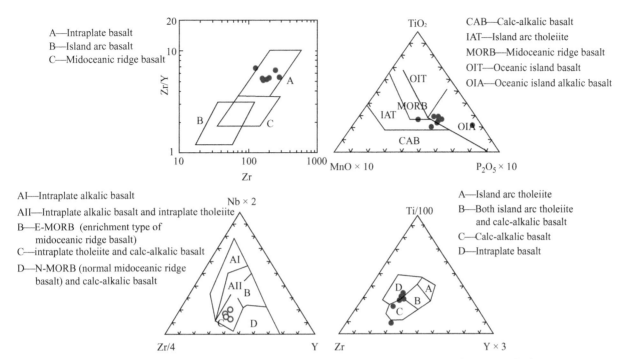

FIGURE 2.12 Tectonic setting discrimination diagram of trace elements in volcanic rock in the Ludong region, Junggar Basin.

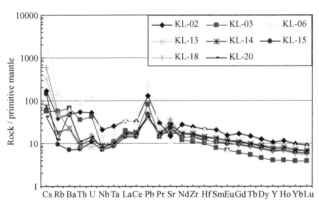

FIGURE 2.13 Primitive mantle standardized trace and rare-earth elements partitioning diagram of volcanic rock in Junggar Basin.

is related to this event. During this time, it formed extensively developed Permian volcanic rock with a distribution area of $13 \times 10^4 \, \text{km}^2$. The Permian volcanic rock in the basin came from a magmatic chamber at the early stage of evolution and was formed in a tectonic setting in the continental plate (Chen et al., 1997).

The Permian volcanic rock is developed in the northern part of the basin, with different lithologies and thicknesses. The lithology mainly consists of basalt, andesite, tuff, volcaniclastic rock, diabase, and intermediate and acidic andesite, dacite, ryolite, granodiorite, granite, and granite porphyry. The basalt and tuff are the most extensively distributed, not only in the west section of the Manjiaer depression and the west section of the Tabei uplift but also in the Tazhong uplift, the Bachu uplift, and the southwest depression. They are present over almost the entire Tazhong area. The Permian volcanic rock in the Tahe oilfield is an alkalic series volcanic rock, and it has two major volcanic rock eruption-effusion cycles. A relatively complete volcanic eruption-effusion cycle appears as a basalt-dacite sequence from bottom to top. Also, it appears as a basalt-volcanic breccia-dacite sequence or volcanic breccia-lava sequence in a few wells. Generally, the vertical distribution pattern is basalt in the lower part and dacite or dacite with rhyolitic structure in the upper part. This conforms to the general rule of volcanic activity and its product evolution, i.e., that volcanic rock is basic at the early stage and evolves toward acidity at the late stage. The volcanic activity has an intermittent eruption characteristic, with low eruption frequency and a short eruption interval. It is primarily a serene effusive eruption, or accompanied by intensive explosive eruption, forming two volcanic eruption patterns—basalt and dacite or simple dacite.

2.3.1.3 Chuanzang Basin Group

The Permian Emeishan basalt in Sichuan Basin, formed in an important magmatic event in the Hercynian, mainly occurs in the three southwest provinces of Sichuan, Tibet, and Yunnan. The Sichuan province is the unique volcanic rock province recognized by international academicians, being a kind of continental effusive basalt related to the mantle plume (Zhang et al., 2005; Xiao (2004); Xu et al., 2001). The volcanic rock in the Sichuan Basin covers an area of $7 \times 10^4 \, \text{km}^2$.

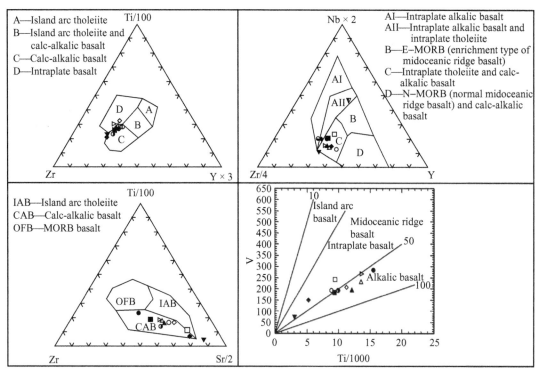

FIGURE 2.14   Trace element discrimination diagram of volcanic rock in Santanghu-Tuha Basin.

We can see from the aforementioned distribution and tectonic setting characteristics of volcanic rock in the western Paleozoic basin group that all magmatic activities in western China basins were affected by the collision between the Eurasian and Indian plates, and closely related to the subduction and dying out of the Paleotethys ocean. Regionally, a great deal of intraplate magmatism developed in the early to late Permian in the northern margin of the eastern Tethys, including Permo-Carboniferous intraplate magmatism in the Junggar Basin and early Permian intraplate volcanism in the Santanghu-Tuha Basin. Permian intraplate volcanic rock developed in both the Tarim Basin and the Emeishan basalt in the Sichuan Basin in the west margin of the Yangtze plate. Therefore, the Paleozoic volcanic rock basin group in western China is closely related to the evolution of the Paleotethys ocean and constitutes the Neopaleozoic magmatism zone at the northern margin of the Paleotethys ocean.

2.3.2   Mesozoic and Cenozoic Volcanic Rock Basin Group

China's continent has been affected by the westward subduction of the Pacific plate and the opening and closing of the Neo-Tethys ocean since the Mesozoic and Cenozoic eras. Magmation is developed, and a northeast-trending Mesozoic active continental margin and a Cenozoic tectonic setting in which intraplate rift dominates were formed.

As a result, the volcanic rock of this period mainly occurs in the eastern China basins, including the Dongbei-Huabei Basin group, Bohai Bay Basin group, and Qingzang Basin group. The volcanic rock was mainly formed in the Jura-Cretaceous and Paleogene (Figure 2.15), with lithologies dominated by intermediate and acidic volcanic rock and basic volcanic rock; other lithologies are also found.

2.3.2.1   Mesozoic Dongbei-Huabei Basin Group

The Dongbei-Huabei Basin group mainly refers to the Songliao Basin in northeast China, where the volcanic rock was formed in the Mesozoic Yanshanian. The strongest magmation in the geologic history of China occurred at Yanshanian. This magmation was stronger in the east than in the west and stronger in the early stage than in the late stage. Yanshanian volcanic rocks occur extensively throughout the area near the Pacific and secondly in the southwest-northwest regions. The volcanic activity of the volcanic rock zone in eastern China intensifies from west to east, and the ejecta components are predominantly intermediate and basic, as well as intermediate and acidic in the west and predominantly acidic in the east. The magma started as intermediate and basic and ended as acidic; as a result, basic volcanic rock rarely occurs in the area, and andesitic and trachyandesitic volcanic rock occurred at the early stage, and trachytic or even phonolitic volcanic rock occurred at the late stage.

FIGURE 2.15 Mesozoic and Cenozoic volcanic basin group in China.

① Mesozoic Dongbei-Huabei Basin group; ② Cenozoic Bohai Bay Basin group; ③ Qingzang Basin group

Passive margin subsiding depression

Foreland

Rift (aulacogen)

Intracontinental (intracraton) depression

Plate movement direction

Mesozoic volcanic rock

Collision suture

Uplift zone

Fault

South China Sea Islands

Songliao Basin is a late Mezozoic rift basin developed on the pre-Mesozoic basement (Gao et al., 2007), and has a "binary stratigraphic formation"; that is, both the volcanic rock and the sedimentary rock were formed in the same basin (Guo et al., 1997). Volcanic rock occurs extensively throughout the basin. The distribution area reaches about 5×10^4 km^2 and can be subdivided into four volcanic eruption cycles at different scales: (1) The first volcanic cycle occurred in the Huoshiling stage, in which the volcanic eruption activity is strong, developing along the vicinity of a discordogenic fault and presenting a fissured linear distribution. The lithology is mainly basic, intermediate and basic, intermediate, intermediate and acidic lava, and partly volcaniclastic rock, among which the intermediate and basic volcanic rock predominates, and the eruptive-effusion facies and volcanic-sedimentary facies are dominant. (2) The second volcanic cycle occurred in the Shahezi stage, in which the volcanic rock is present widely in the whole basin. The lithology is mainly ryolite, dacite, and tuff, with relatively thin beds. (3) The third volcanic cycle occurred in the Yingcheng stage, which is a late Jurassic-early Cretaceous volcanic eruption event that occurred in the Songliao Basin, with the characteristics of large scale, long duration, and extensive distribution. This volcanic eruption can be divided into two stages, which compose a complete volcanic eruption cycle. During the first eruption, the lithology is predominantly intermediate and acidic volcanic rock; during the second eruption, the lithology is predominantly intermediate and basic volcanic rock. There is also an acidic volcanic end member, and the volcanic rock occurs mostly in layers. (4) The fourth volcanic cycle occurred in the Quantou stage, had a much smaller scale and influence, and is only found locally in the basin. The two largest volcanic eruption activities occurred in the deep layer of the basin in the Huoshiling and Yingcheng stages.

Different rock series are formed in different tectonic settings, and this fact is used to work backward from rock types to infer tectonic settings. It is generally agreed that the tholeiite series and the alkali basalt series are the products of an oceanic rise or intraplate continental rift, respectively, while the calc-alkalic basalt series and the shoshonite series are the products of subduction zones at plate margins or the intraplate continental collision zone, respectively. The geochemical analysis on volcanic rock in the Songliao Basin showed that the sample points fall into the region of intraplate volcanic rock (Figure 2.16). The sample points of Huoshiling Formation volcanic rock are closer to the island arc region than that of Yingcheng Formation volcanic rock, which shows that the Huoshiling Formation volcanic rock

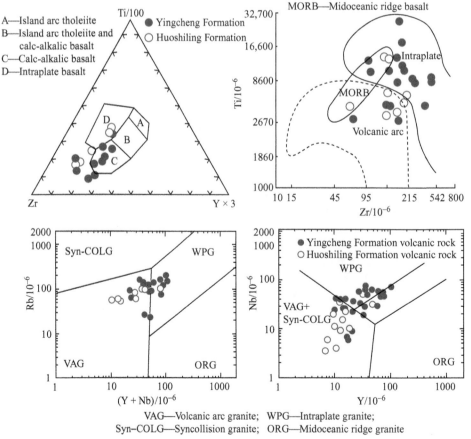

FIGURE 2.16 Tectonic setting discrimination diagram of trace elements of volcanic rocks in Songliao Basin.

was formed at the early stage of rift development and was contaminated by the oceanic crust component of westward-diving Pacific plate, while the Yingcheng Formation volcanic rock was formed at the rift growth phase. Therefore, the volcanic rock in the Songliao Basin was formed in an intraplate extensional environment under the late Jurassic-early Cretaceous compressional background and was related to the underthrust of the Pacific plate toward the passive continental margin of the Eurasian plate and subsequent back-arc extension action. The Songliao Basin is a back-arc rift basin at the active continental margin of the western Pacific.

2.3.2.2 Cenozoic Bohai Bay Basin Group

The Bohai Bay rift basin was affected by the subduction of the western Pacific and the opening and closing of the Neo-Tethys ocean in the Cenozoic, and the volcanic rock was all formed from the magmation during the Himalayan stage. The Cenozoic volcanic rock occurs more in the east than in the west and more in the north than in the south in map view. It is relatively enriched in Kongdian Formation, Member 3 and Member 1 of Shahejie Formation, Dongying Formation, and Guantao Formation; it basically distributes in a north-northeast trend along the faults. The Kongdian Formation volcanic rock was distributed extensively and in a scattered pattern. The magmatic activity was strong at the early stages and weakened to some extent at the late stages. The volcanic rock thickness is more than 1200 m in western Liaohe. The Shahejie Formation volcanic rock is less common in the Liaohe and Huanghua depressions and Dongpu sag and tends to migrate to the basin center on the whole. Volcaniclastic rock is seen in the Huimin sag, while in the Jizhong depression, magmation evidence is absent and the migration toward the basin center is more obvious. The Dongying Formation volcanic rock mainly occurs in the Huanghua depression (and the Bozhong depression), volcaniclastic rock is seen in the Huimin sag, and there is basically no trace of volcanism in the Jizhong depression. As for the Neogene Guantao stage basalt, the fissured eruption predominates, mainly occurring in the Liaohe and Huanghua (and Bozhong) depressions and sparsely in the Huimin sag.

Our understanding of the volcanic rock of the Qingzang Basin group is still limited at present due to intensive latter transformation and a low level of exploration.

The Cenozoic volcanic rock in the Liaohe Basin is primarily trachytic and basaltic volcanic rock, and lesser amounts of diabase, andesite, basaltic andesite, gabbro, trachite, and volcaniclastic rock distributed locally, characterized by the albite-rich alkalic series. It is shown on the trace element diagram that its continental basalt and rare-earth element characteristics also reflect that it is formed in a continental rift environment as the product of continental extensional structure background. The properties and distribution of volcanic rock are controlled by the tectonic setting of the basin formation and its hypomagmatic activity.

To summarize, volcanic rocks are widely distributed in major hydrocarbon-bearing basins of China. Spatially, they occur from the Northern Xinjiang, Tarim, and Chuanzang Basin group in the west to the Dongbei-Huabei and Bohai Bay Basin group in the east. Historically, they occurred in the west basin group in the Paleozoic and in the east basin group in the Mesozoic and Cenozoic, which shows a differential pattern of old volcanic rock in the west and new rock in the east. In the western region, affected by the subduction and dying out of Paleotethys ocean, intraplate magmation developed at the northern margin of the eastern Tethys ocean from early Permian to late Permian, and both the volcanic rock in basins such as Junggar, Santanghu-Tuha, and Tarim and the Emeishan basalt in the Sichuan Basin were all formed by the intraplate magmatism in a rift environment. In the eastern region, under the influence of subduction of the west Pacific and opening and closing of the Neo-Tethys ocean, a back-arc extension rift basin system related to subduction was formed, and the volcanic rocks in basins such as Songliao and Bohai Bay were all formed in this tectonic setting. The difference lies only in that the early volcanic rock was formed at the early stage of rift development, and the late volcanic rock was formed at a relatively mature stage of rift development.

Chapter 3

Characteristics of Volcanic Reservoirs

Considerable attention has been paid to the formation and distribution rules of volcanic reservoirs since volcanic reservoirs were first discovered as a special type of reservoir. Research at various levels has been conducted, including the development characteristics and distribution rules of volcanic rocks; eruption mode and eruption type of volcanic rocks; features of volcanic edifices; different types of volcanic facies and lithologies; characteristics; formation mechanisms; and distribution rules of pore structures in volcanic reservoirs, types and hydrocarbon-bearing properties of volcanic reservoirs, and types and hydrocarbon accumulation conditions of volcanic reservoirs. Current research activities, however, mainly focus on the type and genesis of the reservoir space, diagenetic evolution of the reservoir, and factors controlling reservoir development. Research on the volcanic reservoir formed mechanism is seldom conducted or needs to be strengthened. Purposeful research efforts should be made during hydrocarbon exploration for standardizing the lithology, classifying lithofacies, and naming volcanic rocks, as well as understanding the evolution mechanism of hydrocarbon accumulation, main factors controlling reservoir development, and the distribution of favorable reservoirs.

3.1 CONCEPTS OF THE VOLCANIC RESERVOIR

Volcanic rocks are a series of products formed by volcanism. They are very different from sedimentary rocks in terms of formation conditions, development environments, and distribution rules (Table 3.1). Since the 1960s when volcanic reservoirs were discovered in hydrocarbon-bearing basins across China, scientists conducted a massive amount of research on various aspects of volcanic reservoirs, including rock characteristics, lithofacies combinations, development environments, reservoir space, reservoir capacity, and controlling factors. During the research, however, problems such as chaotic classification and naming, and conceptual ambiguity, for example, were introduced into the literature, because the researchers had different research purposes and objects, they chose different systems of volcanic reservoir classification. This resulted in issues such as the same name having different meanings, different names having the same meaning, and so on. Thus, correlating volcanic rocks across regions and communicating with the fundamental geologic research community were difficult. Unification of volcanic

rock classification and naming is a challenge that urgently needs to be resolved. We must establish a set of volcanic reservoir concepts based on the unified rock and lithofacies classification and naming system in volcanic geologic research, by emphasizing the operability and practicability in lithology and lithofacies study on volcanic reservoirs in hydrocarbon-bearing basins, and by focusing on the relationship between lithofacies and the reservoir's physical properties. At the same time, two requirements must be satisfied simultaneously, i.e. the practicability of hydrocarbon exploration, and communications in the fundamental research field (Zou et al., 2008).

3.1.1 Volcanic Eruption Mode, Type, and Environment

3.1.1.1 Volcanic Eruption Mode

Depending on whether magma intrudes into the earth's crust or erupts onto the surface, magmatism can be divided into two types: magmatic intrusion and volcanism. Volcanism is the process where deep magma erupts onto the surface or emplaces near the surface through a volcanic conduit, thereby forming various types of volcanic debris (Figure 3.1). Volcanism typically includes the following processes: (1) formation and evolution of magma; (2) volcanic eruption from the magma chamber to the surface through a volcanic conduit; (3) transportation, accumulation, and settlement of ejecta in different environments or emplacement of magma near the surface, thus forming various types of volcanics and related structures; and (4) formation of various types of volcanic rocks by the processes of cooling, welding, consolidating, and degassing after settlement of ejecta, as well as geothermal heat, hot springs, gas blowing, and hydrothermal alteration. The rock formed by volcanism is called volcanic rock. The volcanic rock commonly discussed includes two types: eruptive rock and subvolcanic rock. Eruptive rock includes lava formed by condensation of lava flow and volcaniclastic rock composed of volcaniclastic debris. Subvolcanic rock is the superhypabyssal intrusive rock related to volcanism (Qiu et al., 1996).

A volcanic eruption can, according to its energy, velocity, and components, be divided into three basic modes: explosion, effusion, and extrusion. Explosion means violent

TABLE 3.1 Comparison Between Formation Conditions of Magmatic Rock and Sedimentary Rock

	Magmatic rock		Sedimentary rock	
	Volcanic rock	*Intrusive rock*	*Clastic rock*	*Carbonate rock*
Temp/pressure	High temperature, high pressure		Normal temperature, normal pressure	
Environment	About 3 km below surface	Deep in earth crust	Surface or near surface	
Mode	High-temperature melting, extrusion	High-temperature melting, intrusion	Mechanical transport, gravitational differentiation	Biochemistry, autochthonous production
Components	Quartz, feldspar, dark-colored minerals		Quartz, feldspar, debris	Calcite, dolomite
Structure	Vesicular, pillow shaped	Joint	Bedding structure	Biogenic structure
Occurrence	Bedded, blocky	Blocky	Bedded	Bedded, blocky
Compaction	Affecting volcaniclastic rock	Does not affect	Obvious	Obvious
Distribution	Discontinuous		Fairly continuous	

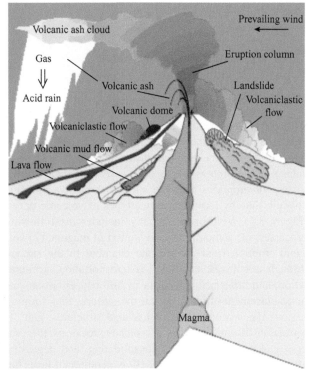

FIGURE 3.1 The volcanism process.

eruption under conditions that usually result in earthquake, landslide, tsunami, thunderbolt, rainstorm, and orogenic deformation. Effusion means fairly quiet overflow of volcanic magma onto the surface, thereby forming lava streams, lava falls, or lava lakes. Extrusion is neither violent eruption nor quiet overflow, but is the process where magma is extruded through relatively small conduits to form lava domes.

Factors determining whether volcanic eruption is explosive or nonexplosive include properties of the magma (such as components, volatile matter content, viscosity, and eruption rate), confining pressure, or the presence of extraneous water (seawater, lake water, and groundwater). Explosive eruption is driven by volatile matter of different origins. Volatile matter can come from exsolution of magma (CO and water) or gasification of extraneous superheated water (seawater, lake water, or groundwater that contacts the magma and becomes gasified). When the surrounding hydrostatic pressure is high enough to suppress explosive expansion of juvenile gas and explosive expansion of superheated seawater in contact with magma (i.e. explosion of magmatic vapor), only nonexplosive eruptions will take place in the form of fissured volcanic and hot spring eruptions. Its driving force is the load of lithostatic pressure and the buoyancy of fluidic magma, rather than the volatile matter pressure.

3.1.1.2 Volcanic Eruption Type

The occurrence of volcanic rocks is mainly related to how magma rises to the surface. Traditionally, eruptions were subdivided into three types: deroofing eruption, fissured eruption, and central eruption (Qiu, 1985). Today, volcanic eruptions are subdivided into three types: central eruption, fissured eruption, and complex eruption, based on modern observations and research on volcanic eruptions and the types of volcanic extrusion conduits.

3.1.1.2.1 Central Eruption

The magma erupts through neck-shaped conduits and usually erupts through the intersections of cross faults. It is most significantly characterized by the formation of a volcanic cone, where the center of the crater sags into a basin shape, which forms because of the volcanic eruption and magma recession.

3.1.1.2.2 Fissured Eruption

The magma gently flows out through fissures. This is also called a linear eruption, which often involves primarily basic lava that has a low-viscosity, high-fluidity magma. Craters are usually arranged linearly, and their deep portions are connected to form wall-shaped conduits. After flowing slowly through fissures, the lava can flow in all directions on the surface, thereby forming lava sheets, lava plateaus, or lava mesas. Their area can be up to hundreds of thousands of square kilometers and their thickness can be up to hundreds of meters or even more than 1000 m. Their thickness and components are stable. Volcaniclastic rock is typically rare. For fissured eruptions, the quantity of eruptive materials is different for magma of different properties, which would also result in very different occurrences, scales, and shapes (Qiu, 1985; Sun, 1985).

3.1.1.2.3 Complex Eruption (Fissured-Central Eruption)

This is the transitional type between central eruption and fissured eruption. Initially, almost all volcanic cones and shield volcanoes are erupted and distributed along faults. Only at a later time will they concentrate at one or several craters. When a volcano reaches a certain scale, radial cracks can be created by local and regional stresses, and craters become arranged linearly along the volcano's limb and base. For the magma erupted under such a background, its lithology can change dramatically from one volcano to another, or even along a simple fissure. For some volcanoes, for example, rhyolite is erupted at high horizons, while basalt is erupted at the base. The property of the eruption varies significantly depending on the differences in the structure of the volcano and the role of groundwater.

3.1.1.3 Volcanic Eruption Environment

The volcanic eruption environment can generally be divided into two types—onshore and underwater—based on the geographic environment, production depth, eruption type, transportation pathway, and rock characteristics of the volcanics. Onshore environments include continents, islands, and coastal areas, where volcanic rock erupts in air (Figure 3.2). Underwater environments include deep seas, shallow seas, lakes, and the bottom of glacier bands (Figure 3.3). The products developed in these two different eruption environments are significantly different (Qiu, 1985):

1. Distribution: onshore volcanic rocks are mostly planar and generally cover a small area; underwater volcanic rocks are usually banded and generally cover a large area.
2. Color: onshore volcanic rocks mostly show primary or oxidation color, where intermediate-basic rocks are mostly purple, red, black, or gray, and intermediate-acidic rocks are mostly light brick red or light yellow white. Underwater volcanic rocks that are basic are mostly blue or green, and the intermediate-acidic rocks are mostly silver gray or off-white.
3. Minerals: onshore volcanic rocks are characterized by having primary high-temperature minerals and volcanic glass, some of which can show dark-colored alkaline minerals and feldspathoids; underwater volcanic rocks are usually characterized by secondary low-temperature minerals and hydrous minerals (hydrated), without showing dark-colored alkaline minerals and feldspathoids.
4. Texture: onshore volcanic rocks feature development of hyaline and spherulitic textures, the spheroids are large in diameter and are mostly radial and concentric net-shaped, and the recrystallized felsite often shows equigranular texture, while underwater volcanic rocks are just the opposite.
5. Structure: onshore volcanic rocks are characterized by vesicular structures, oxidized tops, and columnar joints; underwater volcanic rocks feature fairly high development of pillow and bedding structures and usually show thick quenched edges, irregular cracks, and quench-broken glass debris.
6. Debris: volcanic debris deposits can be divided into three genetic types: fall-down and accumulation of volcaniclastic debris, accumulation of volcaniclastic flow, and accumulation of volcaniclastic surge. Onshore volcanoes feature a high debris content and large grain size as well as moderate development of volcanic bombs, volcanic mud balls, welded volcaniclastic rock, and conglomerate, while underwater volcanoes are just the opposite, but lithification of underwater nonexplosive quenched clastic debris and lithification of autoclastic breccia can also form a massive amount of volcaniclastic rock. Note that all volcaniclastic rocks are not the product of volcanic eruption. We must also remember to distinguish between reconstructed and redeposited distal volcaniclastic rocks.
7. Lava: onshore lava flows in a molten state without any transportation media and typically forms ropy lava, slaggy lava, or block lava; underwater lava and underwater rolling lava both are lava consolidated in water and can easily form pillow lava and vitreous clastic rock.
8. Facies change: onshore volcanic rocks are poorly sorted; in particular, the change of facies such as thickness or grain size along the horizontal direction is significant, while underwater volcanic rocks are well sorted, but their facies change along the vertical direction is more significant (in particular, the change of density is significant).
9. Paragenesis: onshore volcanic rocks contain continental zoolite and freshwater phytolite and feature

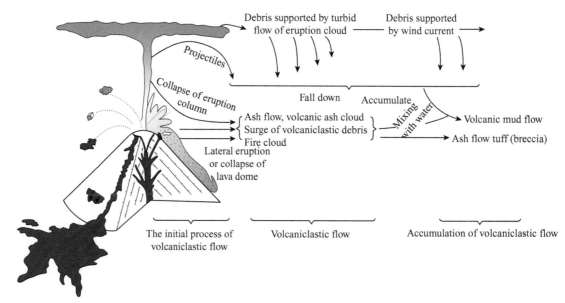

FIGURE 3.2 The genetic process of falling-down volcaniclastic debris of onshore volcaniclastic flow. *Based on Fisher et al. (1984).*

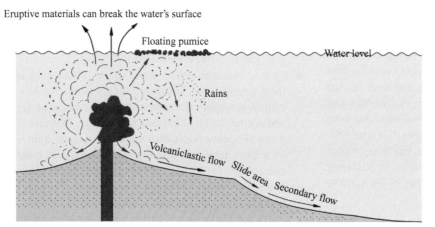

FIGURE 3.3 The genetic process of eruption and falling-down of underwater volcaniclastic flow. *Based on Fisher et al. (1984).*

development of clastic rock, especially development of coarse debris; underwater volcanic rocks contain marine and limnetic fossils and are commonly characterized by siltstone, argillaceous rock, silicalite, and carbonate rock.

10. Occurrence: onshore volcanic rocks often have eruptive unconformable contacts with underlying strata, have a large primary dip angle, and feature moderate development of weathered crusts and scour pits, while underwater volcanic rocks have more conformable contacts with underlying strata.

3.1.1.4 Volcanic Eruption Cycle and Volcanic Sedimentary Structural Sequence

A volcanic eruption is the event that sends volcanic materials (lava, volcaniclastic debris, and volcanic gas) to the surface. For a single volcano, eruption always follows the cycle of starting, intensifying, gradual weakening, and finally ceasing. At present, however, the application of concepts about the longitudinal evolution of volcanic eruption activities among researchers is not unified, and their use of research on volcanic reservoirs in hydrocarbon-bearing basins is even more chaotic. Commonly used terminology includes eruption cycle, volcanic eruption cycle, volcanic cycle, volcanic activity cycle, volcanic rhythm, volcanic sequence, volcanic facies sequence, volcanic stratum, volcanic layer, eruption stage, eruption period, volcanic activity period, volcanic eruption-sedimentary cycle, and volcanic sedimentary sequence. For all this terminology, the definitions provided by different researchers are very different. This is very inconvenient for research activities.

Eruption cycle means "the whole process of volcanic activity going from the initial eruption stage, and moving through the peak stage, declining stage, and dormant stage" (Editorial Committee of "Dictionary of Earth Sciences", 2006). One eruption cycle often includes multiple eruption activities. There is usually a certain time gap between two cycles, which is reflected by structural unconformity, eruptive unconformity, or the existence of certain sedimentary interbeds. They constitute rhythmic eruptions of a volcano, termed the "volcanic rhythm." Volcanic cycle is the changing process of volcanic activity that develops alternately between strong and weak intensity. For a single volcano, it is the process of happening, developing, and dying. These periodic changes all can be included into the category of volcanic cycle. But, this still remains an issue under discussion.

Qiu (1991) concluded that the number of eruptions can be determined according to the typical characteristics of lava or volcaniclastic rock. For example, a thick and highly vesicular red top layer or a blue-black colored, bark-like vitreous top crust is often seen on top of the lava (bed) of each eruption in basic rock, while vitreous rock (perlite, obsidian, pitchstone, or porous lava) is often seen on top of the lava (bed) of each eruption in acidic lava. One or more eruptions will form the volcanic layer. Periodic changes of lithofacies, components, texture, and structure in volcanic rock are also called the "volcanic rhythm." One rhythm is composed of multiple layers of rock and can be as thick as several to a dozen meters. A lithologic interval is generally used as its unit. Volcanic cycle means greater periodic changes created by volcanic eruptions, where the interface between two cycles is quite clearly rhythmic, often having certain discontinuities (unconformity and thick sedimentary interbeds). One cycle usually consists of multiple rhythms and can be as thick as several hundred to several thousand meters. A stratigraphic formation (or group) is generally used as its unit. Cycles not only reflect the process of volcanic activities but also roughly reflect the sequence of structural evolution.

Some scientists also connected volcanism to the research of sequence stratigraphy. The Mesozoic and Cenozoic magmatic cycles in eastern China have three primary evolutionary levels: megacycle (135 Ma), second-order cycle (35 Ma), and third-order cycle (5-13 Ma). Wherein, six second-order cycles correspond to the six major structural cycles of the Yanshan and Himalayan movements. The transformation of every structural cycle from a compression to a rifting extension mechanism is all accompanied by a gradual increase in regional volcanic activities and rock series alkalinity rate, as well as the change of magmatic activity edifice moving through "explosive facies—eruptive facies—eruptive-effusion facies—extrusive facies—intrusive facies." Third-order cycles correspond to the positive sequence of sedimentary basins. Large-scale magmatic activities in rift valleys are accompanied by abrupt tectonic subsidence, changes of paleoclimate, and changes of the source supply rate. According to the research by Lv et al. (2003), for all

basins in eastern China, volcanic rocks can be classified into subfacies such as explosive facies, eruptive facies, eruptive-effusion facies, and intrusive facies. Exploration practices indicated that typical explosive and eruptive facies are mainly distributed at lowstand systems tracts and lacustrine transgressive systems tracts in basins, eruptive-effusion facies appeared during the peak lake-flooding period, and intrusive facies appeared at or after the lacustrine transgressive systems tract and has diachronous characteristics. Yan et al. (2007), by investigating strata containing the volcanic rock series at District 5-8 along the northwest margin of Junggar Basin, concluded that both volcanic rock and sedimentary rock showed bedding structures, the bedding planes of both have isochronism, both are controlled by structural denudation, and their unconformity surfaces are completely comparable. Therefore, the Jiamuhe Formation of District 5-8 can be classified into four third-order sequences based on sequence stratigraphic classification and correlation of strata containing volcanic rock series according to their unconformity surfaces and bedding surfaces. During the fault activity period, magma upwelled through active fault zones to form a catenulate arrangement of multiple craters, where volcanic rocks of eruptive facies and eruptive-effusion facies were distributed radially around the crater. When the supply of source materials remained unchanged, the volcanic eruption stage was an important period when the fan delta changed its development scale, distribution characteristics, and advancing direction. Such sedimentary development characteristics of the Permian system at District 5-8 along the northwest margin of Junggar Basin resulted in frequent interbedding between volcanic rock and clastic rock within the Jiamuhe Formation. This interbedding is also the result of the combined effect of basin-mountain coupling in foreland thrust belts, origin of basin, evolution of orogenic belts, supply of source materials, and change of lake level.

The above discussions reveal that the concepts used by different researchers are vastly different. In hydrocarbon-bearing basins, volcanic strata are often associated or interbedded with sedimentary rocks to constitute strata containing a volcanic rock series. Therefore, third-order classification of volcanic rock series in hydrocarbon-bearing basins is necessary by reference to the sedimentary sequence grade classification system.

(1) Volcanic rhythm: this is the general terminology for products during the whole process of a single volcanic activity through the initial eruption stage, peak stage, decline stage, and dormant stage. The vertical pattern shows eruption-effusion-extrusion, manifesting the characteristics of volcanic activities declining from strong to weak.

(2) Volcanic cycle: this means the total collection of all products and their spatial distribution characteristics

formed within a stage during the process of volcanism. One volcanic cycle can include multiple basic continuous eruptions and can develop multiple volcanic rhythms. The products of volcanic activities of the same cycle constitute the volcanic formation, which reflects the characteristics of the same activity cycle or stage, including volcanic formation thickness, material composition, debris grain size, texture structure, eruption type, transportation way, accumulation environment, and the contact relationship with adjacent horizons in 3D space. They have their own characteristics and correlatability in terms of lithology, lithofacies combination rhythm, petrochemical evolution, and spatial distribution of volcanic edifices because they are controlled by synchronous regional tectonism and magmatic origin. A certain discontinuity exists between two adjacent volcanic cycles. It is often reflected by structural unconformity, eruptive unconformity, or quite thick regional normal sedimentary rock interbeds.

(3) Volcanic sedimentary structural sequence: in strata containing volcanic rock series, volcanic rock and sedimentary rock mostly show bedded occurrence with complicated and variable lithology and an absence of lithologic marker beds. Also, stratigraphic correlation is very difficult, because eruption is instantaneous, staged, and periodic, and there is no positive connection between base level cycle and sedimentary cycle. However, the structural unconformity surface does have isochronous correlatability, because tectonic movement controls volcanism and sedimentation, the volcanic eruption cycle reflects the characteristics of structural environment, and the structural unconformity often results in a gap or discontinuity of compositional evolution of a volcanic rock series and interruption of sedimentation, thereby leading to the formation of basal conglomerate and a broad regional distribution. Volcanic sedimentary structural sequence means the volcanic sedimentary rock series being developed and filled before, during, and after volcanic activities during a certain tectonic activity period. As thick as several hundred to several thousand meters, it often shows eruption and subsidence for multiple episodes and can involve multiple volcanic cycles. Volcanic sedimentary structural sequence can be shortened to volcanic sequence.

As an example, the Huoshiling and Yingcheng Formations at Xujiaweizi in Songliao Basin can be classified into three major magmatic activity stages (Table 3.2). According to isotope dating, the first magmatic activity stage was 147-158 Ma, corresponding to the top member of the Huoshiling Formation; the second magmatic activity stage was 125-135 Ma, corresponding to Member 1 of the Yingcheng Formation; and the third magmatic activity stage was 113-119 Ma, corresponding to Member 3 of the Yingcheng Formation. Dormancy between these three magmatic activity stages was long, and normal sedimentary rocks were formed.

The late Carboniferous Batamayineishan Formation in Junggar Basin is subdivided into three volcanic sequences and eight volcanic cycles according to the data of nearly 400 exploratory wells drilled through the Carboniferous system in the basin and based on seismic and logging analysis. The top of the early Carboniferous series is subdivided into one volcanic sequence and one volcanic cycle because only the top formation of early Carboniferous series was encountered and strata had not been penetrated completely. The early Carboniferous series mainly shows volcaniclastic rock and volcanic sedimentary clastic rock, with acidic rhyolite observed locally. For the late Carboniferous series, the bottom volcanic sequence features development of three volcanic cycles, mainly consisting of volcanic breccia of explosive facies interbedded with lava of eruptive-effusion facies, and shows a lithology that transforms gradually from intermediate basic to intermediate acidic from bottom to top, reflecting the structural setting of intense volcanic eruption, where the strata are equivalent to the volcanic rock assemblage of the lower member of Batamayineishan Formation. The middle volcanic sequence features development of two volcanic cycles and mainly consists of tuffaceous sandstone (glutenite) of the volcanic sedimentary facies, reflecting the structural setting of volcanic eruption moving from weak activity to dormancy, where the strata are equivalent to the sedimentary rock assemblage of the middle member of the Batamayineishan Formation. The top volcanic sequence features development of three volcanic cycles and mainly consists of volcanic breccia of explosive facies interbedded with lava of eruptive-effusion facies, reflecting the structural setting of intense volcanic eruption. Complete cycle series where lithology transitions gradually from intermediate basic to intermediate acidic from bottom to top is observed in the depressed area within the basin. For the uplift area, the cycle series is not revealed completely, because the top of Carboniferous system had been exposed and denuded for a long time, where its strata are equivalent to the volcanic rock assemblage of the upper member of the Batamayineishan Formation.

3.1.1.5 Types of Volcanic Edifices

Terminologies related to volcanic edifice are highly varied, and most of them have been translated from overseas. Some only mention nouns without providing definitions and identification marks. It is very easy to get confused and difficult to apply in practice. Volcanic edifice (or volcanic

TABLE 3.2 Classification of Magmatic Activity Cycles at the Xujiaweizi Rifted Area in Songliao Basin

| Cycle | | Basis | | | |
		Isotopic age (Ma)	Location of sequence under formation	Main lithology	Representative well
Yingcheng stage (K₁yc)	Third magmatic activity cycle	113-119	Conformity coverage over normal clastic rock of Member 2 of Yingcheng Formation, unconformity coverage over Denglouku Formation	Mainly intermediate-acid rocks (andesite, rhyolite)	Songshen3, Songshen2, Songshen1, Weishen4, Fangshen6, Zhuangshenl, Shuangshen10
	Second magmatic activity cycle	125-135	Unconformity coverage over Shahejie Formation (coal measure)	Mainly intermediate-acid rocks (andesite, rhyolite)	
Huoshiling Formation (J₃h)	First magmatic activity cycle	147-158	Unconformity coverage over Mesozoic Formation, located beneath sedimentary clastic rock of Shahejie Formation (coal measure)	Mainly intermediate-basic rocks (basalt, andesite)	Shangshen1, Shengshen101, Zhuangshen1, Shuangshen10

Reproduced from Wang et al. (2007).

apparatus), for example, is defined in the "Dictionary of Earth Sciences" (2006) as: "it is also called volcanic body or edifice and means all kinds of volcanic landscapes formed on the ground surface during volcanic eruption, including volcanic cone, volcanic dome, volcanic crater, caldera, and lava plateau; sometimes it also refers to sub-structures such as volcanic neck, volcanic conduit, etc." Qiu (1991) suggests that volcanic edifice is "the collection of all kinds of lithofacies and structures formed during a certain period, under the same volcanism, and centered on the volcanic conduit, including extrusive facies, volcanic conduit facies, ringed and radial subvolcanic dike, central stock, and caldera, and it typically refers to volcanic cone (including volcanic crater and volcanic conduit) and caldera." Qiu et al. (1996) suggests that volcanic edifice comprises "volcanic conduit and its adjacent deposits and structures of all kinds formed within a certain range of space and time, including crater, volcanic neck, near-crater deposits, and extrusive lava dome or volcanic rock where volcanic neck and volcanic crater are only the basic elements of volcanic edifice." Li et al. (1992) thinks that volcanic edifice is "the general name of all kinds of products and structures formed around the crater of the volcano during a quite long period or a longer geologic time." Chen (2002) defined volcanic edifice as "the general name of volcanic conduit and its adjacent deposits and structures of all kinds formed within a certain range of space and time, including volcanic neck, volcanic crater, and volcanic facies around the crater." Tao et al. (1999) proposed that volcanic edifice is "the complex geologic body constituted by volcanic conduit and products of its related various volcanisms during a certain period, including volcanic crater,

volcanic neck, volcanic cone, and all kinds of volcanic eruptive rock, extrusive rock, and subvolcanic rock." Huang et al. (2007) stressed that volcanic edifice is "the collection of products of all kinds of volcanisms formed during a certain time frame, constituted by volcanic materials from the same eruption source that deposited around the source area, and having a certain shape and syngenetic combination relationship. Modern volcano generally means a certain scale of accumulation body formed by eruptive materials around volcanic conduit. Modern volcanic edifice is usually well preserved, while paleovolcanic edifice is often damaged to different degrees due to weathering, denudation, and tectonic deformation."

The classification of volcanic edifice types is quite chaotic because volcanic edifice is controlled and affected by the property of eruptive materials, the eruption mechanism, and related external factors, resulting in diversified shapes and scales. Some scientists emphasize the eruption mode. For example, Li et al. (1992), according to eruption modes, divided it into fissured volcanic edifice, central volcanic edifice, and complex volcanic edifice. Most researchers stress the occurrence and shape of volcanic rocks. For example, the United States Geological Survey (USGS) suggested that the shape of volcanic rocks be primarily classified into four types, i.e. cinder cone, stratovolcano, shield volcano, and lava dome. Chen (2000) classified the volcanic formation edifices at Yingcheng Formation in Songliao Basin into three types, i.e. stratovolcano, mini shield volcano, and cinder cone volcano. Qiu et al. (1996) classified volcanic edifice into seven types, i.e. shield volcano, flood basalt, cinder cone, stratovolcano, maar crater, caldera, and volcanic dome. Huang et al. (2007), based on modern volcano classification, identified four types of buried volcanic edifices in the Xujiaweizi rifted area of Songliao Basin, where stratovolcanos and lava domes are formed mainly in acidic and acidic plus intermediate-basic volcanic eruption areas, and shield volcanoes and volcaniclastic cones are formed mainly in intermediate-basic volcanic eruption areas. They believe that the types and shapes of local paleovolcanic edifices are controlled by the lithology and lithofacies of volcanic rocks and that adjacent craters and their ejecta of different ages are often interlaced and superimposed to form larger-scale complex volcanic edifices.

In general, although the definitions of volcanic edifice are somewhat different, their contents are basically similar, except some authors stress more the occurrence and shape, while others focus more on the lithofacies combination. Therefore, we believe that the definition of volcanic edifice should stress the concept of time and space and the location near the volcanic conduit, and we recommend that it be classified into three basic types—shield volcano, stratovolcano, and cinder cone volcano (Figure 3.4).

A shield volcano is entirely or almost entirely composed of lava of eruptive-effusion facies, where volcaniclastic debris accounts for about 1%, wide and short domes are formed by magma often through multiple rather than single effusions, and the slope angle is 2-10° and rarely exceeds 15°. The top has flat-floored and steep-walled crater(s). Several shield volcanoes are often arranged directionally. They are controlled by faults and most of them are composed of basalt.

A stratovolcano is composed of interbedded lava and volcaniclastic rock. The lava distributes like "ribs." Some modern stratovolcanos can constitute very tall peaks. The slope angle is less than 35°. The cone top has funnel-shaped crater(s).

A cinder cone volcano is entirely composed of volcaniclastic rock, the original slope angle is less than 30°, and large debris and volcanic bombs are distributed around the crater.

Volcanic edifices in north Xinjiang can, in map view and moving from the crater outward, be classified into near-crater belts, subvolcanic facies belts, medium-from-crater facies belts, and distant-crater belts, primarily comprising volcanic conduit facies-explosive facies. Various facies belts have different rock assemblage types and lithofacies association types (Figure 3.5).

3.1.2   Volcanic Petrography

3.1.2.1   Types of Volcanic Rocks

The rock types of volcanic reservoirs determine the facies types in the said region. The more complicated the rock types, the more complicated the facies types, and therefore the more complicated the sequential relationship and spatial through-going relationship of volcanic crater activities.

Magmatic rock is the rock formed by condensation and consolidation of magma and is generally divided into volcanic rock (eruptive rock) and intrusive rock. Volcanic rock means the rock formed by magma that rises through structural fissures and erupts (effuses or explodes) onto the surface through volcanic conduits (Qiu, 1985; Sun, 1985). Some disagreement and chaotic terminology still remain today regarding the issue of classification, naming, and appraisal of volcanic rock, the hierarchy of which is unclear. Most researchers classify it into two types, namely lava and volcaniclastic rock, where the rock formed after cooling of magma erupted and effused by the volcano is called lava, while the rock formed by (onshore or underwater) accumulation of all kinds of debris erupted by the volcano is called volcaniclastic rock (Qiu, 1985; Sun, 1985). Luo et al. (1996) classified volcanic rock into three major categories, namely, lava, subvolcanic rock, and volcaniclastic rock. Wang et al. (2007), using rock textures and origins, classified deep volcanic rock in Songliao Basin into four major categories, namely, volcanic lava type, volcaniclastic lava type, volcaniclastic rock type, and sedimentary volcaniclastic rock type. We herein suggest that subvolcanic rock not be separately classified into a major rock category, because it can be combined into the lava category as its formation space and time are both related to cogenetic volcanic lava, it has an appearance similar to lava under most circumstances, and its chemical composition, mineral components, texture, and geologic characteristics are very close to those of lava of similar rocks. For discrimination during naming, a "sub" prefix can be added before the rock name.

Determination of mineral components and quantitative statistical calculation of the contents of constituent minerals of volcanic rocks during thin-section analysis are difficult, because their degree of crystallization is poor, crystalline particles are tiny, and they contains a huge amount of cryptocrystalline and vitreous matter. We believe that volcanic reservoirs can be classified into two major categories, namely, lava and clastic rock, according to the "Lava and Volcaniclastic Rock Classification and Naming Scheme" recommended by the Volcanic Rock Classification and Naming Group of Petrology Sub-Committee of Geological Society of China, as well as the People's Republic of China's Oil and Gas Industry Standard (SY/T 5368-2000, Thin-Section Examination of Rock). We recommend using the Chemical Composition Classification and Naming Scheme (Table 3.3) for volcanic lava, which can, based on its SiO_2 content, be further classified into five types, namely, picrite, basalt, andesite, trachyte, and rhyolite. These five types can have transitional types between them, such as basaltic andesite, dacite, and rhyolitic dacite. For

Shield volcano: effusive products mainly consist of lava, the volcano is broad and flat with a slope angle of 2°-0°

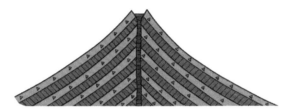

Stratovolcano: volcaniclastic debris and lava are interbedded, explosive facies and effusive facies alternate alternately, and the slope angle is <35°

Cinder cone volcano: primarily composed of pumice and molten slag of explosive facies, with a slope angle of <30°

FIGURE 3.4   Illustration of typical volcanic edifices.

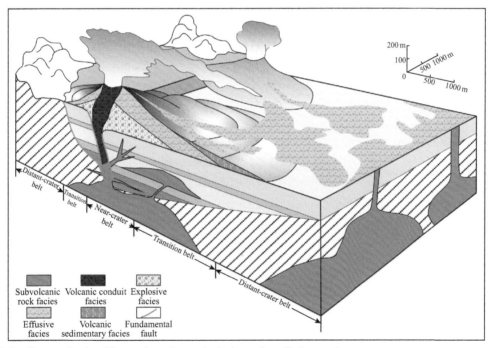

FIGURE 3.5   Carboniferous volcanic edifices and eruption modes in the northern Xinjiang region.

volcaniclastic rock, it is first classified into three types, namely, volcaniclastic lava, volcaniclastic rock, and volcanic sedimentary clastic rock, according to its source, genetic mode, and cementation. It is then subdivided into agglomerate, breccia, and tuff according to its size grade and components. Finally, a detailed name is assigned according to the state of matter, components, and texture of volcaniclastic debris (Table 3.4).

3.1.2.2   Characteristics and Types of Volcanic Facies

Volcanic facies means the collection of the types, characteristics, and accumulation types of volcanic products during volcanism. Definitions given by different researchers are different to a certain extent, such as "overall characteristics of different rocks and rock masses produced under different

TABLE 3.3 Classification of Volcanic Lava (Based on the Volcanic Rock TAS Classification Scheme Recommended by International Union of Geological Sciences, 1989)

Texture category		SiO_2 content (%)	Lithology	Rock name	Characteristic mineral assemblage or debris components
Volcanic lava type	Lava texture	45-52	Basic	Basalt/vesicular-amygdaloidal basalt	Basic plagioclase, pyroxene, olivine
		52-57	Intermediate basic	Basaltic andesite/basaltic trachyandesite	Intermediate-basic plagioclase, pyroxene, hornblende
		52-63	Intermediate	Andesite	Intermediate plagioclase, hornblende, biotite, pyroxene
				Trachyte/trachyandesite	Alkaline feldspar, intermediate plagioclase, hornblende, biotite, pyroxene
		63-69	Intermediate acid	Dacite	Intermediate-acid plagioclase, quartz, alkaline feldspar, biotite, hornblende
		>69	Acidic	Rhyolite	Alkaline feldspar, quartz, acidic plagioclase, biotite, hornblende
	Vitreous texture	Typically >63	Intermediate acid	Perlite/obsidian/pitchstone/pumice (prefixed with rhyolitic/andesitic/basaltic, based on chemical composition)	Frequently seen are quartz and feldspar porphyritic crystals (crystallite); also seen are phenocrysts such as biotite, hornblende, pyroxene, olivine

TABLE 3.4 Classification of Volcaniclastic Rock (Based on the Volcaniclastic Rock Classification Scheme Recommended by International Union of Geological Sciences, 1989)

Category		Volcaniclastic lava	Normal volcaniclastic rock		Volcanic sedimentary clastic rock	
			Welded volcaniclastic rock	*Common volcaniclastic rock*	*Sedimentary volcaniclastic rock*	*Volcaniclastic sedimentary rock*
Content of volcaniclastic debris		10-90%	>90%		90-50%	50-10%
Genetic type		Volcaniclastic lava type	Aerial fall volcaniclastic rock type	Volcaniclastic rock type of volcaniclastic debris (ash) category	Depositional (sedimentary) volcaniclastic rock type	Volcaniclastic sedimentary rock type
Cementation mode		Welding and cementing primarily	Welding primarily	Compaction primarily	Welding and hydrochemical cementing	
Grain size	>64 mm	Agglomerate lava	Welded agglomerate	Agglomerate	Sedimentary agglomerate	Tuffaceous breccia
	2-64 mm	Breccia lava	Welded breccia	Volcanic breccia	Sedimentary volcanic breccia	Tuffaceous breccia
	<2 mm	Tuffaceous lava	Welded tuff	(Crystal vitric) tuff	Sedimentary tuff	Tuffaceous sandstone

genetic conditions of rock masses" (Qiu, 1985); "the rock characteristics that reflect the genetic conditions of volcanic rocks" (Sun, 1985); "the general term of characteristics of products of volcanic activities in certain environments" (Qiu et al., 1996); "the overall collection of characteristics of products of volcanic activities under certain environments and conditions" (Zhao et al., 1999); and "the overall collection of genetic conditions of volcanic rock and lithologic characteristics of volcanic rock formed under such conditions" (Luo et al., 2003). In general, volcanic lithofacies includes a rock's genetic conditions and rock characteristics, is the authentic record of volcanic eruption types, and is also the most essential and most important geologic entity of the products of volcanism. It reflects comprehensive geologic characteristics such as volcanic eruption types, transportation media and ways, accumulation environments, and climate.

Today, Chinese and foreign classifications of volcanic lithofacies seriously lack unification. Sun (1985), using the formation conditions of volcanic rocks, the general mechanism of volcanism, and diagenetic modes, classified six lithofacies, i.e. volcanic conduit facies, subvolcanic facies, extrusive facies, eruptive-effusion facies, explosive facies, and eruptive-sedimentary facies. Using central eruption as an example, Qiu (1985) roughly classified six lithofacies, i.e. eruptive-effusion facies, explosive facies, extrusive facies, volcanic neck facies, subvolcanic facies, and volcanic sedimentary facies. Qiu et al. (1996) classified 11 volcanic facies, i.e. eruptive-effusion facies, fallout facies, volcaniclastic flow facies, surge facies, volcanic mud flow facies, collapse facies, extrusive facies, volcanic crater-neck facies, subvolcanic facies, cryptoexplosive breccia facies, and volcanic eruptive-sedimentary facies. Xie et al. (2002) classified 13 lithofacies, including eruptive-effusion facies, explosive fallout facies, volcaniclastic flow facies, explosive-effusive facies, base surge facies, volcanic mud rock-flow facies, eruptive-sedimentary facies, volcanic neck facies, extrusive facies, subvolcanic facies, cryptoexplosive breccia facies, intrusive facies, and volcanic lake-basin sedimentary facies. Luo et al. (2003), using factors such as the formation conditions of volcanic rocks, the mechanism of volcanism, occurrence, and shapes, classified volcanic lithofacies into volcanic conduit facies, sedimentary volcanic facies, extrusive facies, eruptive-effusion facies, explosive facies, and explosive-sedimentary facies.

Chen (2000) classified volcanic facies in Songliao Basin into explosive fallout facies, eruptive-effusion facies, volcaniclastic flow facies, base surge facies, and eruptive-sedimentary facies, wherein eruptive-effusion facies is further divided into three subfacies, namely, upper, middle, and lower. Wang et al. (2003a,b,c) and Wang (2006) classified volcanic facies into 12 classes in 4 types, i.e. volcanic conduit facies, explosive facies, eruptive-effusion facies, and extrusive facies; and later further classified into 5 facies and 15 subfacies, where the 5 facies include volcanic

conduit facies, explosive facies, eruptive-effusion facies, extrusive facies, and volcanic sedimentary facies. Zhao et al. (1999) classified Mesozoic and Cenozoic volcanic rocks in Liaohe Basin into four kinds, i.e. explosive facies, eruptive-effusion facies, extrusive facies, and volcanic sedimentary facies, and further classified them into eight subfacies. Luo et al. (1996) classified volcanic rocks at Fenghuadian into explosive facies, eruptive-effusion facies, and volcanic sedimentary facies. Wang et al. (2003a,b,c) classified volcanic rocks at the Linshang area, Huimin depression, and Bohai Bay Basin into explosive slump subfacies and subaqueous volcaniclastic flow subfacies. Wang et al. (2005a,b) classified lower Cretaceous volcanic rocks at the Tsagaan Tsav depression into explosive facies, eruptive-effusion facies, volcanic neck facies, and eruptive-sedimentary facies. Wang et al. (1991) classified volcanic rocks at the Abei oil field into three lithofacies, namely, fluid lava, extrusive lava, and volcaniclastic rock. Yan et al. (1996) classified volcanic rocks in Jianghan Basin into eruptive-effusion facies, subvolcanic facies, cryptoexplosive crypto-flow facies, and rock-flow autoclastic tuff facies. Liu (1986) classified Carboniferous volcanic rocks along the northwest margin of Junggar Basin into effusive-lava facies and explosive-volcaniclastic facies. Dun (1995) classified Carboniferous volcanic rocks at the northwest edge area of Junggar Basin into volcanic crater extrusive lava facies, explosive-clastic facies, and subvolcanic facies. Yan et al. (2007) classified Permian volcanic rocks at Jiamuhe Formation along the northwest margin of Junggar Basin into volcanic crater facies, near-crater facies, and distant-crater facies. Luo (2007) classified volcanic rocks at the Malang depression of Santanghu Basin into eruptive-effusion facies and explosive facies. Luo et al. (2006) classified volcanic rocks in the Tahe oil field into volcanic eruption facies, volcanic eruption-effusion facies, and volcanic sedimentary facies.

Obviously, fundamental geologic research on volcanic rocks and geologic research on petroleum are different in their scales. In fundamental geologic research, lithofacies concepts such as explosive facies, eruptive-effusion facies, and extrusive facies are often established based on single magmatic edifices. These three facies represent products of the same magmatic cycle during different stages and can be superimposed vertically. However, in petroleum geology, the classification of volcanic facies serves more to satisfy the requirement of "core-logging-seismic" triple-phase conversion. Its research subject is the products of crypto-cycles and multicycles. It favors using environment-corresponding and seismically identifiable classification schemes, such as the "volcanic conduit-volcanic agglomerate facies, volcanic lava facies, tuff facies" classification scheme. The classifications of volcanic facies given by different authors mentioned above do not strictly differentiate between the naming of environmental facies belts and lithofacies. The commonly used near-crater facies, medium-from-crater facies, and distant-crater facies classification scheme is closer to the definition of environment. Therefore, "near-crater facies" can also be called "near-crater environment," thus endowing it with more geographical environmental characteristics.

The thickness, material composition, debris grain size, texture, and welding degree of volcanic facies often have regular variations vertically (i.e. bottom to top) and laterally (i.e. from the crater to areas far from the crater). Establishing the facies pattern by finding out the variation characteristics and rules of volcanic facies in 3D space is the priority for classifying and studying volcanic facies. Classification of volcanic facies should mainly be based on basic geologic attributes, such as lithology and petrographic makeup, which can be observed and accurately identified through outcrops, cores, or debris. It should also emphasize the operability in basinal volcanic facies research; should focus on the relationship between lithofacies and the reservoir's physical properties; and should take into account the formation conditions of volcanic rocks, the mechanism of volcanism, occurrence, and shapes. The volcanic facies model is a conceptualized and simplified intuitive model that demonstrates the dependence relationship between volcanic facies, is the generalized summary of facies sequence research findings of known sections or wells, and should also offer a guidance role for lithofacies observation and prediction of new sections or wells. The most important role of a volcanic facies model in exploration and development is to constrain and guide seismic-lithofacies interpretations. Like sedimentary rocks and sedimentary facies, the dependence relationship (facies sequence) and variation rule (facies rule) between volcanic facies and subfacies are the important components of knowing and depicting volcanic facies and, moreover, are the basis for building the volcanic facies model, constraining seismic data interpretation, and conducting volcanic reservoir prediction (Wang, 2006).

Existing volcanic facies classification schemes offer limited discussion on diagenetic modes, which are simply the determining factors that decide the change of a reservoir's physical properties during the burial process. Previous volcanic facies classifications only briefly discussed the relationship between lithofacies and a reservoir's physical properties, and this no longer satisfies the actual needs in basinal volcanic facies research designed for reservoir modeling, evaluation, and prediction. We believe that lithofacies classification must follow both the general classification principle and give consideration to its practicability, including the following several aspects: (1) basic form of volcanic eruption, whether explosive, eruptive-effusive, or extrusive; (2) volcanic eruption environment, whether onshore or underwater; (3) mechanism of volcanic eruption and transportation way/accumulation mechanism of volcaniclastic debris, whether fallout, volcaniclastic flow, volcanic surge,

or volcanic mud flow; (4) emplacement mechanism of magma at a certain depth below ground surface; and (5) specific position in volcanic edifice, whether at the volcanic neck or crater. The purpose of such classification is to provide a practical and effective classification scheme for basinal volcanic rock research by restoring its eruption or accumulation environment based on the characteristics of volcanic products, thus allowing the researcher to identify various volcanic facies and subfacies on the scale of profile, core, debris, and thin section by referring to the lithofacies appraisal marks in the classification table, identifying volcanic facies and subfacies by using geophysical data through geofacies-seismic/log facies transformation, and preliminarily evaluating the physical properties of volcanic reservoirs on the basis of lithofacies and subfacies identification. We classified volcanic facies into 4 facies groups, 6 facies, and 10 subfacies (Table 3.5) according to the "lithology-fabric-origin" classification criteria, based on basinal volcanism patterns or eruption/transportation ways, and using the characteristics of volcanic activities and volcanic rock distribution. In creating this classification, we referenced the classification scheme of 5 facies and 15 subfacies that Wang (2006) provided when studying volcanic facies in Songliao Basin. We also conducted a comprehensive review of previous research on volcanic facies in hydrocarbon-bearing basins across China. The said classification scheme is mainly based on basic geologic attributes, such as lithology and petrographic makeup, which can be observed and accurately identified using cores or debris, emphasizes the operability in basinal volcanic facies research, and focuses on the relationship between lithofacies and a reservoir's physical properties (Table 3.5 and Figure 3.6).

3.1.2.2.1 Subvolcanic Rock Facies

This facies is often located near the crater and interfingers with or injects in the form of stocks, dikes, or veins into other lithofacies and surrounding rock. The representative characteristics of this facies are that the degree of crystallization of its rock is higher than that of all other volcanic facies and it has porphyritic to crystalline texture, chilled border structure, platy flow and linear flow structures, and columnar and tabular joints. Its phenocrysts often give rise to the resorption phenomenon due to fluid activities at later stages of magmatic activity. It is mainly reflected by various kinds of subvolcanic rocks, such as diabase, diorite-porphyrite, and granitic porphyry. For example, the volcanic rock near 3900 m in the Fangshen#9 well is subvolcanic facies. Subvolcanic facies could be formed during the same and later stages of volcanic cycle, is formed by synchronous or later magma that intruded into the surrounding rock, then slowly condensed and crystallized, and intersects with other lithofacies and surrounding rock.

TABLE 3.5 Major Characteristics of Volcanic Facies

Facies group	Depth	Location	Facies	Subfacies	Major rock type	Occurrence	Formation mechanism
Volcanic sedimentary facies group	Surface	Far from crater	Volcanic sedimentary facies	Transition subfacies (transitioning with normal sedimentary facies) Eruptive sedimentary subfacies	Sedimentary tuff (volcaniclastic debris 90-50%), tuffaceous sandstone/mudstone (volcaniclastic debris 50-10%) Tuff (volcaniclastic debris >90%)	Bedded and lenticular shaped	Volcanic ash/dust drift, fallout deposits
Volcanic eruption facies group	Surface	Near crater	Extrusive facies		Perlite	Rock dome, cupola, breccia cupola	Magma is extruded onto surface, cooled, and consolidated
			Eruptive-effusion facies	Upper subfacies	Vesicular lava, amygdaloidal lava	Lava flow, lava sheet, lava rope, pillow lava, lava bed, shield volcano	Eruption, effusion
				Lower subfacies	Various basic, intermediate, and acidic lava		
			Explosive facies	Pyroclastic flow subfacies Fallout subfacies	Welded volcaniclastic rock, volcaniclastic rock	Volcaniclastic layer, volcaniclastic cone, aerial fallout accumulation, ash flow accumulation, splashed materials accumulation near crater	Explosion, eruption, fallout
Volcanic conduit facies group	About 0.5 km from surface	Crater	Volcanic conduit facies	Volcanic crater subfacies Volcanic neck subfacies	Collapsed welded volcanic breccia (cogenetic breccia, heterogenetic breccia) Cataclastic lava	Steep occurrence, with concentric ringed single-eruption neck, multieruption neck	Extrusion-intrusion
Subvolcanic facies group	Near surface <3 km	Bottom of crater	Subvolcanic facies		Basic, intermediate, acidic hypabyssal intrusive	Stock, dike, apophysis, laccolith	Intruded into near surface

(Reproduced and modified from Qiu (1991) and Wang (2006).

3.1.2.2.2 Volcanic Conduit Facies

Volcanic conduit facies refers to the entire magma guidance and transport system from the magma chamber to the crater top. Located at the base of the entire volcanic edifice, the volcanic conduit facies is the assemblage of various volcanic rock types stagnated and backfilled in the volcanic conduit, while magma migrates upward to reach the ground surface. Its lithology is mainly volcanic lava and welded

volcanic breccia. It is formed by volcanic materials filling up the volcanic conduit, while volcanic activity approaches a stop, when magma condensation is relatively slow and the degree of crystallization improves. The volcanic conduit facies can be divided into volcanic crater subfacies and volcanic neck subfacies. It can be formed during the entire process of the volcanic cycle, but what remains is primarily the product of later activities.

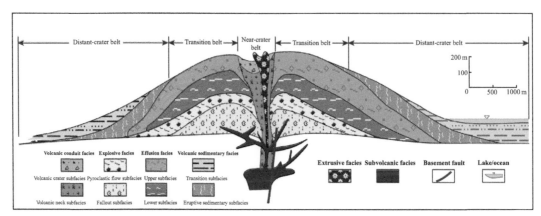

FIGURE 3.6 Volcanic facies models.

3.1.2.2.3 Explosive Facies

Explosive facies is formed through on-surface accumulation of volcaniclastic debris produced by intense volcanic explosions. Its morphology is cone shaped, and it is formed during the early and later stages of volcanism. This facies is the most widely distributed volcanic facies as well as the volcanic rock type that has many different structural types, and can easily get interbedded with normal sedimentary rocks. It can be divided into two subfacies, namely, fallout subfacies and hot clastic flow subfacies.

3.1.2.2.4 Eruptive-Effusion Facies

Eruptive-effusion facies is formed during the middle stage of the volcanic eruption cycle. It is the highly movable lava formed on the surface, while low-viscosity or volatile-constituent-saturated magma gently effuses out of the crater. Its distribution is highly affected by paleotopography. It is typically distributed at and near the lava cone. Eruptive-effusion facies is observed in all acidic, intermediate, and basic volcanic rocks. This facies is typically divided into a lower subfacies and upper subfacies.

3.1.2.2.5 Extrusive Facies

Primarily present in acidic rocks, the extrusive facies is formed during the later stage of a volcanic eruption cycle. A volcanic massif of extrusive facies (vitreous) is formed near the crater when high-viscosity magma is extruded by internal force to flow onto the surface, gets quenched in water, or gets cooled rapidly in air after the caldera-volcanic lake system is formed. The perlite, obsidian, and pitchstone types in eastern China areas where Mesozoic acidic rocks are developed are all volcanic rocks of extrusive facies. Extrusive facies massifs are primarily dome shaped. Lava domes are as high as several scores to several hundred meters, with diameters ranging from several hundred to several thousand meters. In general, based on characteristics such as lithology and texture, it can be further classified into a central belt (dense blocky zone) subfacies, transition belt (high-fracture-density zone) subfacies, and

marginal belt (marginal autoclastic-genetic breccia zone) subfacies.

3.1.2.2.6 Volcanic Sedimentary Facies

This is formed via accumulation of volcaniclastic materials created by denudation, transportation, and sedimentation of volcanic eruptive materials falling into water or underwater and being vibrated by water. It is located at the base of the volcanic cone and can also be distributed near or even far from the volcanic cone. Volcanic sedimentary facies is often associated with volcanic rocks and may appear during all stages of volcanic activity. It can alternate laterally or interbed with other volcanic facies, and its distribution range and scale are far greater than those of other volcanic facies. The volcanic sedimentary facies assemblage is formed primarily in hollow zones between volcanic uplifts during the volcanic eruption process, especially during the dormancy period of volcanic activity. Its lithology features a huge amount of hydrochemical cements contained between volcaniclastic debris particles and shows certain bedding or graded structures. In addition, the rock may contain a slight amount of terrigenous clastic debris or may sometimes show a gradational relationship with mudstone, wherein its lithology primarily features sedimentary rocks containing volcaniclastic debris. The clastic constituents are mainly volcanic debris, tuffaceous debris, crystal fragments, and vitric fragments. Volcanic sedimentary facies is mainly formed in alluvial fan and intermontane fluvial sedimentary environments.

3.1.3 Volcanic Hydrocarbon Reservoir Geology

3.1.3.1 Research in Volcanic Hydrocarbon Reservoir Geology

Volcanic hydrocarbon reservoir geology is a comprehensive geologic discipline that studies the developmental structural environment, genetic type, characteristics, formation, evolution, geometric shape, distribution rules, reservoir research method and description technique, and reservoir evaluation and prediction of volcanic reservoirs in hydrocarbon-bearing basins. Research in volcanic hydrocarbon reservoir geology must provide the scientific basis for volcanic hydrocarbon exploration by focusing more on crossing and integrating multiple disciplines, i.e. combining theories such as structural geology, petroleum geology, reservoir geology, volcanology, volcanic geology, petrology, and reservoir physics, and applying techniques and methods such as seismic, nonseismic, logging, mathematic geology, and computer technology.

The volcanic reservoir itself is characterized by significant changes of scale and physical properties and high heterogeneity; and its rock texture, mineral composition, genesis, reservoir space formation and evolution, reservoir flow unit, reservoir effectiveness and sensitivity, hydrocarbon accumulation process, and main control factors are all obviously different from those of a sedimentary reservoir. All vesicular structures developed in volcanic rock itself are affected by filling during volcanism and its subsequent diagenetic alteration process. Volcanic reservoirs today have all undergone later transformation. The task of volcanic hydrocarbon reservoir geology is to provide the theoretical basis for driving volcanic hydrocarbon exploration and development and to promote advances of fundamental research in volcanic petroleum geology in sedimentary basins by looking deeply into the characteristics of volcanic reservoirs, such as their macroscopic throughgoing, internal texture, reservoir parameters distribution, and pore structure. In addition, the dynamic change characteristics of reservoir parameters during development of volcanic oil and gas fields, ascertaining the main control factors of reservoir development, predicting effective reservoirs and evaluating their distribution, and creating the volcanic reservoir development evaluation technique are also important. Current research includes the following areas:

1. Structural environment of volcanic rock development: including the regional structural framework, basin nature, and major tectonic-volcanic events of sedimentary basins containing volcanic rocks.
2. Volcanic cycle sequence and lithofacies paleogeography: including volcanic development stage, cycle classification and correlation, volcanic developmental paleogeomorphology, development environment, and spatial distribution pattern.
3. Reservoir lithologic-lithofacies characteristics and geologic model: including volcanic lithology, eruption pattern, volcanic petrologic characteristics, geologic characteristics of volcanic facies and individual subfacies, lithofacies model, and the distribution of volcanic edifices, volcanic structures, and volcanic massifs.
4. Reservoir space of reservoirs: including the development characteristics, assemblage and abundance of primary pores, secondary pores, and fractures, and the connection status of various types of pores.
5. Reservoir physical properties, heterogeneity, and effectiveness: including porosity, permeability, fluid saturation, throat type, pore-throat configuration relationship, and the spatial distribution characteristics of these parameters.
6. Diagenesis and evolution: including types of diagenesis, classification of diagenetic stages, diagenetic evolution pattern, the characteristics of individual evolution stages, and the relationship among them.
7. Reservoir development pattern: including how actions such as volcanism, hydrothermal alteration, and dissolution transformation control the forming and evolution characteristics of pores, reservoir development pattern, physical simulation, and numerical simulation of reservoir formation.
8. Reservoir geologic modeling: including geologic elements of modeling, classification of reservoir geologic models, modeling methods, and modeling effect verification.
9. Reservoir prediction technique: including geologic methods, logging, and gravity-magnetic-electrical-seismic techniques for volcanic reservoir prediction.
10. Comprehensive reservoir evaluation: including single-well reservoir evaluation, regional reservoir evaluation, development reservoir evaluation, reservoir sensitivity evaluation, and reservoir behavior evaluation.

The advent of volcanic hydrocarbon reservoir geology is the result of convergence between development trends of science, exploration, and development in petroleum geology. As a discipline, volcanic hydrocarbon reservoir geology as established will no doubt help drive the establishment of the theoretical system and methodology for volcanic reservoirs, help drive further deepening of volcanic hydrocarbon reservoir research and further increase the exploration and development success rate, and also help train high-level professional technical experts and theory innovators in related fields. The advance of volcanic hydrocarbon reservoir geology as a comprehensive discipline involving multiple subjects and multiple technologies requires collaboration and cooperation of multiple disciplines. At the same time, it can also promote the advancement of other related disciplines. Therefore, research in volcanic hydrocarbon reservoir geology is significantly important to further drive hydrocarbon exploration and development in volcanic rocks.

3.1.3.2 Main Areas and Key Points of Research on Volcanic Reservoirs in Sedimentary Basins Across China

China boasts a massive distribution area of volcanic rocks. The total area is $215.7 \times 10^4 \, \text{km}^2$, where the predicted payable exploration area is $39 \times 10^4 \, \text{km}^2$. Today, volcanic reservoirs have been discovered in many basins, including Songliao, Bohai Bay, Hailar, Erlian, Subei, Junggar, Santanghu, Tarim, and Sichuan. The proven original oil in place (OOIP) is about 6×10^8 tons, and the proven original gas in place (OGIP) is about $4700 \times 10^8 \, m^3$. Volcanic hydrocarbon discovery and development of different extents have been made in almost all hydrocarbon-bearing basins. Deep layers in Songliao Basin and the Carboniferous system in northern Xinjiang are the new areas of major exploration efforts in recent years where hydrocarbon exploration has made important breakthroughs and advances. They are one of the important succession areas for future onshore hydrocarbon development across China.

Throughout geologic history, China has been a country with many volcanoes. Volcanic rocks of different ages are widely developed. Volcanic rocks of marine facies and continental-oceanic interaction facies mostly appeared before the Triassic Period and are mainly distributed in western China, where weathered crust volcanic reservoirs are developed and large-scale stratigraphic hydrocarbon reservoirs are formed primarily. Volcanic rocks of continental facies mostly appeared after the Triassic Period and are mainly distributed in eastern China where lithologic hydrocarbon reservoirs exist primarily. Here, primary volcanic reservoirs are formed because of weak tectonism. There are many types of volcanic rocks across China, including those ranging from femic to acidic to alkaline volcanic rocks, where intermediate and acidic rocks have the widest distribution. Paleozoic volcanic rocks are significantly different from Mesozoic and Cenozoic volcanic rocks (Table 3.6).

3.1.3.2.1 Lithologic Characteristics and Distribution of Volcanic Rocks

Through many years of research, the lithologic characteristics of volcanic rocks in Chinese sedimentary basins have become clear through wide-ranging discussions in a huge amount of literature. The key is that system concepts such as classification and naming of volcanic rocks await to be unified. Logically determining the types of volcanic rocks, volcanic lithofacies, and the petrologic characteristics of volcanic rocks to facilitate correlation in hydrocarbon reservoir research becomes necessary. By providing unified rock and

TABLE 3.6 Comparison of Characteristics Between Volcanic Reservoirs and Hydrocarbon Reservoirs of Eastern and Western China

	Eastern China	Western China
Age	Mesozoic Era, Cenozoic Era	Paleozoic Era
Distribution	Through-going mostly along NEE or NE direction	Through-going along NWW direction
Rock series and structural setting	Ranging primarily from alkali-rich calc-alkalic volcanic rocks at activated zones in continental plate of Mesozoic Era to alkaline series volcanic rocks appearing during end of late Cretaceous Epoch-Cenozoic Era, reflecting the structural setting of first compressing and then extending	Primarily calc-alkalic volcanic rocks formed at former island arc or active continental margin orogenic belts
Development environment	Continental facies primarily	Marine facies and continental-oceanic interaction facies primarily
Lithology	Intermediate acidic, with appearance of alkaline rocks	Intermediate basic
Lithofacies	Explosive facies primarily, eruptive-effusion facies secondarily	Eruptive-effusion facies primarily, explosive facies secondarily
Volcanic edifice	Stratovolcano	Stratovolcano, lava dome
Later transformation	Weak	Intense
Reservoir	Primary type	Transformation and superimposition type
Hydrocarbon reservoir type	Lithologic type	Lithologic and stratigraphic type
Reservoir forming combination	Proximal	Proximal, distal
Keys in reservoir formation	Lithology, lithofacies, volcanic edifice	Volcanic cycle, leaching, and dissolution mechanism

lithofacies classification and naming based on the contents of volcanic geologic research, and by creating a set of fairly systematic volcanic reservoir concepts system, we can thereby resolve the problem of chaotic classification of volcanic lithology and lithofacies.

3.1.3.2.2 Diagenesis Types and Diagenetic Evolutionary Series in Volcanic Reservoirs

The research on the evolution mechanism for forming volcanic reservoirs is the core issue in reservoir research. During the formation process, the volcanic reservoir underwent diverse changes, such as volcanism, hydrothermal alteration after the magmatic stage, weathering-leaching transformation during the eruption dormancy period, diagenetic transformation during the burial period, elevated denudation-leaching, and tectonism that generated various impacts on reservoir development. As far as the current situation is concerned, much research has been conducted regarding the types and impacts of diagenesis. However, how does a particular type of diagenesis affect reservoir development? Through what means and in what form does it transform reservoir texture and structure? What are the chemical thermodynamic conditions required for various diagenetic changes to happen? All these specific questions require more in-depth research.

3.1.3.2.3 Genetic Mechanisms of Weathered Crust, Primary, and Fractured Volcanic Reservoirs

Hydrocarbon exploration and development practices in China indicate that three major types of volcanic reservoirs are developed primarily, namely, the weathered crust type, primary type, and fractured type. Unlike sedimentary reservoirs, the volcanic reservoir is the result of the combined action of both internal and external factors. Internal factors include lithology, lithofacies, and eruption environment. External factors include tectonism and diagenesis. Internal factors set the foundation of reservoir formation and distribution and determine the development degree of reservoir space. External factors can significantly transform the reservoir's accumulation and seepage properties. Packing action is the main reason that causes decreased reservoir performance of volcanic reservoirs. Dissolution is the key for improving reservoir performance of volcanic reservoirs. Tectonism also helps improve the reservoir.

Among the discovered volcanic reservoirs, weathered crust reservoirs are usually excellent. For example, such reservoirs were discovered in the Junggar and Santanghu Basins, where the Ma17 and Ma19 wells show good prospect of exploration. Today, however, research on weathered crust volcanic reservoirs is fairly limited. Many issues still remain unclear, such as their vertical zonality,

the intensity of diagenetic transformation of individual zones, and their reservoir performance.

Primary volcanic reservoirs have also been discovered continuously in recent years, including the volcanic reservoir in the Carboniferous Haerjiawu Formation in the Tiao 16 well at the Tiaohu depression in Santanghu Basin, and the primary volcanic reservoir in the Beisantai area in Junggar Basin. Their development is not only related to later transformations but also depends on their original volcanic eruption types and facies belts. The key is how to accurately evaluate and predict volcanic facies.

The effort of accurately predicting the distribution rule of natural fractures and building 3D quantitative models in fractured volcanic reservoirs is still in the exploration stage, although great advances have been made. Today, borehole fracture identification techniques and methods are becoming increasingly mature and very effective, including core, outcrop fracture observation and description, various sidewall imaging techniques, and conventional logging fracture interpretation methods. The key is to predict the spatial distribution of fractures.

3.1.3.2.4 Volcanic Reservoir Distribution Prediction Technique

The supportive exploration technologies available at CNPC (China National Petroleum Corporation), such as seismic volcanic reservoir prediction techniques and large-sized fracturing equipment, are improving continuously and preliminarily form an exploration series for volcanic reservoirs. This series includes a four-step method for predicting volcanic reservoirs: Step 1, regional prediction of volcanic rocks, primarily using high-precision gravity-magnetic-electrical and 3D seismic techniques; Step 2, identification of volcanic rock targets; Step 3, prediction of volcanic reservoirs; and Step 4, prediction of volcanic rock fluids. We are presently tackling key problems continuously and have made advances in volcanic rock lithology prediction, target identification, reservoir prediction, and fluid detection techniques.

3.1.3.2.5 Geologic Modeling and Development Reservoir Evaluation of Volcanic Reservoirs

For volcanic reservoirs, development evaluation is quite difficult because the lithology and lithofacies are complicated, the spatial through-going and internal texture of reservoirs are more complicated, pores, cavities, and fractures are developed, many types of storage-seepage combinations exist, and the accumulation and permeation abilities are highly different. Therefore, the geologic attribute model must be built by finely depicting the shape, internal texture, and attributes of volcanic reservoirs, thereby ascertaining the texture characterization and distribution rule of volcanic reservoirs. A system framework for evaluating volcanic reservoirs must be established by evaluating their petrographic characteristics, lithofacies characteristics, reservoir space types and characteristics, and reservoir classification, thereby guiding exploration and development practices.

3.2 DEVELOPMENT ENVIRONMENT AND TYPES OF VOLCANIC RESERVOIRS

3.2.1 Characteristics of Volcanic Eruption in Sedimentary Basins

China has a wide distribution of volcanic rocks, which include two types: marine facies and continental facies. The latter is further divided into two kinds: above water and underwater. Volcanic rocks of marine facies are mainly distributed in pre-Triassic strata, while volcanic rocks of continental facies are mostly distributed in post-Triassic strata. Rock types mainly include basic and acidic rocks, with a minor amount of intermediate and alkaline rocks. Mesozoic and Cenozoic volcanic rocks are mainly distributed in eastern China and Tibet and are a component of the grandiose West Pacific volcanic rock belt along the north-northeast direction. In eastern Chinese regions, for example, the area of Yanshanian volcanic rocks is more than $50 \times 10^4 \, \text{km}^2$, and the area of the volcanic rock belt at Great Xing'an Mountains exceeds $100 \times 10^4 \, \text{km}^2$. Such a massive area is rarely seen among geotectonic units worldwide. The explosion in the form of volcaniclastic flow is the strongest, followed by magmatic fallout, with only minimal eruptive-effusion activity. For the southeastern coastal volcanic rock belt, volcanic eruption is absolutely dominant, where the volume of volcanic eruption products is about $48 \times 10^4 \, \text{km}^3$, accounting for approximately 80% of the volume of eruptive materials ($60 \times 10^4 \, \text{km}^3$). In addition to fallout eruption, the volcanic eruption mode is primarily Plinian-type eruption; i.e. tall eruption columns are formed, which then collapse to become high-temperature (above 600 °C) volcaniclastic flow that flows rapidly on the surface and finally becomes wide-coverage, nonwelded to welded tuff. Yanshanian volcanic eruptions in eastern China are all central eruptions; 153 paleovolcanoes have been restored along the southeast coast, including 102 calderas (Tao et al., 1996).

Volcanic rocks are widely distributed in hydrocarbon-bearing basins across China, where four stages of volcanic rocks are developed primarily: Precambrian, Paleozoic, Mesozoic, and Cenozoic. Sinian volcanic rocks are mainly intermediate basic, which are observed in the Tarim Basin. Paleozoic volcanic rocks are mainly upper Paleozoic and are primarily developed in western Chinese regions, such as the Permian System in Tarim Basin, the Carboniferous Batamayineishan Formation and Permian Jiamuhe Formation in Junggar Basin, and the Carboniferous-Permian System in Tuha and Santanghu Basins. Mesozoic volcanic

rocks are primarily intermediate basic and intermediate acidic. Cenozoic volcanic rocks are primarily basalt. For example, deep layers in Songliao Basin mainly feature development of volcanic rocks of the Jurassic Huoshiling Formation and Cretaceous Yingcheng Formation; Bohai Bay Basin mainly features development of Cenozoic volcanic rocks. Onshore hydrocarbon-bearing basins across China mainly involve central eruptions, with some regions featuring development of fissured eruptions.

Individual edifices of volcanic rocks in the Yingcheng Formation in Songliao Basin are formed primarily through central eruptions, where their entirety shows catenulate planar distribution because they are controlled by regional faults (Huang et al., 2007) (Figure 3.7). However, some researchers also concluded that fissured eruption and central eruption are both developed in the volcanic rocks of the Yingcheng Formation, their lateral thickness varies significantly, their volcanic facies mainly comprises eruptive-effusion and volcanic sedimentary facies, and volcanic cones are often developed. The volcanic rocks of the Huoshiling Formation mainly feature fissured eruption mode, the range of their lateral distribution is wide, their thickness variation is relatively uniform, stratovolcanic edifices are developed mostly, and their volcanic facies is mostly the eruptive-effusion facies. There are two types of eruption conduits for volcanic rocks in Songliao Basin. The fissured eruption often features a streak-shaped distribution along a single direction, a wide distribution range of lava, a large lava thickness near the gap that gradually thins toward both sides, primarily volcanic lava, low percentage of volcaniclastic debris, and development primarily at the slop region west of the fault depression. The central eruption features quite intense volcanic eruptions that formed different sizes of volcanic cones. This type of volcanic rock consists of volcaniclastic rock and lava and has a high percentage of volcanic breccia and volcanic ash.

The Mesozoic volcanic activity in Erlian Basin is a continental facies eruption. After early and middle Jurassic sedimentation, as the Yanshan movement that swept eastern China intensified, the earth's crust had undergone strongly fault depression, and magma erupted or overflowed via fissured and fissure-central modes along the NE-NNE fault or at the intersection between two groups of faults. The late Jurassic volcanic rocks of the Xing'an Mountain group are, from bottom to top, composed of acidic, intermediate basic, and acidic volcanic lava and tuff and feature fissured eruption. The Aershan Formation of the lower Cretaceous Bayanhua Group shows basalt, basaltic andesite, and tuff, and features fissured-central eruption; intermediate-basic magma was also erupted during the early part of the upper Cretaceous Epoch and is characterized by fissured eruption. Cenozoic volcanic eruptions mainly occurred during the Himalayan movement phase. The gray-black and dust-colored basalt of the Neogene Baogedawula Formation

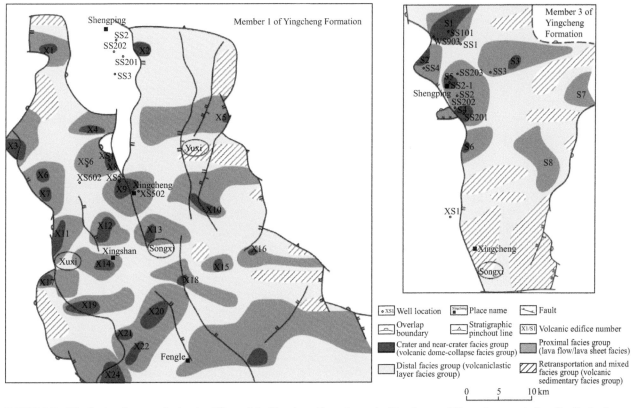

FIGURE 3.7   Distribution diagram of volcanic edifices of the Yingcheng Formation at the Xingcheng and Shengping areas in Songliao Basin. *Reproduced from Huang et al. (2007).*

was formed via fissured eruptions, and the Abag basalt of the Pleistocene Epoch was formed via central intermittent eruptions (Yu, 1988).

Since the Paleogene Period, multiple stages of tectonic activities at Liaohe depression have all been accompanied by different degrees of magmatic eruptions. Volcanic rocks are distributed along faults; faults controlling magmatic activities mainly align along the northeast direction, followed by the northwest and near east-west directions. Near the spillway, volcanic rocks mostly appear as dikes or lava cones; far from the spillway, they appear as lava sheets interlayered with sand or shale of various strata and members. Paleogene-Neogene volcanic activities were most intense during the depositional stage of the Fangshenpao Formation, including 12 eruptions during 4 stages. Profiles show that volcanic rocks are developed in all kinds of lithologic sections. For example, Long 13 well is drilled in thick sandstone, Re 3 well is drilled in thick argillaceous rock, Niu4 well is drilled in carbonaceous mudstone, and Liao 3 well is drilled in interbedded sandstone and mudstone. They are interbedded with sedimentary rocks and have conformable contacts with both upper and lower sedimentary rocks. Core observation shows that no baking or metamorphosing phenomenon is present in the upper and lower layers of volcanic rocks. For example, the siltstone

above the basalt at 2726 m in the Yu 17 well was not metamorphosed, and the layer above the basalt at 2111.66 m in the Re 21 well is a simple fossil bed, where gastropod fossils were well preserved. Cenozoic volcanic rocks in the Liaohe fault depression are the products of multiple underwater effusions that occurred simultaneously with sedimentation. The distribution of volcanic rocks is not controlled by the sedimentary environment and lithology because they were intermittently erupted underwater and effused through faults over multiple episodes (Liu, 2001).

The Nanpu depression was undergoing the early rift valley evolution stage during the early Cretaceous Epoch, when both tectonism and volcanic activities were very intense. Its volcanic rocks primarily feature dotted eruption or upward intrusion and are mainly located near fundamental major faults and boundary faults. Since the Paleogene Period, multiple tectonic activities occurring at the Nanpu depression were all accompanied by magmatic eruptions of different intensities; in particular, lower Paleogene rocks are the most well developed. Eruptions of volcanic rocks are divided into five stages according to the eruption intensity and spatiotemporal distribution rule, where volcanic rocks of individual stages are all regularly distributed along faults. Deep major faults along the east-west and northeast directions are the principal faults that control magmatic activities at Nanpu depression, followed by faults along the northwest direction. The mode of magmatic activities features central eruption and fissured eruption (Tan and Tian, 2001).

The volcanic rocks at the Dongying depression mainly feature lava flow and lava sheet, where multiple stages of volcanic activities created composite lava sheets. The facies are mainly eruptive-effusion, with minor explosive and extrusive facies; the eruption environment features continental facies, with some underwater eruption as well. Volcanic lava sheets mostly show isometric shapes with primarily isolated distribution, presumably from a central eruption. A few lava sheets are controlled by faults; show linear arrangement; and feature large thickness, large area, and smooth occurrence. For example, 352 present-day wells encountered volcanic rocks within the range of nearly 1000 km^2 west of the Huimin depression. Volcanic rocks are all highly developed from Member 3 of the Shahejie Formation to Guantao Formation. Yuhuangmiao is the area with the most significant development of Huixi volcanic rock, where the single-well average thickness is 443.7 m among all wells drilled therein and each well encountered volcanic rocks 53.6 times on average. Well Block Xia 5 features the strongest volcanic activity, where volcanic rocks were encountered for a total of 188 times, with an accumulated thickness of 1028 m. Within the Dongying depression, volcanic activities on the east and west sides are very different. On the west side of the central uplift zone, the depositional stage of Member 2 of the Shahejie Formation underwent the most intense volcanic activity. Only sporadic distribution of basalt comes from the depositional stage of the Dongying to Guantao Formations, and no volcanic rock is observed from other periods. On the east side, volcanic activities were well developed during the depositional stage of Member 3 of both the Shahejie Formation and Guantao Formation, where rock types are complicated, comprising volcanic rocks, subvolcanic rocks, and volcaniclastic rocks. Underwater effusive eruptions were predominant, accompanied by explosive eruptions and volcanic rock intrusions. The place where faults merge is the eruption center with the most intense volcanic activity. Research indicates that Huixi had many volcanic eruption centers during the depositional stage of Member 3 of the Shahejie Formation to the Guantao Formation, which were distributed at the place where two or more groups of faults merged.

The Minqiao volcanic rocks at the Gaoyou depression in north Jiangsu Province were formed by onshore central effusions and later onshore eruptions, where they were formed by magma that flowed into water and accumulated underwater. The closer to the crater, the thicker the volcanic rock.

From the Cretaceous Period to Paleogene Period in Jianghan Basin, volcanic rocks were formed mainly by underwater eruptions; however, onshore-erupted volcanic rocks are developed at structural highs such as Jinjiachang of the Jiangling depression.

In the Santanghu Basin, current research concluded that the eruption environments of the Kalagang and Haerjiawu Formations are also different. The Kalagang Formation is formed mainly by onshore volcanic eruptions, while the Haerjiawu Formation as a whole demonstrates the characteristics of underwater sedimentation. Therefore, two different types of volcanic reservoirs were created (Table 3.7).

The Permian volcanic activities at the Tahe area in Tarim Basin have intermittent eruption characteristics, characterized by alternation between long-term slow effusions and short-term rapid eruptions. The eruption mode mainly consists of fairly quiet effusive eruptions, occasionally accompanied by fairly intense explosive eruptions, thereby forming a volcanic eruption cycle ranging from basalt to dacite or simple dacite.

For the Carboniferous Formation at the northwest margin of Junggar Basin, the base is volcanic breccia, the middle is volcanic rock, the top is glutenite, and the volcanic eruption is primarily fissured-central type. In the breccia lava widely distributed across the Shixi area in the middle of Junggar Basin, the percentage of brown and liver-brown volcanic rock is high, which features onshore (especially during atmospheric precipitation while erupting) or shallow underwater eruption. At the Wucaiwan depression in the east, the base consists of Neopaleozoic-Carboniferous volcanic rock (alternating lava and volcaniclastic rock) primarily, and the color is generally dark and mostly green gray. It includes very little breccia lava and welded breccia laminated with thin layers of mudstone and sandstone. Its sedimentary formation contains fossils of marine facies. Volcanic activities as a whole present the characteristics of being relatively weak during the lower Carboniferous Epoch and being relatively intense during the upper Carboniferous Epoch, showing intermittent continental volcanic eruptions. The volcanic rock is formed in an epicontinental volcanic sedimentary environment and is characterized by deep-water eruption (i.e. erupted in deep water). From west to east, the volcanic eruption environment tends to transition from above water to underwater (Yu et al., 2004).

3.2.2 Types of Volcanic Rocks in Hydrocarbon-Bearing Basins

There are many rock types in volcanic reservoirs in China's hydrocarbon-bearing basins. The lava type includes basalt, andesite, dacite, rhyolite, and trachyte; the volcaniclastic rock type includes agglomerate, volcanic breccia, tuff, and welded volcaniclastic rock. Mesozoic volcanic reservoirs in eastern China were mostly formed from the late Jurassic to the early Cretaceous epochs, and all lithologies

TABLE 3.7 Comparison Between Characteristics of Onshore and Underwater Volcanic Eruptions

Type	Onshore eruption	Underwater eruption	
		Onshore eruption-underwater sedimentation	Underwater eruption-underwater sedimentation
Lithology	Including all kinds of lava, clastic lava, volcaniclastic rock, and sedimentary volcaniclastic rock	Volcaniclastic rock and volcanic lava, where volcaniclastic rock mostly comprises layers containing crystal fragments, vitric fragments, and sedimentary tuff, commonly with all kinds of clay minerals	
Color	Oxidation color predominantly, showing pale red color	Reduced color predominantly, showing pea green color	
Texture structure	Joints mostly	Appearance of lava is perfectly round, presumably pillow lava with orbicular structure; easy to form graded bedding, horizontal bedding, and deformation bedding	
Occurrence	Showing unconformable contact with underlying formation, commonly with paleoweathered crust and scour pits	Showing conformable, pseudoconformable, or erosional contact with underlying formation, weathered crust is not developed	
Alteration characteristics	Commonly with olivine phenocryst iddingsitation and chloritization	Albitization, chloritization, with spilite-quartz keratophyre formed	
Reservoir space	Primary vesicles, secondary pores, contraction fractures, structural fractures (fractures-pores predominantly)	Blast fractures, interpillow fractures-pores, later structural fractures, acidic water-dissolved fractures/cavities (pores-fractures predominantly)	
Example	Ma17 well	Ma29 well	

from basic to acidic are developed, but acidic lithology is predominant. Cenozoic volcanic reservoirs in eastern China include volcanic reservoirs at the Jiangling depression in Jianghan Basin, volcanic reservoirs at various depressions in Jiyang sag, and volcanic reservoirs at the eastern depression in Liaohe sag. All lithologies from acidic to basic are developed, but intermediate-basic lithology is predominant. Volcanic rocks in western Chinese Basins are intermediate basic predominantly.

The Cretaceous Xing'an Mountains group in Hailar Basin can be divided into three sections from bottom to top: (1) the bottom is the intermediate-acidic volcanic rock section, which mainly comprises a set of intermediate-acidic lava, volcaniclastic rock, dusty-yellow rhyolitic porphyry, trachyte, and green-gray tuff; (2) the middle is the intermediate-acidic volcanic rock plus a laminated coal layer section; it is pseudoconformed on top of the intermediate-acidic volcanic rock section, its lithology features purple-sage andesite, andesitic basalt, and laminated coal bed, and this layer is 0-1.5 m thick; and (3) the top is the intermediate-basic volcanic rock section, whose lithology features thick black-gray to black basalt laminated with thin layers of black mudstone.

The types of volcanic rocks in Songliao Basin are diverse. Being intermediate basic to acidic predominantly, they include 12 types, namely, rhyolite, andesite, dacite, basalt, basaltic andesite, trachyandesite, rhyolitic breccia

tuff, rhyolitic volcanic breccia, dacitic volcanic breccia, basaltic andesitic volcanic breccia, andesitic crystal tuff, and sedimentary volcanic breccia (Figure 3.8). Rhyolite has the highest thickness frequency, which accounts for 70.46%; followed by andesite and dacite, accounting for 6.15% and 5.42%, respectively; basalt, for 3.43%; trachyandesite, for 3.04%; rhyolitic breccia tuff, for 2.80%; dacitic volcanic breccia, for 2.72%; basaltic andesite, for 1.73%; rhyolitic volcanic breccia and basaltic andesitic volcanic breccia, each for 1.49%; andesitic crystal tuff, for 0.86%; and sedimentary volcanic breccia, for 0.40%. The intermediate-acidic volcanic rock accounts for 86% of the sample total, while basic volcanic rock accounts for 14% of the sample total (Figure 3.9) (Zheng et al., 2007).

Erlian Basin features development of autoclastic brecciated andesite, vesicular-amygdaloidal lava, blocky lava, tuff, breccia, and reservoir rock.

The volcanic rocks at the Tsagaan Tsav depression in Yingen Basin include intermediate-basic basalt, trachyandesite, and andesite, with a minor amount of tuff, welded breccia, and diabase.

Bohai Bay Basin features basalt, trachyte, and diabase. For example, Cenozoic volcanic rocks in Liaohe Basin feature an assemblage of basalt-andesite and trachyandesite-trachyte, including types such as trachyte, trachyandesite, trachybasalt, basalt, basaltic andesite, and andesite. Mesozoic volcanic rocks include andesite, while Paleogene

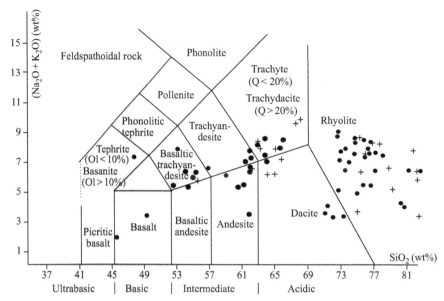

FIGURE 3.8 Total alkaline and SiO_2 (TAS) diagram for deep volcanic rocks in Songliao Basin. Q, quartz; Ol, olivine.

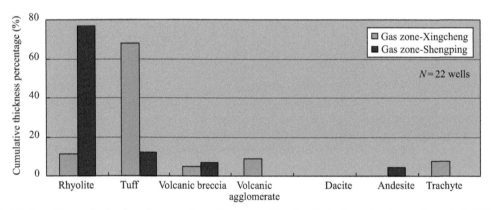

FIGURE 3.9 Distribution of types of volcanic rocks across the northern area of Songliao Basin. *Reproduced from Zheng et al. (2007).*

volcanic rocks include basalt and trachyte. The Jurassic Formation in the Jizhong depression features dark-mauve and gray andesite laminated with tuff, and its top features basalt, andesitic breccia, and volcaniclastic sandstone. Its Cretaceous Formation features mottled volcanic breccia at the bottom, while the top features gray tuffaceous glutenite, sandstone, and andesitic breccia. The Dongying depression features widespread development of basic volcanic rock, subvolcanic rock, and volcaniclastic rock. Main rock types include olivine basalt, basalt, basaltic porphyrite, tuff, and volcanic breccia. Volcanic rocks in the Fenghuadian area in Huanghua depression include pantellerite, dacitic rhyolite, rhyolite, and rhyodacite. The Nanpu depression features basic volcaniclastic rock, intermediate volcaniclastic rock, and basalt. The Gaoyou depression features gray-black, green-gray, and purple-sage basalt. The types of Cretaceous-Paleogene volcanic rocks in the Jianghan Basin mainly include quartz tholeiite, olivine

tholeiite, and basaltic porphyrite and, secondarily, include diabase and volcaniclastic rock.

Permian volcanic rocks in the Sichuan Basin include plagiobasalt, tuff, and tuffaceous breccia.

The types of Permian volcanic rocks in Tarim Basin are fairly simple, where the lava type includes basalt and dacite. Dacite is predominant, accounting for 80.3% of the total thickness of volcanic rocks, followed by brecciated dacite and a minor amount of breccia basalt, brecciated tuffaceous dacite, brecciated tuffaceous basalt, tuffaceous breccia, volcaniclastic breccia, crystal-vitric tuff, crystal-lithic tuff, crystal tuff, sedimentary tuff, sedimentary volcanic breccia, conglomeratic tuffaceous mudstone, and conglomeratic tuffaceous siltstone (Yang et al., 2004).

Carboniferous volcanic rocks in the northern Xinjiang region are intermediate basic predominantly, including basalt, basaltic andesite, and andesite, with also minor acidic rhyolite, characterized by medium-low potassium

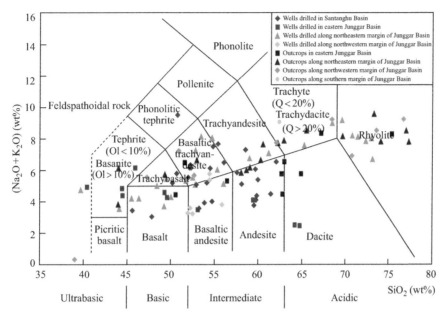

FIGURE 3.10 TAS diagram of Carboniferous volcanic rocks in the northern Xinjiang region. Q, quartz; Ol, olivine. *Base map reproduced as per recommendation by the Igneous Rock Taxonomy Sub-Committee of International Union of Geological Sciences (1989).*

FIGURE 3.11 Potassium contents of Carboniferous volcanic rocks in northern Xinjiang region (categorized based on GB/T 17412.1-1998).

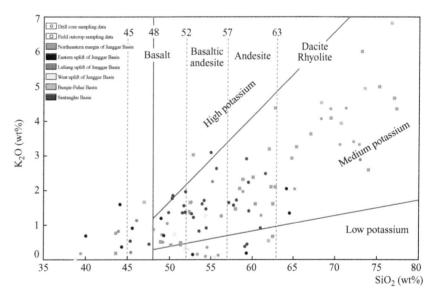

(Figures 3.10 and 3.11). The types of volcanic rocks include lava predominantly, followed by volcaniclastic lava, volcaniclastic rock, and sedimentary volcanic rock. The Ludong-Wucaiwan area in Junggar Basin features basalt, andesite, dacite, rhyolite, volcanic breccia, and tuff; the Carboniferous lithology at the northwestern edge features andesite, basalt, andebasalt, volcanic breccia, tuffaceous breccia, welded breccia, tuff, and agglomerate. Major types of Carboniferous volcanic rocks in the Santanghu Basin include basalt, andesite, dacite, rhyolite, tuff, volcanic breccia, diabase, and gabbro, showing quite complicated rock types.

3.3 CHARACTERISTICS OF VOLCANIC RESERVOIRS IN HYDROCARBON-BEARING BASINS

The characteristics of volcanic rock are the key parameters of volcanic reservoirs. The earliest and most in-depth research on the formation mechanism of volcanic reservoirs occurred in Japan, but these are mostly descriptive. Volcanic reservoirs in Japan are distributed in the deep seabed and are composed of Miocene rhyolite erupted on Paleogene base rock. Rhyolites include lava, pillow breccia, and vitreous clasolite, and they possess primary and secondary pore space. Pore types in volcanic reservoirs mainly include vesicular, intercrystalline, pumice, perlite, intracrystalline, and microcrystalline types. Primary pores originate from rapid cooling, breaking and flaking, breccia lithification, and crystallization of erupted magma on the seafloor. Secondary pores are generated by hydrothermalism as well as by subsequent volcanism and tectonic movement. Macrofractures develop in lava and pillow breccia, huge cavities occur in vitreous clasolite, lava, and pillow breccia, and microfractures have significant impact on permeability. For example, based on the analysis of 380 core samples from 150 m deep in the Nanchanggang gas field, their porosity ranges from 10% to 20%, and there is no obvious difference in lithofacies; however, permeability varies from 5 to 100×10^{-3} μm^2 of pillow breccia, through 1 to 20×10^{-3} μm^2 of lava, to vitreous clasolite less than 1×10^{-3} μm^2.

Recently, overseas research on volcanic reservoirs has focused on their development mechanism with the increasing discoveries of volcanic oil and gas fields. Sruoga et al. (2004) and Sruoga and Nora (2007)) systematically analyzed volcanic reservoirs of the Austral and Neuquen Basins and believed that both porosity and permeability of volcanic reservoirs depend on the characteristics of source rock as well as late diagenesis. And volcanic reservoirs could be divided into primary and secondary processes (Table 3.8). The primary process involves clinkering, crystal dissolution after the magmatic stage, gas escape, and cataclasis, and the secondary process involves hydrothermal alteration and fracturing. The primary process may lead to high porosity and low permeability, and the secondary process may play the role of reducing porosity. Both porosity and permeability tend to increase under the action of dissolution and cataclasis.

Liu (1986) was one of the earliest domestic researchers focusing on volcanic reservoirs. Based on the analysis of volcanic reservoirs of the Gu-65 well in the No. 1 zone of Karamay oil field, he believed that one crater exists near the Gu-65 well area, and volcanic rock is divided into effusive-lava facies and explosive-volcaniclastic rock facies. Secondary minerals were generated from early-stage chlorite and calcite, through middle-stage zeolites to late-stage calcite. Volcanic reservoirs have eight types of pores, of which the most favorable ones are intercrystalline dissolved pores and dissolved pores. Volcanic reservoirs also include four types of pore assemblages, of which the most favorable is the assemblage of gravel-edge fractures, intercrystalline pores, intergravel dissolved pores, and intercrystalline dissolved pores. Pores of volcanic reservoirs are related to dissolution, and their formation depends on fractures generated by tectonic movement. Plenty of secondary pores form on the premise of secondary change.

Recently, research on volcanic reservoirs has made great progress with renewed oil and gas exploration in the deep layers of Songliao Basin and the upper Paleozoic

TABLE 3.8 Primary and Secondary Processes of Volcanic Reservoirs and Produced Pores

Genesis	Process		Pore type
Primary	Clinkering		Inner hole of vitric fragments and pumice
	Crystal dissolution after the magmatic stage		Intracrystalline cribriporal or moldic pores
	Gas escape		Vesicle and gas conduits
	Cataclasis	Cataclasis of fluid	Interbedded fractures or tensional fractures
		Autoclastic brecciation	Intragranular pores
		Cataclasis of phenocrysts	Crystalline cataclase pores
Secondary	Alteration	Dissolution and precipitation of secondary minerals	Spongiform pores to caves
		Precipitation of secondary minerals in pores	Drusy
		Migration of secondary minerals	Secondary cribriporal or moldic pores
	Fracture	Tectonism	Structural fractures
		Quenching	Quenching fractures
		Hot-quenching brecciation	Drusy breccia pores

volcanic rock of northern Xinjiang. This research has focused on the development environment of volcanic reservoirs, lithologic and lithofacies characteristics, reservoir properties, main controlling factors of reservoir development, and reservoir assessment and prediction.

To date, researchers Huang et al. (2009) and Li and Wang (2008) have arrived at the following basic conclusions in the formation of volcanic reservoirs:

(1) Types and genesis of reservoir space: includes primary and secondary reservoir space, which are further divided into primary pores, primary fractures, secondary pores, and secondary fractures.

(2) Diagenetic evolution of reservoirs: volcanic reservoirs generate and evolve in stages. Generally, it includes a

volcanism stage and late diagenesis or alteration stage, although the division varies with researchers. The volcanism stage involves volcanism during volcanic eruption stage as well as hydrothermal alteration after magmatic stage, when volcanics erupt to the surface to form vesicles, condensed contraction fractures, and volcanic interbreccia pores. The late diagenesis stage involves weathering and leaching alteration during the volcanic dormancy period, and alteration during burial diagenesis, uplift denudation, and structural faulting. During this stage, dissolved pores, cavities, and fractures are formed, as well as filling of calcite, dolomite, and zeolites under the action of burial compaction, dissolution, leaching, filling, and faulting.

(3) Controlling factors of reservoir development: lithology, eruption environment, diagenesis, and tectonic disruption all impact the formation, development, blocking, and reformation of volcanic reservoir space.

3.3.1 Volcanic Reservoir Space

Pores, cavities, and fractures in the volcanic reservoir are the spaces and channels where oil and gas accumulate and migrate through. As a result, research on pore structure is an important topic of volcanic reservoirs.

The structure of the volcanic reservoir is under the control of magma properties and changes in the crystallization process and physicochemical conditions. In particular, either temperature drop or exsolution of volatile constituents may lead to crystallization and finally form volcanic rock with a certain structure and texture. The physicochemical properties of magma (including chemical composition, temperature, and content and type of volatile constituents), together with the physicochemical environment of its *in situ* location (including temperature and pressure), have an impact on mineral constituents and structure of rocks. Therefore, the same magma may solidify to form volcanic rock with completely different mineral constituents and structure under various physicochemical environments. Rapid pressure drop may lead to exsolution of the volatile constituents, where exsolved gas rises and accumulates to form abundant vesicles above the lava flow. After complete cooling, the vesicles may subsequently form intercrystalline pores and contraction pores, which are predominantly primary pores in volcanic rock. Primary pores in lava are favorable for the formation of deep hydrocarbon reservoirs because they are not vulnerable to the impact of compaction. Volcaniclastic rock is a transitional type between volcanic rock and sedimentary rock. Large rock masses and plastic debris contain vesicles, and volcaniclastic rock undergoes compaction. Generally speaking, volcanic reservoirs have complex pore structures and mostly are dual-porosity reservoirs (including pores and fractures). Fractures and different assemblages of fractures

and pores exist in the reservoirs. Therefore, volcanic reservoirs have diversified physical properties and high heterogeneity.

Based on the data, observations, and analyses of cores, rock thin sections, and impregnated sections of volcanic reservoirs in Chinese hydrocarbon-bearing basins, and in combination with the formation and evolution mechanism of volcanic reservoirs, reservoir space is divided into primary pores, secondary pores, and fracture pores (Table 3.9). Primary pores involve vesicles formed by volcanic eruption onto the surface, as well as residual pores, intercrystalline micropores, and volcanic interbreccia pores filled with amygdaloid. Secondary pores are products of a volcanic massif suffering from weathering and denudation, and they include devitrified pores of pyromeride, dissolved pores of feldspar (product of phenocrysts, crystal fragments, microlites, and devitrification), volcanic ash and clay minerals, vesicles filled with carbonate amygdaloid, fractures filled with carbonate veins, and dissolved pores in carbonate minerals generated by rock carbonatization. Epidiagenesis, physical weathering, and leaching and dissolution by atmospheric water may lead to dissolved pores, cavities, and fractures. Explosive fractures and contraction fractures are generated by volcanism and diagenesis, but relatively large faults are developed in volcanic massifs by structural stress, and tectonic fractures associated with these faults make up the dominant seepage channels and reservoir space in volcanic rock. Weathering denudation and dissolution, together with structural stress, may denude and destroy a volcanic massif. Even if volcanic rock is covered with an overlying formation, water or an organic acid solution may flow into the volcanic massif, resulting in dissolved pores and fractures.

3.3.1.1 Primary Pores

Primary pores refer to the pores and fractures generated in volcanic rock during magma intrusion, eruption, cooling, and crystallization, prior to diagenesis, and preserved to this day. Primary pores may be divided into primary vesicles, intergranular pores, and intercrystalline pores (Feng et al., 2008; Wang et al., 2003a,b,c).

3.3.1.1.1 Vesicles

Vesicles in volcanic rock are an important component of good reservoirs. They are primary pores formed in the process of volcanic eruption. For eruptive rocks condensed and consolidated on the surface, rapid cooling results in the exsolution of volatile components or the escape of excess gas. The exsolved gas rises, accumulates, and swells to form numerous vesicles in lava, especially in the upper lava flow. As a result, a certain amount of vesicles exist in consolidated volcanic rock. These vesicles are filled with minerals to form amygdaloids, and in the case of incompletely filled

TABLE 3.9 Space Type and Characteristics of Volcanic Reservoirs

Reservoir space type		Rock type	Genesis	Characteristics	Oil- or gas-bearing properties
Primary pore	Vesicle	Andesite, basalt, breccia, and breccia lava	Gas expansion and effusion during diagenetic processes	Mostly distributed in the top and bottom of rock-flow with various sizes and shapes	Vesicles connected with fractures or cavities have relatively good oil- or gas-bearing properties
	Intergranular pore	Volcanic breccia, agglomerate, and volcanic sedimentary rock	Residual pores between clastic particles by diagenetic compaction	Mostly occur in volcaniclastic rock	Good oil- or gas-bearing properties
	Intercrystalline and intracrystalline pore	Basalt, andesite, and autoclastic brecciated lava	Rock-forming mineral framework	Mostly distributed in the middle of rock flow with relatively small pores	Mostly are free of oil
	Condensed contraction pore	Basalt	Volume contraction of magma due to condensation	No definite directionality and irregular shape	Contraction joints connected with vesicles are filled with oil or gas
Secondary pore	Devitrified pore	Pyromeride	Devitrification of rhyolitic glass	Micropore with good connectivity	Favorable gas reservoir space
	Dissolved pore of feldspar	Various rock types	Dissolution of feldspar developing along cleavage crack	Irregular pore shape	Dominant reservoir space
	Dissolved pore of volcanic ash	Tuff, welded tuff, and volcanic breccia	Dissolution of volcanic ash	Abundant small pores with good connectivity	Capable of forming favorable reservoirs
	Dissolved pore of carbonate rock	Various rock types	Dissolution of calcite and siderite	Relatively large pores	Good oil- or gas-bearing properties
	Dissolved cavity	Basalt, andesite, breccia lava, and breccia	Weathering, leaching, and dissolution	Develop along fractures, autoclasis clastic belts, and structural highs	Good oil- or gas-bearing properties
Fracture	Explosive fracture	Autoclastic brecciated lava and subvolcanic rock	Autoclasis or cryptoexplosion	Resumable	Realtively good oil- or gas-bearing properties
	Contraction fracture	Basalt, andesite, and autoclastic brecciated lava	Magmas upwell to destroy upper lava during the cooling contraction and condensation	Columnar joints in open form, and planar fissuring with few dip offset	Common to relatively good oil- or gas-bearing properties
	Structural fracture	Various rock types	Tectonic stress	Well developed near faults, relatively straight, and most are high-angle fractures	Related to the timing of tectonic movement
	Weathered fracture	Various rock types	Various weathering	Connected with dissolved pores, fractures and cavities, and structural fractures	No reservoir significance

amygdaloids, they form residual pores. Some pore space may be preserved in the amygdaloids, or some amygdaloids that fill primary vesicles may be dissolved by late fluid action to form intra-amygdaloid pores. These types of pores have irregular shapes, various sizes, and uneven distribution. The pores are associated with the growth of quartz, albite, and carbonate in the vesicles. Vesicles in volcanic rock or intra-amygdaloid pores mainly occur in pyromeride and rhyolite, such as the vesicles and residual pores in the pyromeride of the Shengshen202, Shengshen203 and Xushen9 wells in Songliao Basin. Flow structures are developed in pyromeride, where vesicles show zonal distribution and form vesicle belts locally. Pyromeride with underdeveloped quartz and albite cystals in the vesicle margin of pyromeride may form primary ultra-large pores. In addition, vesicles or residual vesicles also exist in the plastic deformation debris of welded tuff. When volcanic rock masses in agglomerate and volcanic rubble in volcanic breccias are pyromeride, rhyolite, or welded tuff, vesicles or intra-amygdaloid pores also exist. For example, in Bohai Bay Basin, primary vesicles are common in breccia or basalt of volcanic breccia. Vesicles account for more than 50% of the total volume of breccia and average 20-30%, while they are less than 10% in basalt. Vesicles may be round, elliptical, or irregular shaped and are mostly isolated in the rock. Fractures (structural and contraction fractures) developed in rock generally cut through the vesicles and connect isolated vesicles. Occasionally, vesicles are filled with carbonate minerals and gypsum and form amygdaloidal structures (Vernik, 1990).

Based on their genesis, shape, and attitude, vesicles in volcanic rock are classified into escape pores, expansion pores, explosion pores, and creep pores.

Primary pores may form favorable fracture-vesicle type reservoirs with the alteration of fractures. For example, high-production layers in the volcanic reservoir in Member 3 of the Shahejie Formation in Binnan of Shandong include vesicle-enriched volcanic rock with the alteration of weathered and structural fractures, where vesicles are the dominant reservoir space. In addition, vesicles are well developed in the volcanic reservoir of Daxing'anling of Erlian Basin.

3.3.1.1.2 Intergranular Pores and Volcanic Interbreccia Pores

Intergranular pores are common in volcaniclastic rock, especially in volcanic breccia. They have similar pore characteristics to intergranular pores in clastic rock. With shallow burial depth, poor diagenesis, and incomplete filling, primary interbreccia pores in the Paleogene volcanic rock of Jiyang sag developed in Member 1 of the Shahejie Formation and volcaniclastic rock of Dongying Formation and appear as residual interbreccia pores. Primary

interbreccia pores in the lower Liaohe sag predominantly developed in the lower Submember-3 of the Shahejie Formation, as well as autoclastic lava. Abundant interbreccia pores are developed in the volcanic reservoirs of Yingen and Songliao Basins. These pore types are supported by volcanic breccia facies and mainly occur in volcanic explosive facies.

3.3.1.1.3 Condensed Contraction Pores

Condensed contraction pores are found in basalt of volcanic reservoirs in the Jinjiachang structure of Jiangling sag. They are pores and fractures formed by eruptive magma on the surface as a result of volume contraction. These pore types, with no directionality and an irregular shape, are mostly associated with vesicles. They intersect other fractures to connect with vesicles and other reservoir spaces. If they connect with vesicles, they may contain disseminated or be semifilled with crude oil.

3.3.1.1.4 Intercrystalline and Intracrystalline Pores

Both intercrystalline and intracrystalline pores are generated by mineral crystallization. These pore types are commonly found within phenocrysts, especially within matrix microlites. The higher the degree of crystallization, the more developed these pore types are. The pores are also found in varieties of volcanic rock. Generally, intercrystalline pores have relatively small scale, and the diameter width is less than 0.05 mm. Although they are small, they play a great role in improving rock porosity and permeability. For example, some tiny intercrystalline pores in feldspar crystal particles in the volcanic reservoirs of the Jinjiachang structure of Jiangling sag have relatively small size (generally 10-15 μm in diameter), but most of them have good oil-bearing properties.

3.3.1.2 Secondary Pores

These pore types refer to the pores or cavities resulting from minerals or filling minerals between grains, crystalline particles, or primary pores that have been dissolved or carried by water under the action of hydrothermal alteration, dissolution, or weathering. Dissolution may occur in unstable components in volcanic rock, such as the Paleogene basalt reservoir of the Binnan area, whose reservoir spaces are mainly dissolution pores with dissolved components of amygdaloid, basic volcanic glass, and dark mineral phenocryst.

3.3.1.2.1 Micropores Generated by Devitrified Vitreous Rhyolitic Glass

Pyromeride has fluidal, vesicular, and porphyritic structure. Its matrix appears as a spherolite texture, and phenocrysts include quartz and alkali feldspar. Micropores in the center of a spherolite are the products of devitrified rhyolitic glass. Rhyolitic glass may form feldspar and quartz after devitrification and volume contraction. That is, rhyolitic glass has lower density than feldspar and quartz, and crystal minerals generated by devitrification of rhyolitic glass may shrink to form micropores. In addition, feldspar generated by devitrification of rhyolitic glass may dissolve to form a great quantity of secondary pores under the action of acid fluid. Feldspar particles generated by devitrification of rhyolitic glass are relatively fine and may also form micropores. Given that the rock is completely composed of spherical particles generated by devitrification of rhyolitic glass and quartz content is neglected, if calculated with the density of orthoclase (lowest density), the density of rhyolitic glass is about 2.36 g cm^{-3}, while the average density of orthoclase, albite, and quartz is 2.59, 2.615, and 2.65 g cm^{-3}, respectively. All products are denser than the rhyolitic glass. Devitrified feldspar particles of 1 kg rhyolitic glass may form 37.63 cm^3 pores. The comparison and estimation of rhyolitic glass in pyromeride before and after devitrification show that devitrification of rhyolitic glass may produce a large quantity of micropores. Therefore, devitrification of rhyolitic glass in pyromeride is important to the genesis of pore space (Zhao et al., 2004).

3.3.1.2.2 Dissolved Pores in Feldspar Minerals

These pore types include dissolved micropores in feldspar phenocrysts, microlites, and crystal fragments, which are developed in volcanic rock in large quantities. Feldspar is partly dissolved by an acidic medium to form dissolved pores in feldspar with irregular shapes. Generally, dissolved pores in feldspar are developed along cleavage fractures, and the main body of the feldspar grain is still preserved. This type of dissolved pore is the dominant reservoir space in volcanic rock.

Impregnated thin sections show that microcrystalline feldspar generated by devitrification of porphyritic crystal feldspar in rhyolite and microcrystalline feldspar and acid glass in the matrix may dissolve to form pores, and especially the central spherical particles tend to dissolve. Although some pores are tiny, well-developed micropores connected to each other will form favorable gas reservoir space.

Petrographic analyses give evidence of dissolved pores in feldspar. Dissolved pores in feldspar involves the dissolution of phenocrysts, crystal fragments, microlites, and devitrified feldspars. Acid fluid in rock fractures reacts with the feldspar to dissolve it. When the reaction reaches balance, the feldspar components may be dissolved. Inorganic acid reacts with albite to form kaolinite, quartz, and potassium and sodium ions. These ions are carried away by fluid to form pores. In addition, the acidity of the solution has obvious impact on the dissolution of feldspar.

An increasing acidity (H$^+$ concentration) is favorable for the dissolution of feldspar. This explains why feldspar tends to dissolve in an acid medium and transform into kaolinite and quartz and form secondary pores favorable for hydrocarbon accumulation. Thermodynamics calculations show that the solubility of feldspar is related to temperature and pressure, and it decreases with rising temperature and increases with rising pressure. Besides the above-mentioned temperature, pressure, and fluid acidity, feldspar composition also has an important impact on its solubility. Feldspar with various compositions dissolves in a sequence of anorthite, albite, and potash feldspar. In consideration of the dissolution temperature, thermodynamics tendency, and volume contraction, the large-scale dissolution of potash feldspar and albite in rock is the dominant factor that forms secondary pores in reservoirs. A quantity of feldspar in volcanic rock, along with the introduction of acidic fluid, is essential to form a reservoir. Dissolved pores cannot form if an acidic medium does not react with feldspar, and acidic fluid does not permeate a rock without pores such as vesicles and fractures.

In conclusion, primary vesicles, fluidal structure, fissures, and micropores generated by devitrification of rhyolitic glass in volcanic rock are prerequisites for late-stage acid fluid entering volcanic rock, and they are favorable for the dissolution of feldspar. Inorganic acid reacts with potash feldspar and albite to form kaolinite, quartz, potassium, and sodium ions. The ions are carried away by fluid, and kaolinite and quartz provide filling materials for fractures in volcanic rock, which are generally filled with quartz and clay minerals. The occurrence of kaolinite in fractures in reservoir rock means that kaolinite does not accumulate in primary pores or plug pores. Instead, it is driven by seepage water to migrate to other places and deposits, where it is carried away by a reaction product. This contributes to the reaction mentioned above and the formation of pores.

3.3.1.2.3 Dissolved Pores in Volcanic Ash

Dissolved pores in volcanic ash are an important type of reservoir space. They mainly occur in rhyolitic tuff, rhyolitic welded tuff, and rhyolitic volcanic breccia. Generally, they occur as a quantity of micropores with good connectivity and, as a result, also create a favorable reservoir. When volcanic ash is extensively dissolved, the pores form large dissolved cavities, which are favorable for reservoir formation.

Volatiles in magma have an impact on its eruption mode, as a volcano strongly explodes when volatiles accumulate. Volcanic ash or dust is the product of magma bursting as a result of strong expansion of volatiles near the surface or as a result of gas release in high-viscosity rhyolitic magma. Volcanic ash or dust refers to semiplastic vitric fragments with granularity less than 2 mm. Vitric fragments are the product of vesicle walls bursting in vesicular magma, which does not solidify completely during eruption. Semiplastic vitric fragments are unstable materials that may devitrify to finally form fine quartz and feldspar and release some space when the physical condition changes. In addition, semiplastic vitric fragments and fine quartz and feldspar may undergo dissolution or alteration because of their large surface area under the action of fluid.

Both the alteration of volcanic ash and dissolution and the alteration of feldspar may form clay minerals, which are hydrous aluminosilicates with complex chemical components. Besides kalium and natrium, they also contain magnesium and ferrum. The minerals have various textures. Feldspar has a framework silicate texture, and clay minerals have a layer silicate texture. Despite all this, clay minerals may dissolve to shrink their volume and increase porosity under the action of acid fluid. They have a dissolution mechanism similar to feldspar, and acid may dissolve aluminumsilicate minerals.

3.3.1.2.4 Dissolved Pores in Silicate

Calcite and siderite in volcanic rock are developed after the formation of volcanic rock. They plug early-stage pores in reservoir rock and decrease reservoir capacity. However, the dissolution of carbonate minerals, especially calcite and siderite, may greatly improve reservoir capacity. An abundance of carbonate minerals may deposit to plug pores under alkaline environments, and conversely, they react with acid fluids and dissolve to form pores under acidic conditions. The dissolution of carbonate minerals is closely related to fluid properties, temperature, CO_2, salinity, pH value, and pressure. The transition between calcite and siderite is closely related to Ca^{2+}/Fe^{2+} concentration, temperature, and pressure of fluid in pores. Ca^{2+}, Fe^{2+}, and HCO_3^- dissolving in water may increase the porosity.

3.3.1.2.5 Dissolved Cavities

Surface freshwater or groundwater seeps downward and dissolves minerals along structural fractures to form dissolved pores, dissolved fractures, and dissolved cavities. Dissolved cavities refer to dissolved space larger than 2 mm, which are mainly developed in concentrated fracture zones, fracture intersection positions, or fracture breaking zones. Dissolved cavities are well developed in some rock cores, accounting for 30-40% of core volume. Dissolved cavities have irregular shapes and generally are 10-50 mm in size. Most cavities are filled with zeolites and calcite in the margin and center, respectively. Some cavities are only filled partially or unfilled. During core sampling, it is common to find crude oil emerging from dissolved cavities. For example, dissolved pores in the Bohai Bay Basin have relatively large diameter, generally larger than 5 mm. This is observed in cores, such as

diabase cores in the Xia-39, Xia-382, and Luo-152 wells, where dissolved pores have diameters up to 2 cm as a result of simultaneous dissolution of pyroxene phenocrysts and plagioclase matrix.

3.3.1.3 Fractures

Fractures occur in rock under the action of tectonic stress. Some early-stage fractures may be filled, while unfilled late-stage fractures occur as vertical and lateral fractures. Some fractures transect interconnected vesicles or matrix-dissolved pores. Fractures connect isolated pores or cavities and improve reservoir space in volcanic rock, which is favorable for hydrocarbon migration and accumulation. Core and impregnated thin sections show that locally developed fractures and micropores may connect formerly isolated pores to form effective reservoir space and provide the basis for forming secondary pores in rock, although they are not the dominant reservoir space in volcanic reservoirs.

3.3.1.3.1 Blast Fractures

Early-stage mineral crystals and clasts in surface magmas burst to form micropores within particles under the action of explosive power. This fracture type commonly occurs in volcanic autoclastic lava.

3.3.1.3.2 Condensed Contraction Fractures

During magma crystallization or condensation, tensile stress resulting from thermal loss or cooling contraction of melt mass may lead to breaking of the rock matrix to form contraction fractures, which are also called joint fractures. Contraction fractures are classified into vertical and horizontal joints in terms of attitude. Vertical joints are oriented near-vertically and mostly have an angle of inclination of 80°, near-parallel fractures, and long extension. For example, fractures in the cores of the Luo-151 well extend up to 2 m. In diabase cores from the Shang741 well, this type of fracture terminates in overlying mudstone. Most contraction fractures are open fractures. During the development of contraction fractures, volcanic rock cores are laminated, and the laminate thickness is related to the degree of development of the fractures. Horizontal joints are oriented near-horizontally, and they have relatively poor continuity and shallow fracture width. Fractures in volcaniclastic rock occur in volcaniclastic particles and between the particles. Contraction fractures occurring in volcaniclastic particles have the following shape characteristics: (1) Contraction fractures center around vesicles and radially disperse and gradually die out or narrow down. The long extension of some fractures may connect with isolated vesicles and improve porosity and peameability. (2) Contraction fractures occur in filling materials (amygdaloids) of vesicles, and they have an irregular polygonal structure with mud crack shapes. (3) Contraction fractures have no relationship to vesicles or amygdaloids. They are isolated and have irregular shapes, and individual fractures are wide in the middle and shallow in both ends. Contraction fractures among volcaniclastic particles mainly occur between volcaniclastic particles and filling materials. Because of diagenetic dehydration and contraction, volcaniclastic particles and filling materials occur in the positions that are similar to particle-attaching fractures in clastic rock.

3.3.1.3.3 Structural Fractures

These fracture types are a series of fractures generally related to regional tectonic stress. Structural fractures usually cut vesicles and dissolved pores, thereby connecting multiple pores or cavities so as to improve the physical properties of volcanic reservoirs. Generally, they occur in volcanic rocks in groups, where deep cutting of structural fractures resulting from long extension, especially those with widths of 0.1-0.5 mm, may connect vesicles and dissolved pores to form important seepage channels, which preserves a small quantity of crude oil. Structural fractures are classified as fully filled extension fractures, semifilled extension fractures, and unfilled fractures in terms of filling status. According to their genesis, fractures are divided into three types:

(1) Structural arched fractures: volcanic rocks in one area usually form when magma erupts at various stages. For example, the volcaniclastic rock in Member 1 and diabase in Member 3 of the Shahejie Formation in the Shang741 well area experienced at least two stages of formation. Each stage of magma eruption and invasion may lead to upward arched tension. Under the action of arched tension, early-stage intrusive rocks with dense and fragile lithologies may form a series of arched fractures, most of which are low-angle open fractures with dips less than 60°. This type of fracture extends for only a short distance, generally 0.3 m or so; vertically, they have a radial distribution (disperse upward and converge downward). In addition, this fracture type is also developed in metamudstone overlying diabase where low-angle open fractures are developed. Both of them have similar characteristics.

(2) Structural shear fractures: the structural fractures generated by regional structural shear stress have a strike identical to regional faults. For example, the structural fractures in the volcanic rock of Member 3 of the Shahejie Formation in the Shang741 well area mostly occur along an east-west direction and correspond with subsidiary faults in the regional Linyi fault zone. These types of fractures generally are high-angle inclined fractures with large dips (mostly greater than 65°), smooth fracture faces, and long fracture extensions. Multiple fractures commonly occurring in cores are arranged like a brush, where they converge downward, bend, and then spread or disperse upward. While the fracture inclinations are identical, their dips are different.

(3) Structural microfractures: these are mainly seen under the microscope. Generally, this pore type with an irregular shape is observed cutting through mineral crystals.

3.3.1.3.4 Weathered Fractures

Volcanic rocks may burst to the surface and dissolve to form weathered fracture zones under various external agencies. Weathered fractures in outcrop have no directionality, and they gradually diminish from the surface of the rock matrix, while fractures are developed in rock as a whole. Weathered fractures are commonly interconnected with dissolved pores, dissolved fractures, and structural fractures and cut rock into fragments of various sizes. This fracture type is sometimes filled with mauve iron mud materials and has little reservoir capacity. However, weathered fractures provide conditions for later-stage stuctural fractures or deep buried hydrothermal dissolved pores. For example, the Carboniferous weathered fractures in the northwest margin of the Karamay oil field are developed near the weathered and denuded top surface of volcanic rock. These types of irregularly shaped fractures may form favorable reservoir space when they are not completely filled. For example, the section from 471 to 486 m of top volcanic rock in 95,704 well belongs to fractured reservoir.

3.3.1.3.5 Dissolved Fractures

Dissolved fractures are primarily developed along structural fractures and mineral cleavage fractures or developed along former microfractures that were filled and then dissolved during weathering. During dissolution, fractures widen to form a special type of fracture with an irregular margin.

Core and petrographic thin-section observations show that the most common secondary reservoir space in Santang Basin involves dissolved pores, unfilled or semifilled vesicles, and dissolved fractures. Dissolution pores are common in rock, including dissolved olivines, plagioclases, pyroxene phenocrysts, and matrix minerals. In addition, mineral filling vesicles are soluted and flushed to form dissolved pores. Microfractures are not common in volcanic reservoirs in this area, only occurring in some individual intervals. Available data show that microfractures occur in all well intervals with concentrated shows of oil and gas (Ma17 and Tangcan 3 wells). Fractures observed under the microscope account for 75% of the total area of thin sections in the oil-bearing layer of the Kalagang Formation in Ma17 well, and the fractures are filled with laumontites. In thin section, we see that fractures in the Carboniferous of the Tangcan 3 well are completely filled with calcite and quartz. We have observed that thin sections in well intervals without shows of oil and gas are zones where dissolved fractures are undeveloped.

3.3.1.4 Reservoir Space Assemblage

The quality of volcanic reservoirs depends on the degree of development of reservoir space. Volcanic reservoir space is diverse. Generally, vesicles and dissolved pores in volcanic reservoirs contain plentiful crude oil, while structural fractures and weathered fractures mainly connect vesicles, dissolved pores, and other reservoir space and act as pathways during the process of hydrocarbon migration. They can be oil reservoir space in themselves, but their reservoir capacity is small. Generally, all types of reservoir space exist as assemblages rather than independently. Moveover, reservoir space assemblages vary with reservoir intervals. Besides volcaniclastic rock, primary pores such as intercrystalline and intracrystalline pores, contraction cavities, intergranular pores, and vesicles developed in other volcanic rocks are also dispersed, and they are incapable of forming networks and accumulation and permeation space. Only if volcanic massifs undergo tectonic, weathering, hydrothermal, and condensation processes can they form varieties of pores and fractures which, together with cavities, interconnect to form hydrocarbon reservoir space.

Wang et al. (2003a,b,c) classified the reservoir space of volcanic reservoirs in the Xujiaweizi fault depression of Songliao Basin into 3 categories and 12 basic genetic types (Table 3.10). Each type of reservoir space has its own genesis and fabric characteristics, which can be identified by core or debris observation and microfabric analysis. The lithology of volcanic rock has little relationship to its petrophysics, as an identical lithology may have favorable or unfavorable petrophysics. For example, the porosity of tuff, rhyolite, and andesite is 0.7-8.8%, 0.8-9.8%, and 0.2-8.6%, respectively. Reservoir properties are influenced by lithofacies of volcanic rock and location of the volcanic edifice. Generally, porosity is relatively low in subvolcanic facies (for example, the porosity of porphyrite in the Shengshen3 well is 0.3-0.8%); it is relatively high in the upper subfacies of eruptive rhyolite as well as fallout tuff (for example, the porosity of eruptive rhyolite in the Fangshen701 well is 8.4%). Layered volcanic edifices have better physical properties than shield volcanoes, and volcanic edifices with multiple volcanic facies have better reservoir properties than individual volcanic facies.

Volcanic reservoir space assemblages invlove pore-fracture assemblages, pure fractured reservoirs, assemblages of matrix-dissolved pores and fractures, and assemblages of porphyrocrystic dissolved pores and fractures. Reservoir space varies with the specific volcanic reservoir. For example, the volcanic reservoir space in the Shengshen201

well is the assemblage of primary vesicle or amygdaloid inner pore and fracture. Former vesicles with various sizes in volcanic rock are isolated and disconnected. Fractures in volcanic rock induced by tectonic movement connect the primary vesicles to form pore structure in the reservoir. Volcanic reservoirs in the Wang 903 well include two types of pore structures: one is the fracture-fracture assemblage, comprising individual fractures, parallel fractures, intersecting fractures, and irregular network fractures, and the other is the pore-fracture structure, comprising porphyrocrystic dissolved pores and fractures, which has the same genesis as the assemblage of matrix-dissolved pores and fractures. Volcanic reservoirs have their own characteristics of pore space and assemblages. Vesicles vary in diameter from several microns to tens of centimeters, and they are irregularly shaped and have different fracture widths. These vesicles are partially connected after diagenetic alteration. Epigenetic alteration has obvious impact on the characteristics of volcanic reservoirs as well as the reservoir space assemblage.

For example, the Meso-Cenozoic volcanic reservoir in Liaohe Basin has dissolved pores dominant in the reservoir, and the reservoir space is the assemblage of vesicles, structural fractures, dissolved fractures, and dissolved pores. This assemblage is mainly developed in the vesicle-enriched basalt and partial trachyte. When fractures are dominant, it is the assemblage of structural fractures, dissolved fractures, and dissolved pores, which is developed in trachyte. When intercrystalline pores are dominant, it is the assemblage of intercrystalline pores, intramicrolite pores, dissolved fractures, and primary vesicles, which is developed in tight volcanic rock and belongs to unfavorable reservoir space. When intragranular pores are dominant, it is the assemblage of intragranular pores, structural fractures, and matrix-dissolved pores, as well as the assemblage of intragranular pores, vesicles, structural fractures, and matrix-dissolved pores, which occur in welded volcanic breccia and belong to moderate-quality reservoir space. When intergranular pores are dominant, it is the assemblage of intergranular pores, primary pores, structural fractures, and dissolved pores, which is common in breccia lava or other volcaniclastic rock. During diagenesis, pore types and assemblages may change; however, reservoirs with identical lithologies in the same area may have similar or identical reservoir space.

The lithology of volcanic reservoirs in the Beipu area of Huanghua sag includes basalt, basaltic andesite, and diabase. Generally, reservoir space is classified into pores and fractures, and they are divided into primary pores and secondary pores in terms of genesis.

The Carboniferous volcanic reservoir space of the Junggar-Santanghu Basin mainly appears as pores and fractures, and it can be divided into primary and secondary pore space (Table 3.11). Generally, secondary reservoir space is

TABLE 3.10 Types and Characteristics of Volcanic Reservoir Space in the Xujiaweizi Fault Depression of Songliao Basin

Genetic type	Pore type	Genesis	Characteristics	Distribution
Primary pore	Primary vesicle	Volcanics containing plentiful gaseous-liquid inclusions erupt onto surface to form late-stage unfilled vesicles in the upper flow units	Circular, elliptical, linear, and irregular shape, unequal size and uniform distribution	Common in rhyolite and basalt
	Lithophysae vesicle	Volcanics containing plentiful gaseous-liquid inclusions erupt onto surface to form large vesicles in the upper flow unit, and the vesicles filled with hydrothermal material condense and contract to form fractures along pore walls	Larger than common vesicles, 4-6 cm in diameter, circular and elliptical shape dominantly, and highest distribution density	Common in rhyolite
	Intra-amygdaloid pore	Amygdaloids are vesicles not completely filled with minerals, and pores among the amygdaloids are called intra-amygdaloid pores	Rectangular, polygonal, or angular irregular shape	Common in rhyolite
	Interparticle or intercrystalline pore	Residual pores that are generated by volcaniclastic particles of diagenetic compaction	Irregular shape, commonly occurring along the clast margins, with good connectivity	Common in volcaniclastic rock
	Matrix fracture and mineral cleavage fracture	Formed by bursting of mineral phenocrysts caused by rapid cooling and pressure release during magmatic eruption	Irregular or cleavage-like crystal surface	Common in porphyritic volcanic rock
Secondary pore	Intracrystalline dissolved pore	Pores are generated by dissolution of common feldspar, quartz, olivine, and picrite phenocrysts in porphyritic igneous rock	Irregular shape, and completely dissolved minerals exhibiting ghost image of former crystal	Common in rhyolite and basalt
	Matrix-dissolved pore	Volcanic matrix commonly consists of microcrystalline feldspar and a vitreous mass. Vitreous mass devitrifies to form chlorite and siliceous materials, and microcrystalline feldspar dissolves to connect fractures	Connective to a certian extent in spite of small pores	Developed in rhyolite
	Interbreccia pore in fault breccia	Structural fractures form in consolidated volcanic rock associated with strong tectonic activity in volcanic conduit facies and appear as multiple structural activities when they are not completely filled with late-stage hydrothermal fluid	Irregular shape in fault breccia, good connectivity, and high pore-throat coordination number	Common in volcanic conduit facies
Fracture	Primary contraction fracture	Stress difference within inner matrix leads to uneven contraction caused by rapid cooling during magmatic eruption	Columnar, lamellate, or spherical joints	Common in rhyolite, perlite, andesite, and basalt
	Structural fracture	Caused by tectonic stress, promotes diagenesis of volcanic rocks	Some early-stage fractures are filled, while late-stage fractures are not filled. Horizontal, vertical, and intersecting distribution; some fractures connect vesicles and matrix-dissolved pores by transecting	Common in rhyolite and andesite
	Residual filled structural fracture	Fractures incompletely filled with late-stage hydrothermal fluid after the diagenesis of volcanic rock	Drusy vug in irregular shape	Volcanic structural belt
	Filling-dissolution structural fracture	Fracture is developed in rocks under the action of structural stress, and the filled fracture after diagenesis dissolves to form effective reservoir space	Remaining former fracture shape	Common in rhyolite, andesite, and basalt

Modified from Wang et al. (2003a,b,c).

TABLE 3.11 Type and Genesis of Volcanic Reservoir Space in the Junggar-Santanghu Basin

Type		Genesis	Characteristics	Corresponding lithology	Typical wells
Primary pore	Primary vesicle	Products of gas expansion and effluence during diagenetic processes	Mostly occur at the top and bottom of rock flow, various sizes and shapes	Volcanic breccia and lava	Shinan-1, Cai6, Cai27, Cai201, Cai202, Cai25, and Dixi10
	Residual vesicle	Pores remaining when vesicles are not completely filled with secondary minerals	Semifilled pores	Basalt, volcanic breccia	Cai201, Cai202, Cai25, and Dixi10
	Intergranular pore	Residual pores among clastic particles after diagenetic compaction	Common in volcaniclastic rock	Volcanic breccia, agglomerate, and volcanic sedimentary rock	Dixi5
	Intercrystalline and intracrystalline pore	Pores in the framework of rock-forming minerals, most porphyrocrystic minerals (such as pyroxene and plagioclase) have cleavage, existing intracrystalline pores	Mostly occur in the middle of rock flow and have small size	Lava and volcaniclastic rock	Dixi10, Cai25, and Cai203
Secondary pore	Phenocryst dissolved pore	Phenocrysts dissolve to form pores under fluid action, and these types of pores are generally developed along cleavage fractures	Irregular shape, intercrystalline pores dominantly	Andesite	Cai25 and Dixi10
	Amygdaloid dissolved pore	Filling materials in vesicles undergo metasomatic dissolution to form dissolved pores	Irregular shape and relatively poor connectivity	Lava	Cai25, Ma19, and Ma17
	Matrix inner dissolved pore	Devitrification of glass in matrix or dissolution of microlitic feldspar	Small pores are dissolved pores with certain connectivity	Tuff, andesite, and basalt	Cai25, Shinan-12, and Ma36
	Interbreccia dissolved pore	Epigenesis, such as weathering, leaching, and dissolution	Developed along fractures, autoclasolite belts, and structural highs	Basalt, andesite, and breccia	Dixi18, Dixi14, Dinan-1, Dixi5, and Niudong9-8
Fracture	Primary condensed contraction fracture	Magma condenses and crystallizes to form contraction microfractures	Columnar joint, open type, planar cracking, and few dip offset	Volcanic breccia, andesite, and trachyte	Cai202, Dixi10, and Shinan-1
	Structural fracture	Volcanic rocks undergo tectonic stress to form microfractures	Developed near faults, relatively smooth and high-angle fracture predominantly	Basalt and andesite	Cai25, Dixi10, Dixi8, and Di-8
	Weathered fracture	Interconnected with dissolved pores, fractures, and structural fractures, and cut rock into fragments of various sizes	Connected with dissolved pores and cavities, and structural fractures	Volcaniclastic rock and volcanic breccia	Common in the area
	Dissolved fracture	Leaching and dissolution		Amygdaloidal andesite and volcanic breccia	

superposed above primary reservoir space. This greatly improves the petrophysics of the reservoirs and leads to complications of reservoir space types. As a whole, secondary pores and fractures are well developed in weathered, denudated, and fractured zones.

The lithology of Permian volcanic reservoirs in the Tahe area of Tarim Basin is primarily dacite, and secondarily basalt, tuff, and volcanic breccia. Volcanic breccia is the most favorable reservoir rock, and secondarily is dacite, followed by basalt. Tuff is the worst reservoir rock. Both the upper and lower parts of volcanic rock are favorable reservoirs. Reservoir space is composite, mostly being fracture-pore types. Reservoir space is divided into primary and secondary pores (Table 3.12).

3.3.2 Diagenesis and Pore Evolution of Volcanic Reservoirs

Volcanic rocks experience a series of diagenetic changes after their eruption and accumulation, which have important impact on reservoir capacity. Deep high-temperature magma may migrate upward along the vulnerable position of structures under the action of volatile components and geologic stress. During the diagenetic process from magma eruption to consolidation, vesicles and cavities are generated in volcanic rock because of changes in physicochemical conditions. Magmatic eruptions disrupt strata and form multiple

stages of first-order minor faults and fractures, which further form fractures and intercrystalline pores during the process of condensation. Diagenesis of volcanic rock includes compaction, filling, hydrothermal alteration, and dissolution, which may play favorable or unfavorable roles in the formation of reservoirs (Feng et al., 2008; Wang et al., 2003a,b,c; Yan et al., 1996).

3.3.2.1 Diagenesis

3.3.2.1.1 Filling

Filling is common in basalt and volcaniclastic rock. Filling of authigenic chlorite and illite is most common in volcanic rock and occurs in vesicles, fractures, and intercrystalline pores. Generally, intercrystalline pores and vesicles are filled with pompon-like or cancellate chlorite, and the minerals grow from the inner vesicle wall to the center. In addition, we have seen that vesicles and interbreccia pores are filled with calcite and zeolites. Filling is unfavorable for the development of volcanic reservoirs, as it reduces porosity and permeability.

3.3.2.1.2 Compaction

Compaction is unfavorable for the fomation, preservation, and development of volcanic reservoirs. It has the most significant impact on volcaniclastic rock. Shallow burial depth and poor compaction are favorable for the preservation of

TABLE 3.12 Pore types and Characteristics of Permian Volcanic Reservoirs in the Tahe Area of Tarim Basin

Pore type			Genesis	Pore assemblage characteristics
Primary		Vesicle	Gas expansion and escape	Most vesicles are isolated and connected by fractures and dissolved fractures
		Interbreccia pore or fracture	Area between volcanic breccia fragments	Connected with pores or fractures
		Intercrystalline pore or fracture Condensed contraction fracture	Crystallization Condensation and contraction of magma	Partially connected with dissolved pores or fractures
		Intracrystalline pore or fracture	Crystallization	
Secondary	Dissolved	Intergranular dissolved pore	Dissolution and leaching	Connected with dissolved pores
		Phenocryst Matrix-dissolved pore or interstitial material dissolved pore Dissolved fracture	Dissolution	Connected with pores or fractures
	Structural	Structural fracture	Tectonic stress	Connected with pores or fractures
		Weathered fracture	Weathering and denudation	Connected with dissolved pores or dissolved fractures

Modified from Luo et al. (2003).

primary reservoir space. For example, the volcanic reservoirs in the Paleogene strata of Jiyang sag, where primary pores are dominant, are found in Member 1 of the Shahejie and Dongying Formations.

3.3.2.1.3 Hydrothermal Alteration

Vesicles in volcanic rock are not reservoir space directly. Vesicles are filled with such minerals as chlorite, zeolites, and calcite initally; then the minerals are dissolved by groundwater. Only after the void space is connected with fractures can hydrocarbon accumuate. This explains the favorable effect of secondary changes on pores. Chloritization, calcite replacement, and zeolitization are common diagenetic alteration processes.

3.3.2.1.3.1 Chloritization Chloritization is common in volcanic matrices, as well as feldspar phenocrysts or matrix plagioclase microlite. Authigenic chlorite occurs in volcanic vesicles and behaves as free growth from the margin to the center. Chrysanthemum chlorite with good crystallization may plug pores. Chloritization of plagioclase appears as a replacement of plagioclase, which may form tawny vermiform or foveolate matter. It can be carried away to form vermiform micropores on the surface of plagioclase by late-stage groundwater dissolution. Thin sections show that chlorite in vesicles form an extremely thin ctenoid traps filled with celestite and calcite. This shows that the formation of authigenic chlorite was earlier than that of calcite. Chloritization of dark minerals is more intensive. Chloritization may lead to expansion of microscopic cracks and cleavage fractures due to the dissolution of matrix and feldspar, but on the other hand filling by newly precipitated chlorite may reduce reservoir space. During the process of chloritization, the combined action of fluid participation and material carrying improve the porosity.

3.3.2.1.3.2 Calcite Replacement Calcite replacement is a common phenomenon in deep volcanic rock and volcaniclastic rock. Calcite may replace phenocrysts in lava and crystal fragments in volcaniclastic rock, and they also replace all types of debris and matrix. Extensive replacement may make the former rock texture unidentifiable or form replaced carbonate rocks. Based on an analysis of the attitude of calcite, this type of replacement occurs subsequent to the chloritization and dissolution of feldspar. For example, chloritization seen in thin sections of one well is under the control of primary fabric, while the distribution of ferrocalcite cuts primary fabric. In the crystal tuff of the Shengshen3 well, ferrocalcite is replaced by residual dissolved crystal feldspar, and some ferrocalcites are preserved completely while others are dissolved. Calcite replacement is an important mark of the frequent movement of local fluids. During the process of replacement, former pores enlarge or

form new pores due to the disappearance of replaced minerals. Some pores are filled with newly precipitated ferrocalcite, while it is soluble. As a result, replacement often improves the porosity. Early-forming calcite has relatively good crystal form because of the large growth space available. After calcite formation, dissolved pores and cavities may develop because of groundwater dissolution. Late-forming calcite has poor crystal form because of limited growth space, and the crystals appear as filling structures with undeveloped pores. Besides the filling calcite in fractures, calcitization of plagioclase is also found, from extremely fine star shapes to uneven porphyritic shapes, and even complete calcitization of whole particles. After the calcitization, an abundance of dissolved micropores are generated in plagioclase by dissolution.

3.3.2.1.3.3 Zeolitization Zeolitization is a special epigenetic change of volcanics. Zeolites are most common as a fracture filling and occur in column or needle shapes. It is observed in thin sections that zeolites fill in residual pores of chlorite, or zeolite veins cut through chlorite and calcite veins. This shows that the formation of zeolites was subsequent to chlorite and calcite. Pores in zeolites are mainly intercrystalline dissolved pores and intercrystalline pores, whose development determines the quality of the pore structure. Both intercrystalline dissolved pores and intercrystalline pores are developed, and pore structure is better in areas where zeolites are formed. The larger the zeolite crystals, the larger the formed intercrystalline dissolved pores. As a result, zeolitization is extremely favorable for the development of pores.

3.3.2.1.4 Weathering and Leaching

Volcanic rocks always experience weathering and leaching to some extent. Volcanic rocks are primarily formed in the surface environment, and they subside underground to constitute basin-filling sequences by differential vertical movement. For most volcanic rocks, their degree of pore development is closely related to leaching (dissolution, oxygenation, hydration, and carbonatization of minerals), which not only breaks the rock but also obviously changes the chemical composition of rock. Dissolution may carry away soluble materials in rock to enlarge pores and improve permeability. In weathered zones, weathering crust reservoirs reflect the most important impact of weathering dissolution on volcanic reservoirs, and they are often developed in the top of the volcanic massif. For example, in the purple andesitic tuff at the top of the Yingcheng Formation of Shengshen2 well, the former tight, explosive, tuffaceous lava facies is loose and appears as tofu pulp in cores because of weathering and leaching, with porosity greater than 15% and relatively high permeability. Therefore, leaching is not only a common geological phenomenon in volcanic rock but also an important factor affecting volcanic reservoir capacity.

Two regional denudation surfaces exist between the Neogene Guanghuasi and Paleogene Jinghezhen Formations, as well as between the Paleogene Qianjiang and Jingsha Formations in Jianghan Basin. In addition, there also exist local denudation surfaces between the Jingsha and Xingouzui Formations and between Shashi and upper Cretaceous Yuyang Formations in the high position of positive structures. Deeply buried paleodenudation surfaces represent the location of ancient surfaces, and paleohydrogeological conditions below the denudation surface are the same as the groundwater zone below the modern-day surface. The dissolution of soluble rock and minerals decreases with increasing depth. Therefore, the surface porosity of vesicles and dissolved pores in volcanic rocks diminishes downward below the denudation surface, and dissolution does not take place when certain critical depths are reached. With the rising of the earth's crust, the denudation depth increases as well as the critical depth of groundwater dissolution. For example, within the total interval from 836 to 962 m of the Hai-12 well in the Tonghaikou Uplift, Qianjiang sag, Jianghai Basin, the surface porosity of carbonate minerals in basalt is 0.3-1% from 836 to 860 m, 0.2-0.3% from 860 to 930 m, and less than 0.2% from 930 to 962 m. This demonstrates that the depth of 860 m may be the local denudation surface between the Shashi and Yuyang Formations. Surface porosity varying with depth reflects its tendency to diminish when it is distant from a denudation surface. The distance between the top surface of a volcanic rock and an unconformity surface is an important factor that dominates the development of dissolved reservoir space by weathering and leaching.

3.3.2.1.5 Dissolution

Dissolution is another important factor dominating the reservoir capacity of volcanic rocks. Unstable components in volcanic rocks may dissolve to form secondary pores under the dissolving action of acidic water. Internal factors favorable for the formation of secondary pores in volcanic rocks are as follows: (1) tight and fragile lithology tends to form structural fractures under tectonic stress; (2) primary pores or fractures with vesicles and contraction fractures may act as conduits for acidic water when they are connected with structural fractures; (3) abundant unstable components, such as dark mineral crystals and basic volcanic glass, may provide a material basis for the formation of dissolved pores; and (4) both organic and inorganic acid are sufficient to dissolve volcanic rocks. Generally, most volcanic rocks occur near faults, where formation water is acid because of the seepage of atmospheric water along fault surfaces. This type of water may penetrate downward 2-3 km

or more. The penetration depth is related to the density of the pore water and pore pressure. For volcanic rocks at the surface weathering vadose zone, dissolved pores are relatively well developed, which are the dominant volcanic reservoir space.

For example, dissolution of crystal feldspar exists in crystal tuff in the Yingcheng Formation of the Xujiaweizi fault depression, Songliao Basin. In addition, some dissolution is found in matrix plagioclase microlites. Typical dissolution takes place in crystal tuffs in the interval from 2597.0 to 2601.61 m of Shengshen3 well, where crystal tuffs are caesious color and contain about 25% crystal fragments, of which feldspars are dominant. Crystal feldspars dissolve along cleavage to form intracrystalline fenestral dissolved pores. Data by line measurements show that surface porosity of dissolved pores in crystal feldspars ranges from 1.8% to 0.65%, and averages 1.01%. Based on the analysis of the electron microprobe, dissolved crystal feldspars are potash feldspar or albite, and undissolved feldspars are albite. This illustrates that albitization is subsequent to dissolution, and dissolution of crystal potash feldspars precedes albitization. Dissolution carries off the dissolved material, and the total effect is porosity increase.

3.3.2.2 Division of Reservoir Diagenesis Stages

Volcanic reservoir diagenesis may be divided into a diagenesis stage and epidiagenesis stage. On this basis, diagenesis is divided into various periods (Table 3.13 and Figure 3.12). Analyses by various means show that Carboniferous volcanic reservoirs in the northern Xinjiang experienced the long-term combined action of various diagenetic events, and their complex evolution process involved at least two stages of filling and dissolution. Diagenesis varies with basins, zones, and reservoirs.

3.3.2.2.1 Stage of Diagenesis

3.3.2.2.1.1 *Magmatic Crystallization Period* Primary intercrystalline pores are generated in minerals by crystallization of magmatic minerals in this period. However, residual magmas fill in intercrystalline pores and crystallize, which may diminish intercrystalline pores. Moreover, volatile components in magma escape to form vesicles. Compared to other diagenetic stages of volcanic rock, the reservoir space generated in this period is the most favorable for forming reservoirs.

3.3.2.2.1.2 *Consolidation Period* Magmas consolidate and form a variety of contraction fractures with the contraction of rock volume at the late stage of diagenesis. This may increase rock reservoir space and improve rock permeability. This stage is favorable for forming reservoirs.

Primary pores in reservoirs are mainly formed at this stage of diagenesis. A large number of effusive rocks are

TABLE 3.13 Division of Diagenesis Stages of Volcanic Rock

Diagenesis		Diagenetic process	Genesis	Diagenetic products	Type of reservoir space	Hydrocarbon reservoir
Stage	*Period*					
Stage of diagenesis	Magmatic crystallization period	Crystallization differentiation	Magmatic differentiation and fractional crystallization	Volcanic rocks with various crystalline substances and minerals	Contraction fracture, vesicle, joint, cleavage, blast fracture, intergranular pore	Primary pore
	Consolidation period	Consolidation	Gravity differentiation, fallout sedimentation, gas expansion, and consolidation	Volcanic cone, rock quilt, pillow lava, lava bed, shield volcano, lava dome, cupola, breccia cupola, stock, dike, apophysis, laccolith, tabular mass, and lensoid body		
Stage of epidiagenesis	Hydrothermalism period	Filling and alteration	Deep hydrothermal fluid ascending to surface or near surface	Amygdaloid generated from deposits in vesicles (chlorite, lotrite, and zeolites), alteration (olivine replaced by serpentine and iddingsite, and pyroxene and feldspar replaced by chlorite and lotrite)	Intercrystalline micropore in clay minerals, and inner pore of amygdaloid	Fill in primary pore
	Epigenetic transformation period	Weathering and leaching	Thermal expansion and cold contraction, and weathering and leaching of rock	Fractures and spherical weathering, leaching of atmospheric water, and seepage of near-surface water	Weathered fracture, leaching dissolved pore or fracture	Weathered crust
		Tectonism	Tectonic stress	Fractures with various orientations under different rock forces	High-angle fracture, subhorizontal fractures, and net-like fracture	Improving permeability and forming hydrocarbon migration channel
		Dissolution	Burial compaction, dissolution by formation water, and organic acid	Dissolution of phenocryst, matrix, volcanic debris and fractures, improving connectivity	Matrix-dissolved pore, porphyritic crystal dissolved pore, intergranular dissolved pore, and dissolved fracture	Hydrocarbon entrance to form hydrocarbon reservoir

developed in northern Xinjiang, where magmas erupted to the surface and formed vesicles by the escape of volatile components. Primary pores, including fine intercrystalline pores formed between phenocrysts, between microlites, and between microlites and phenocrysts by magmatic crystallization, interbreccia pores created by eruption and explosion, and contraction fractures created by magma condensation, are generated at this stage. These primary pores provide the basis for subsequent development of reservoirs. Vertical profiles show that active volcanic rocks are characterized by an assemblage of large lava and tuff or an assemblage of volcanic breccia, lava, and clasolite.

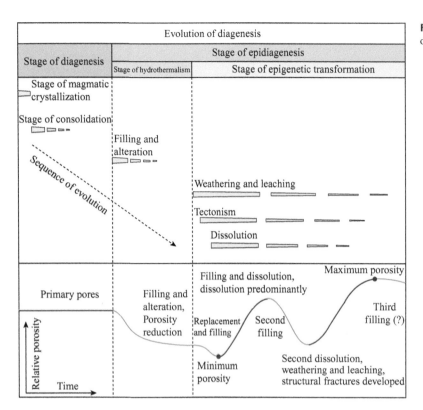

FIGURE 3.12 Evolution processes and their impact on pores during the diagenesis of volcanic rocks.

3.3.2.2.2 Stage of Epidiagenesis

Volcanic reservoir spaces experience an extremely complex evolution process, including formation, development, plugging, and reformation. After the formation of volcanic rock, the degree of development of reservoir space is dominated by epidiagenetic processes, such as tectogenesis, weathering and leaching, and fluid action.

3.3.2.2.2.1 *Hydrothermalism Period* Hydrothermalism is frequent at the end of volcanism. Core observations and thin-section analyses show that dark, rock-forming minerals in volcanic rock are altered by hydrothermalism. For example, pyroxene and hornblende alter to chlorite; basic plagioclase alters to kaolinite, sericite, and chlorite; olivine is replaced by iddingsite; chlorite alters to zeolites and carbonate minerals; and tuff matrix is replaced by carbonate and laumonite. Accompanying this alteration and mineral transformation, minerals (such as chlorite, zeolites, calcite, and quartz) carried by hydrothermal fluids may crystallize, separate out, and fill reservoir space under appropriate conditions. This greatly reduces the reservoir capacity of volcanic rock. Scanning electron microscope (SEM) observations show that pores filled with minerals occur more frequently than other pores in most samples. Pores are filled with minerals to varying

degrees, and filling is the first reason for the decreasing reservoir capacity of Carboniferous volcanic reservoirs in the northern Xinjiang area.

Reservoir space is reduced by quartz filling available pore space. Chlorite has two crystal habits: amygdaloid lining and amygdaloid inner filling, which occur in dense basalt and vesicular (amygdaloid) basalt. There are two stages of calcite filling. Early-stage calcite fills in vesicles and associates with chlorite and zeolites and late-stage calcite mainly fills in fractures and the centers of amygdaloids. Zeolites are dominant filling materials. Thin section, SEM, and electron microprobe results show that most zeolites are radially packed or packed in clusters in the centers of pores, and the others occur in a pore-lining habit similar to horse's teeth. Zeolites that fill fractures are predominantly the horse-teeth habit. Compositional analyses show that zeolites developed in volcanic reservoirs are dominated by laumontite as well as small quantities of heulandite and mordenite. The filling sequence of secondary minerals in primary pores are most easily observed by SEM and impregnated thin sections. Early-stage pores are filled with chlorite, which surrounds the inner pore as a liner. Late-stage pores are futher filled with zeolites, and they are filled with calcite and chlorite once again when the zeolites fillings are incomplete or the zeolites are dissolved. Research has shown that chlorite is commonly associated with zeolites, and they have basically similar distribution characteristics but occur in different sequences. During the stage of alteration, the formation of chlorite precedes that of zeolites, while chlorite generated in vesicles and cavities by precipitation and crystallization is obviously subsequent to zeolites. During the burial evolution of lavas, the upper vesicle development zones are filled with chlorite, then they evolved downward into a transitional zone where zeolites and chlorite are mixed, and the lowest vesicle development zones are filled with zeolites. During the late-stage alteration process, both zeolites and chlorite are subjected to alteration to some degree under the action of weathering, leaching, and dissolution. When weathering is extensive, filling materials dissolve completely, and no chlorite is found in rocks and only a small amount of secondary zeolite filling remains. When weathering is moderate to relatively minor, undissolved or poorly dissolved chlorite is found in pores, as well as undissolved zeolites.

3.3.2.2.2.2 Epigenetic Transformation Period
After eruption and consolidation, the Carboniferous volcanic reservoirs in the northern Xinjiang area experienced uplift and denudation during the late Carboniferous, long-term weathering and leaching, Mesozoic and Cenozoic multistage tectonism, deep burial compaction, and complex transformations by formation fluid.

3.3.2.2.2.2.1 Epigenesis (Primarily Weathering and Leaching)
Vertically, mutiple stages of volcanic eruption cycles are developed in Carboniferous volcanic reservoirs in the northern Xinjiang area, where depositional breaks are developed. Multiple sets of weathered and leached surfaces were formed in the Carboniferous System. Especially, Carboniferous volcanic rocks formed in the Hercynian underwent intense compression fracturing and long-term weathering denudation, forming large-scale unconformity surfaces. Young strata at various horizons directly overlie the unconformable surfaces made up of Carboniferous volcanic rock weathered crust. This explains the characteristics of inconsistent stages for overlying cap rocks of Carboniferous volcanic reservoirs. The longer the amount of time between basement and overlying cap rock, the longer the duration that basement is exposed to the surface, and hence, the longer it is exposed to weathering and leaching. As a result, the pores and fractures in rocks are relatively well developed.

Currently, most Carboniferous volcanic reservoirs in the northern Xinjiang area are developed in the lower weathered crust. Core analysis statistics show that weathering and leaching can obviously improve the porosity of reservoirs. The higher the weathering intensity, the more well-developed pores are. Average primary porosity is 7.6% for basalt without being altered, 8.7% for basalt with minor alteration, and 15.3% for basalt with extensive alteration. Accounting for material loss as well as Al_2O_3 enrichment, weathering intensity can be determined by the weathering index CIA $(CIA = Al_2O_3/(Al_2O_3 + Na_2O + K_2O + CaO) \times 100\%)$, which has good results in comparison with core analysis. The higher the weathering index, the greater the weathering intensity and the more obviously rocks are improved.

Weathering and leaching have the following three characteristics: (1) dissolution space exists at the top and base of multiple effusive cycles. Dissolution started from the top by weathering and leaching, forming alternating zones of dissolved fractures and tight lava. This phenomenon is especially obvious in the continuous cored intervals in the Ma19 well. Rocks appear as a large set of slags only if weathering is complete or structural fractures connect with the volcanic massif. (2) With the combination of weathering/leaching and faults, leached fluid penetrated 500 m below the weathered crust. For example, the dissection of Niudong reservoirs in Santanghu Basin showed that leaching fluids seeped downward along faults, developing a superior, large-scale reservoir network comprising dissolved pores and fractures. (3) Leaching also promotes initial zeolite formation. Zeolites are stable in a high pH (9.1-9.9) environment, but the seepage of surface water may carry away dissolved matter and alter the pH value, which may speed up the dissolution of zeolites. Zeolites are the primary materials that fill in pores and fractures in the Carboniferous volcanic rock of the northern Xinjiang area, and dissolved pores that once held zeolites are predominant reservoir space. As a result, the dissolution of zeolites has significant impact on the improvement of reservoir capacity. Analyses of volcanic reservoirs in the top of the upper Carboniferous Series in the north Xinjiang area show that the dissolution of top zeolites occurs via surface water leaching. Therefore, the discoveries of effective volcanic reservoirs will focus on the leached areas below weathered crusts.

3.3.2.2.2.2.2 Tectonism
The northern Xinjiang area has experienced multistage tectonic movement since the late Paleozoic Era, leading to well-developed fractures. During faulting, abundant fractures and fractured zones were generated in volcanic rocks. These fractures connected isolated vesicles and improved the effective porosity and reservoir capacity of volcanic reservoirs, such as in the Shinan-1 and Cai27 wells. In addition, the formation of amygdaloids that filled vesicles or dissolved pores is dominated by dissolution leaching, permeation, and precipitation by the fault-controlled surface water.

3.3.2.2.2.2.3 Dissolution
Secondary changes can occur in volcanic minerals generated under high-temperature and high-pressure environments to form stable hydrated minerals. The secondary changes plug pores resulting from mineral expansion but may also provide conditions for late-stage dissolution. Carboniferous volcanic rocks in the north Xinjiang area underwent long-term dissolution by formation water and organic acid during their deep

burial. This is the predominant formation mechanism for reservoir space in the area. The secondary dissolution of early-stage filling and altered materials is predominant in intermediate-basic basalt and andesite. For example, the partial dissolution of early-stage chlorite and calcite is common in the Shinan-3, Shinan-4, and Madong-2 wells. Dissolution of hornblende, feldspar phenocrysts, and matrix is predominant in andesite. For example, the dissolution of plagioclase phenocrysts and dissolved pores and dissolved enlarged fractures in matrix are common in the Carboniferous andesite in the Shixi-1 well area of the Shixi oil field. Alkaline and strongly alkaline trachyte and phonolite are sensitive to acidic environments because of their alkaline lithology. Abundant alkaline feldspar phenocrysts and matrix may dissolve when a neutral pH environment changes from alkaline to acidic. For example, many dissolved pores are found in pollenite and trachyandesite cores of the Xiayan-2 and Shidong-8 wells. Acidic fluids are divided into inorganic and organic acids, which are active in various areas independently or jointly. Inorganic acids are dominant near later volcanic eruption areas and discordogenic faults; organic acids have stronger dissolution effects near hydrocarbon source areas. Widespread dissolution is related to the development of faults. The dissolution mechanism varies with minerals (especially dissolved feldspar) sensitive to dissolution by organic or inorganic acids.

The above-mentioned analyses show that diagenesis has a double impact on volcanic reservoirs, and the matching relationship between filling and dissolution in each diagenetic stage, combined with the diagenetic intensity, may directly influence late reservoir reconstruction. During the diagenetic process, weathering and leaching are crucial for improving reservoirs, while the degree of weathering has obvious impact on reservoir improvement. In addition, the development of reservoirs is also dominated by lithfacies and rock types. The combined effect of above-mentioned factors is essential for the formation of favorable volcanic reservoirs. With the increasing discoveries of weathering-crust-type volcanic reservoirs, the exploration of favorable volcanic reservoirs will focus on primary volcanic massifs to search for volcanic rocks that coincide with hydrocarbon generation and expulsion of organic matter in time and space.

3.3.2.3 Evolution History of Reservoir Space

Volcanic rocks have very complex reservoir space as a result of their complicated secondary changes. Volcanic rocks and their reservoir space experience a long-term and complex evolution subsequent to the formation of the volcanic massif. In summary, the following nine stages of evolution are involved (Xiong et al., 1998).

3.3.2.3.1 Primary Pore Formation Stage

This stage encompasses the volcanic eruption to magmatic condensation to volcanic massif. Pyromagma erupted to the surface or invaded shallow layers along deep faults or volcanic conduits and later rapidly cooled because of the sudden drop in temperature and pressure. During the process, gas contained in intermediate or basic magma effused from the lava flow to form vesicles at the top and bottom of the rock flow. In addition, the magma may form condensed contraction fractures during the condensation and crystallization or cryptoexplosive fractures in subvolcanic rocks.

3.3.2.3.2 Postmagmatic Hydrothermal Mineral Filling Stage

Following the magmatic stage, secondary minerals, the products of crystallization of high-salinity hydrothermal solution in vesicles and fractures, fill vesicles to form amygdaloid structures and fill diagenetic fractures. Filling minerals are mainly chlorite and zeolites. Chlorite occurs in envelope shapes, needle shapes (filling vesicle edge), or chrysanthemum shapes (in vesicle center). Early-stage zeolites (stage I) are flesh colored and fill isolated vesicles in column, sheet, and radial shapes (generally chlorite at the edge). Currently, some pores and fractures are partially filled to form residual vesicles, which provide favorable channels for late dissolution.

3.3.2.3.3 Surface and Near-Surface Weathering and Leaching Stage

Following the formation of effusive rocks, surface or near-surface massifs enter the stage of weathering and leaching dissolution. During each volcanic dormancy period, surface or near-surface volcanic rocks were exposed to weathering and leaching dissolution that enlarged and formed dissolved pores or fractures along vesicles, matrix intercrystalline pores, and primary fractures with the influence of atmospheric water or surface water. In addition, plagioclase and pyroxene in andesite and basalt dissolve to form intracrystalline dissolved pores, and the microphenocryst matrix is dissolved to form micritic matrix-dissolved pores. Thin sections show that a layer of black argitan exists at the edge of dissolved vesicles, which indicates the occurrence of weathering and leaching in surface and near-surface oxidizing environments.

Another type of weathering and leaching occurs during the epidiagenesis stage, namely, near an unconformity surface. Strata tens to more than 100 ms below the surface tend to experience weathering and leaching to form dissolved pores, fractures, and cavities. For example, large amounts of weathered and leached pores were formed in diabases and dolomicrites formerly without reservoir capacity below the weathered surface of the upper Permian top of the Tangcan 3 well, leading to the development of hydrocarbon reservoirs.

3.3.2.3.4 Burial Pore-Filling Stage

Volcanic massifs enter the burial environment after surface or near-surface weathering dissolution. During the burial of volcanic massifs, some pores and fractures are filled with secondary minerals (such as zeolites and calcite) caused by variations in the physicochemical environment. A layer of black iron mud film is common in this type of dissolved vesicle. This demonstrates that the filled vesicles were exposed to an oxidizing environment.

3.3.2.3.5 Early Structural Fracture Formation Stage

Tectonic movement leads to rock breaking to form structural fractures. Generally speaking, this stage of fracturing is widely distributed, the fractures are 0.1-0.15 m wide, and the fractured belt is developed on the top.

3.3.2.3.6 Dissolved Cavity Development Stage

The formation of structural fractures and fractured belts provides conditions for late dissolution. Surface water penetrates along fractures and dissolves their walls to form dissolved fractures. Numerous large or small cavities are generated in the fracture intersection area, fractured zones, and pore development area.

3.3.2.3.7 Dissolved Cavity-Filling Stage

Following the formation of dissolved cavities and fractures, they are completely or partially filled with zeolites and calcite because of changes in the water-body properties in the diagenetic environment. These stage I zeolites are white and have drusy crystal habits at the edge of dissolved cavities. There are also medium to coarse automorphic crystals in the middle and calcite filled in the center. Some dissolved fractures and cavities are partially filled or unfilled to result in residual pores, vugs, and dissolved cavities of various sizes.

3.3.2.3.8 Late Structural Fracture Formation Stage

Following the formation and filling of dissolved pores, structural fractures are generated in massifs by late tectonic movement. This stage of fractures—mostly microfractures—may form microfracture networks in rocks, which cut the rock matrix, amygdaloids, and filling materials and connect with early-formed pores and cavities. This greatly contributes to the permeability of existing volcanic reservoirs.

3.3.2.3.9 Late Dissolution-Filling Stage

Dissolution and filling by groundwater occurs along microfractures and pores. Microfractures are filled with calcite and fine-grained quartz and mostly are open fractures.

In conclusion, volcanic rocks undergo very complex diagenetic changes and show strong heterogeneity. Numerous vesicles and contraction fractures are generated during the stage of cooling diagenesis. Tectonic movement may lead to structural fractures, and weathering, leaching, and burial dissolution may lead to numerous dissolved pores, cavities, and fractures. The pores, cavities, and fractures are further destroyed by various fillings, resulting in very complicated residual reservoir space.

3.3.3 Types of Volcanic Reservoirs

Volcanic rocks occur in most Chinese hydrocarbon-bearing basins, having wide distribution and thickness ranges. Volcanic reservoirs are relatively complex and are classified into various types in terms of their genetic features (Table 3.14). Each reservoir type obviously varies in occurrence, distribution pattern, pore type, porosity, and permeability.

TABLE 3.14 Genesis and Types of Volcanic Reservoirs

Control action	Reservoir space	Reservoir type	Distribution and attitude	Reservoir category
Volcanism	Primary	Volcanic lava Subvolcanic rocks Volcaniclastic rock	Eruptive-effusion facies, bedded Shallow intrusive facies, tubular Exposive facies, heaped and annular	Lava-type reservoir Volcaniclastic-type reservoir
Diagenesis	Secondary	Weathered crust karst	Inner buried hill reservoirs, up to 300 m	
		Buried karst Altered type	Corrosion of acidic fluid, depth unlimited Sill, stock, and alteration zone	Dissolved-type reservoir Fractured reservoir
Tectonism	Fracture	Fractured type	Structural high and fracture zone	

3.3.3.1 Lava-Type Reservoir

These types of reservoirs comprise eruptive-effusion facies lava, such as basalt, andesite, and rhyolite, where vesicles and fractures are the predominant reservoir space. Typical examples include the andesite reservoir of the Jiufotang Formation in Longwantong sag of the Liaohe oil field, basalt reservoir of Member 1 of the Kongdian Formation in Shenjiapu block of the Zaoyuan oil field, and andesite and basalt reservoirs of the Permian system in the Santanghu Basin.

The development of vesicles and fractures in lava is dominated by lithology, lithofacies, and secondary changes. The development of vesicles is restricted by the lava and petrographic properties. Generally, vesicles are prone to develop in basic and intermediate-basic lava (such as basalt and andesite), especially in the eruptive-effusion facies of late-stage volcanism, where vesicles mostly centralize in the upper of lava strata.

Fractures in lava involve two types of geneses: structural fractures and diagenetic joints. Fractures are well developed near fractured zones, where they are characterized by a certain directionality, long extension, deep incision, relatively straight fracture face, and good connectivity. Joints are dominated by lithology and stratum thickness. They are well developed in thick basic rocks of the early magmatic eruption stage, while they are undeveloped in acidic rocks of middle to late stage. Columnar joints are relatively large joint systems in lava, and they occur in whole lava bodies and effectively increase the connectivity of reservoir space. Therefore, research on formation conditions and the associated relationship of vesicles and fractures are crucial for the distribution rules of lava-type reservoirs, which will provide a basis for exploration and development of this type of hydrocarbon reservoir.

Commercial oil and gas flow was obtained in andesite reservoirs of the Jiufotang Formation in Longwantong sag of the Liaohe oil field, where hydrocarbons mainly accumulate in the vesicles and fractures of lava. The development of vesicles is dominated by facies belts, while reservoir permeability depends on the development degree of fractures. Based on their relationship, andesite reservoirs of the Jiufotang Formation are divided into three types of reservoir zones: (1) multivesicle and multifracture zones, (2) few-vesicle and multifracture zones, and (3) multivesicle and few fracture zones. Reservoir porosity is closely related to the degree of vesicle development, and porosity is relatively high in vesicle development belts. Porosity averages 16.0% and 22.0% for the above-mentioned zones 1 and 3, respectively, while it is 3.1% for zone 2. Permeability is dominated by fractures, averaging 7.6×10^{-3} μm^2 in fracture development belts and only 1.0×10^{-3} μm^2 in unfractured zones. As a result, the organic assemblage of vesicles and fractures is critical for the lava reservoir.

Volcanic rocks from the Cretaceous to Paleogene Periods are dominated by subaqueous volcanic rocks in Jianghan Basin. Though shallow-water environments are favorable for the subaqueous eruption of basalt, water-body pressure will obstruct the effusion of volatile constituents, leading to a low degree of vesicle development in subaqueous volcanic rocks. Subaerial volcanic rocks, such as the autoclastic tuff of subaerial rhyolite rock flow formed in structural highs such as the Jinjiachang of Jiangling sag, may form a favorable hydrocarbon reservoir belt because of their high porosity (up to 25%) even in acidic volcanic rock where vesicles are undeveloped. In terms of the lava location, vesicles are much better developed near effusive conduits compared to those apart from the conduits. For example, the vesicle belt of the basalt top has a thickness up to 4-5 m and porosity as high as 37% in the Xiannvmiao stone quarry of Balingshan, Jiangling, where the basalt magma effusive belt of the Lingxi large fault is located. When original magma effuses from underground, its volatile constituents escape to form vesicles because of the sudden pressure drop and porosity decreases rapidly away from the magma effusion channels. In the Bachang stone quarry that occurs 300 m to the east of the above-mentioned stone quarry, top vesicle belts have a thickness of less than 2 m and porosity of 15.6%. Viewed from their location, vesicles are developed at the edge of a single-cycle effusive lava, especially at the top, while they are undeveloped in the center. When single-cycle effusive lava has little volume, the tight zone in the middle is narrow or absent and vesicle development degree is much higher (Yan et al., 1996).

3.3.3.2 Volcaniclastic Reservoir

Clastic reservoirs in volcanic rocks, in general terms, belong to the volcaniclastic reservoir type, and they are not further divided according to their genesis. Actually, volcaniclastic rocks can be classified into normal volcaniclastic rocks and volcaniclastic sedimentary rocks based on their genesis. The former is the product of clast accumulation resulting from volcanic eruption, and the latter is the product of volcanics through weathering, transportation, and sedimentation. With the exception of clastic particles made up of volcanic debris, volcaniclastic rock has the same formation process as normal sedimentary rock.

3.3.3.2.1 Normal Volcaniclastic Reservoir

An example of a volcanic breccia reservoir is Member 1 of the Shahejie Formation in the Shanghe oil region of the Linpan oil field, Jiyang depression. A normal volcaniclastic reservoir that is 800 m wide (east-west) and 1000 m long (south-north) is developed in Member 1 of the Shahejie Formation of the Shang741 well area, and it has a basic cone shape. The normal volcaniclastic reservoir belongs to a volcaniclastic cone by the sixth stage subaqueous volcanic explosion, which contains thin lava, shallow-water micrite, and microcrystalline carbonate rock, silty dolomicrite, tuffaceous dolomite, and mudstone. Based on the difference in lithology and thickness change of volcanic rocks, the volcaniclastic cone is divided into three facies belts from the center to the margin according to their eruption features: namely, crater/near-crater facies, middle-distance volcano slope facies, and long-distance volcano slope facies. The crater/near-crater facies lies near the crater, exceeds 100 m in thickness, and comprises volcanic lava and volcanic breccia. The middle-distance volcano slope facies, as the main body of the slope, occurs in the middle of volcanic slope, is more than 50 m thick, and comprises volcanic tuff. The long-distance volcano slope facies is situated on the exterior margin of the slope, is less than 50 m thick, and comprises tuff interbedded with dolomite and tuffaceous siltstone.

Primary pores are the predominant reservoir space in volcaniclastic cones, including interbreccia pores in breccia, intergranular pores and vesicles in tuff and lava clastic particles, pores in dolomite, dissolved pores and cavities in effusive rock and dolomite by leaching, and crisscross fracture systems resulting from fault activity and regional activity. Near west-east high-angle microfractures are predominant, and they significantly improve reservoir permeability.

Testing results show that volcaniclastic reservoirs generally have relatively high porosity, averaging 14.7% for volcanic lava, 26.6% for breccia, 25.0% for tuff, and 17.8% for tuffaceous sandstone. However, they have significant variations in permeability, which are 1.8×10^{-3}, 386×10^{-3}, 19.6×10^{-3}, and 14×10^{-3} μm^2, respectively. As a result, volcanic breccias are the most favorable reservoirs due to their porosity and permeability. The reservoirs mainly occur in middle-distance volcano slope subfacies and are the main oil-bearing series in the area. Long-distance volcano slope facies are unfavorable reservoir facies since they are composed of relatively low-porosity and low-permeability tuff, dolomite, and tuffaceous siltstone with minor thickness and scale. Crater/near-crater facies are reservoir facies with moderate reservoir capacity (Cao et al., 1999).

3.3.3.2.2 Volcaniclastic Sedimentary Reservoirs

Volcaniclastic sedimentary rocks are the products of volcanic rocks by weathering, transportation, and redeposition and mostly occur near the interaction zone of volcanic cones and lacustrine water bodies or around volcanic islands. Volcaniclastic sedimentary rocks are composed of synchronous volcanic debris and a certain amount of terrigenous debris, and their components are dominated by volcanic explosion materials mixed with a small amount of terrigenous debris

(such as sand and mud). Generally, volcanic debris accounts for 30% or more of the bulk composition.

Volcaniclastic sedimentary reservoirs are developed in the Shang58-5 and 58-4 wells, and a set of inequigranular volcaniclastic sedimentary rocks are developed in Member 1 of the Shahejie Formation, where commercial oil flow is developed. Core observations show that the clastics are composed of volcanics and minor terrigenous sand and mud. Volcaniclastic sedimentary rocks have good sorting, moderate psephicity, and well-developed graded beds; their particle cement is dominated by calcite with a notable lack of argillaceous interstitial matter. Oil can be observed in millimeter-order pores with the naked eye. As a result, volcaniclastic sedimentary rocks have satisfactory porosity and permeability and are one of the favorable reservoirs in the area (Wang et al., 2003a,b,c).

Generally, volcaniclastic sedimentary rocks are formed by repeated wave transformation of lacustrine or marine volcaniclastic materials, volcanic massifs in lake and beach environments, or island-like volcanic massifs. Horizontally, they exhibit zonal, fan, or half-ring distribution with stable thickness. They are often near hydrocarbon source rocks and are favorable hydrocarbon accumulation zones. As a result, emphasis will be placed on this type of reservoir.

3.3.3.3 Dissolved-Type Reservoir

Among dissolved-type reservoirs, weathered crust reservoirs are the best developed.

Crushing and dissolution often occur in lava exposed at the surface (especially basic lava) to form reservoir space (such as fractures and dissolved pores) by supergenesis of the weathered zone. Weathering and leaching type reservoirs are typically developed in Carboniferous volcanic reservoirs in the northwest margin of Junggar Basin, Malang sag of Santanghu Basin, and Bohai Bay Basin. For example, weathering and leaching of volcanic rocks are observed in the interval from 1019.0 to 1026.0 m in the Gao-41-1 well and the interval from 998.2 to 1006.0 m in the Gao-41 well. Neutron logging interpretation and lab analysis show that volcanic reservoirs in weathering and leaching zones have porosity up to 15-20% and an interval transit time of 275 $\mu s\,m^{-1}$. Therefore, a series of fractures and dissolved pores formed by weathering and leaching are favorable for the formation of large-scale reservoirs.

3.3.3.3.1 Paleozoic Volcanic Weathered Crust Reservoir of Santanghu Basin

The reservoir space assemblage of volcanic reservoirs in the Malang sag of Santanghu Basin is mainly weathering and leaching pore and fracture type. The reservoir space of the Kalagang Formation in Ma17 well belongs to this type.

Generally, this type of reservoir space has a porosity of 3.9-7.5% and permeability less than $0.05 \times 10^{-3} \mu m^2$. However, maximum and minimum porosities are 18.4% and 3.4%, respectively, and average porosity is 8% in the interval from 1515 to 1548 m. Thin-section observations show that dissolved pores and fractures are developed in rocks and good reservoirs are formed by long-term weathering and leaching.

3.3.3.3.1.1 Development of Weathered-Crust-Type Reservoirs

Regional geological research shows that structural unconformities are developed in the top of the Kalagang Formation in the north Malang sag of Santanghu Basin. Local uplifts were generated in the middle to late Triassic Epoch in a compressional structure environment, and weathered crusts were formed by the physical and chemical weathering of intermediate-basic volcanic rocks of the Kalagang Formation in a temperate to semiarid climate. Prior to the sedimentation of the Triassic Epoch, water bodies were mainly present in the western Malang sag, and volcanic rocks in the eastern uplift area were denuded. Favorable conditions for the formation of weathered crusts include (1) long-term, stable, low-relief uplift area in the eastern Malang sag during the Triassic-Jurassic depositional stage; (2) surface water penetration into the subsurface along opened faults causes leaching to occur in easily weathered rocks by groundwater flow because of the pressure difference. Seismic profile interpretation shows that multiple reverse thrusts are developed in the eastern weathered crust area of the Malang sag, and volcanic massifs are prone to form vertical fractures under the stress. (3) Vesicles and amygdaloidal structures developed in volcanic massifs vertically and horizontally accelerate weathering and leaching and enlarge the distribution of weathered crust; and (4) numerous columnar joints or partings formed by rapid condensation of basaltic lava are favorable for weathering and leaching of parent materials.

3.3.3.3.1.2 Formation of Weathered-Crust-Type Reservoirs

Volcanic rocks of the Kalagang Formation in Malang sag experienced two stages of weathering and leaching. The first stage occurs in the volcanic dormancy period, when thin sedimentary rocks (fine sandstone, pelitic siltstone, and mudstone) are developed at the top and bottom of volcanic rocks. This provides material basis for dissolution. Secondary dissolution occurs by freshwater leaching (associated with sedimentary rocks) of the top and bottom of volcanic rocks. Leaching dissolution is strongest at the top of volcanic rocks where numerous vesicles and amygdaloids are developed. The second stage of weathering and leaching occurs after the sedimentation of late Paleozoic strata. Regional uplifting led to Paleozoic

volcanic rock being exposed at the surface, which had significant filling by altered minerals (such as zeolites). Most zeolites have been completely dissolved. The volcanic rocks exposed to weathering and leaching eventually formed regional unconformities between Mesozoic and Paleozoic strata.

Volcanic rocks of the Kalagang Formation in Malang sag finally formed volcanic weathered-crust-type reservoirs during the two stages of leaching dissolution.

3.3.3.3.1.3 Development Models of Volcanic Weathered-Crust-Type Reservoirs

Weathering involves physical, chemical, and organic weathering. Physical weathering refers to rock disintegration, which does not lead to obvious changes in chemical composition. Chemical weathering refers to decomposition of rocks and minerals under the action of hypergene water, oxygen, and carbon dioxide, and it involves the interaction of lithosphere, atmosphere, hydrosphere, and biosphere, as well as the geochemical cycle of elements. Organic weathering refers to physical or chemical destruction by biological activity in rocks. Based on the weathering intensity, weathered-crust-type reservoirs are classified into five zones: unweathered (parent rock), disintegration, leaching, hydrolization, and decomposed product zones, whose petrographic characteristics are shown in Table 3.15.

Diagenesis of weathered zones varies because of the vertical zonation of volcanic weathered-crust-type reservoirs.

3.3.3.3.1.3.1 Disintegration Zone

Filling dissolution is obvious in this zone, which is composed of basic basalt containing a high percentage of dark minerals as well as ferrum and magensium elements. With poor stability and ease of alteration, the minerals pyroxene chlorite, iddingsite olivine, zeolites, and basic plagioclase are altered into pumpellyite. These altered minerals fill vesicles and fractures in volcanic rocks and reduce reservoir capacity to some degree.

3.3.3.3.1.3.2 Leaching Zone

This zone is dominated by weathering and leaching, structural fragmentation, and hydrothermal alteration, and it creates the most favorable reservoirs in weathered crusts because of the following three types of diagenesis.

(1) Leaching: it leads to broken rock and a change in rock composition. When atmospheric precipitation seeps into the subsurface along weathered fractures and structural fractures, it dissolves the readily soluble components in tight rocks to form vertical or horizontal dissolved pores, cavities, and fractures in the leaching and hydrolization zones. These irregular pores, cavities, and fractures are widely distributed to form pore and cavity development networks, which improve reservoir space and create favorable hydrocarbon

TABLE 3.15 Predominant Petrographic Characteristics of Volcanic Weathered Crust Zones

Weathered zones	Geological characteristics	Rock integrity	Mechanical properties	Knocking sound	Typical core photos
Final decomposed product zone	Rocks lose their color completely and lose their luster; rock fabrics are completely destroyed and disintegrate, and decompose into loose clay- or sand-like materials; volume varies sharply, not transported, original structure evidence remains. With the exception of quartz, other minerals change into secondary minerals by weathering alteration	Broken	Rapid disintegration when encountering water	Dumb	
Hydrolization zone	Most rocks lose their color and locally retain original color; rock fabrics are mostly destroyed and a fraction of rocks disintegrate and decompose into clay. Most rocks appear as discontinuous frameworks or block rocks, and weathered fractures are developed	Slightly broken	Easy to break with hands	Dumb	
Leaching	Rocks or fracture faces lose their color, while rock color remains fresh in fractures. Weathered fractures are developed and fracture walls undergo strong weathering	Poor to fair integrity	Easy to break by knocking	Slightly dumb	
Disintegration	Rocks or fracture faces fade slightly. No change occurs in rock fabric. Most fractures are closed or filled with silica, and weathering alteration only occurs along large fractures	Relatively integral to integral	Can be broken along fractures	Ringing	
Unweathered	Natural rock color	Integral	Difficult to break	Ringing	

reservoirs. The higher the rock index of alteration (CIA), the higher the weathering intensity, and the more obviously rocks are improved. The CIA of volcanic rocks is up to 54% in the Malang sag of Santanghu Basin, where the weathering and leaching zone of the Ma19 well has a porosity as high as 30.85%.

(2) Structural fragmentation: ductility and compression resistance vary with rocks. The higher the rock porosity, the better the rock ductility and compression resistance, and the more undeveloped fractures are. Volcanic rocks with low porosity tend to form fractures because of low ductility and low compression resistance. When the rocks undergo tectonic stress, they may form fractures. The fractures, together with fractures filled with late secondary minerals, may enlarge porosity and improve effective reservoir space by dissolution because of the inhomogeneous degree of filling. Structural fragmentation is developed in the Malang sag, where structural fractures are found in volcanic cores in the Ma19, Ma17, and Niudong9-10 wells. Generally, the fractures are 0.2-20 mm wide, and they are partially or completely filled with zeolites, chlorite, calcite, and quartz (photos E and K in Figure 3.13).

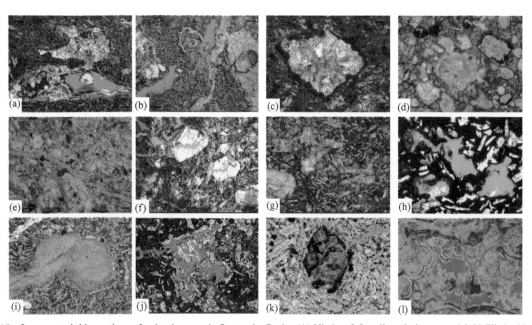

FIGURE 3.13 Impregnated thin sections of volcanic cores in Santanghu Basin: (A) Niudong9-8 well, andesite, amygdaloid filled with laumontite, residual pores, single polars, 1620.54 m; (B) Niudong9-8 well, basalt, dissolution of matrix, and pyroxene phenocrysts, crossed polars, 1643.74 m; (C) Niudong9-8 well, basalt, intra-amygdaloid laumontite intercrystalline pores, single polars, 1646.77 m; (D) Niudong9-8 well, volcanic breccia, vesicles and filling materials in volcanic breccia, single polars, 1541.73 m; (E) Niudong9-8 well, tuff, plagioclase phenocryst replaced by zeolites, developed fractures, single polars, 1550.1 m; (F) Ma17 well, basalt, plagioclase phenocryst replaced by zeolites, developed fractures, single polars, 2372.65 m; (G) Niudong9-8 well, basalt, matrix-dissolved pores, single polars, 1635.00 m; (H) Niudong9-10 well, agglomerate, amygdaloid dissolved pore, laumontite being dissolved residues, single polars, 1430.88 m; (I) Niudong9-10 well, andesite, amygdaloid filled with celadonite, prehnite in the core, single polars, 1464.31 m; (J) Ma19 well, basalt, intra-amygdaloid dissolved micropore, residues of laumontite and analcime being dissolved, single polars, 1537.83 m; (K) Niudong9-8 well, basalt, fracture cutting olivine, associated with dissolution, single polars, 1508.53 m; (L) Niudong9-8 well, andesite, amygdaloid filled with laumontite, residual pores connected with fractures, single polars, 1620.54 m.

(3) Hydrothermal alteration: zeolites are the predominant alteration products in the leaching zone. Microscopic identification by thin section and electron probe shows that pores in volcanic rocks of the Malang sag have a radial distribution of zeolites, which fill the margins of amygdaloids, and they evolve into favorable pores and fractures by long-term alteration, such as shown in photos A, C, E, F, J, and L in Figure 3.13. A few pores are filled with celadonites, which are dissolved along the margin to form microfractures, such as photo I in Figure 3.13. A few zeolites are replaced by olivine and iddingsite, and well-developed fractures cut through the whole olivine, such as photo K in Figure 3.13. When weathering is extensive, pore-filling materials are dissolved completely; when weathering is moderate or relatively minor, undissolved or poorly dissolved chlorite and zeolites with partially dissolved pores are found in cores.

3.3.3.3.1.3.3   Hydrolization Zone This zone is dominated by alteration. With the influence of hypergenesis, minerals undergo alteration to partially form clay minerals and ferric oxide, and eventually the combination of hydrargillite, kiesel, and ferric oxide. The combination is subjected to argillization, chloritization, carbonatization, and saussuritization. The minerals are also subjected to alteration, such as olivine serpentinization, iddingsitation, plagioclase kaolinization, pyroxene chloritization, and biotitization. These altered minerals are dissolved or hydrolyzed to form numerous secondary pores. This greatly improves the reservoir capacity of the hydrolization zone.

3.3.3.3.1.3.4   Final Decomposed Product Zone This zone, the top surface of the weathered crust, is dominated by weathering and structural fragmentation. Rocks are mechanically destroyed on the surface by temperature changes and hydrolytic action. When weathering is extensive, minerals are subjected to argillic alteration and finally decompose into clay and carbonate minerals.

The above-mentioned analyses show that diagenesis has a double impact on volcanic reservoirs. It not only enhances reservoir space but may also decrease reservoir capacity. During the diagenetic process, weathering and leaching are critical for reservoir improvement, and the degree of weathering can improve reservoirs significantly.

3.3.3.3.1.4   Vertical Development Model for Volcanic Weathered-Crust-Type Reservoirs in Malang Sag
Geochemical element analyses have been conducted for the Ma17, Ma19, Niudong9-8, Niudong9-10, and Tangcan

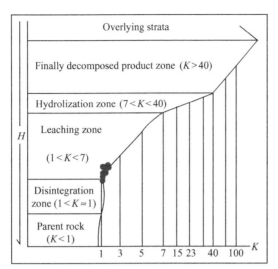

FIGURE 3.14 Relationship between distribution coefficient K and depth H for zoning of effusive rock weathered crust.

3 wells. The distribution coefficient of zoning for weathered crusts is calculated by the following equation:

$$K = Fe_2O_3 + Al_2O_3 + TiO_2/MgO + CaO + NaO_2$$

where oxidates refer to the percentage of oxidates in samples.

Figure 3.14 shows that most K values of samples fall into the leaching zone and only a few values in the disintegration zone. Namely, volcanic weathering and leaching zones are well developed in the area.

Analysis and testing were conducted for the Ma19 and Niudong9-10 wells according to the rock weathering index $CIA = (Al_2O_3/(Al_2O_3 + Na_2O + K_2O + CaO) \times 100\%)$, and in combination with lithologic log data, the weathered crust of Malang sag is vertically divided into the following zones (Figure 3.15).

3.3.3.3.1.4.1 Unweathered Zone (Parent Rock) The rock is gray-brown basic basalt, occurs far from the top of the weathered crust, and is free from or only slightly affected by weathering. No fractures, pores, or cavities are developed in the rock.

3.3.3.3.1.4.2 Disintegration Zone The rock is gray-brown basic basalt, and predominantly composed of plagioclase and pyroxene and secondarily cryptocrystalline with porphyritic structure. The phenocrysts are dominated by plagioclase and pyroxene minerals. Locally, a small quantity of vesicles are found in cores, and microfractures are relatively well developed. Reservoir capacity is relatively poor because the fractures and vesicles are filled with laumontite. The disintegration zone is 14 m deep in the interval from 1543.0 to 1557.0 m in the Ma17 well, 26 m deep in the interval from 1556.0 to 1582 m in the Niudong9-10 well, and 14 m deep in the interval from 1566.0 to 1580.0 m in the Ma19 well, where shows of oil and gas are found.

3.3.3.3.1.4.3 Leaching Zone This zone is one of the most developed of the weathered crust, and it is also the important component of volcanic weathered crusts with good reservoir capacity. The lithology is basic basalt and intermediate andesite. The basalt is brown, composed of feldspar and pyroxene, cryptocrystalline with porphyritic structure, whose phenocrysts are predominantly plagioclase and pyroxene; vesicles or amygdaloidal structures are filled with laumontite, and obvious dissolution is observed under the microscope. The andesite is dominated by plagioclase with porphyritic structure, whose phenocrysts are plagioclase and dark minerals where vesicles or amygdaloidal structures are filled with laumontite. Core observations show that microfractures are developed in a criss-cross pattern; they are primarily vertical fractures and secondarily oblique fractures; horizontal fractures are found occasionally and partially filled with laumontite. When the parent rock is exposed to the surface and undergoes weathering and leaching in a relatively strong hypergenesis environment, secondary change occurs, with the alteration gradually attenuating from deep layers to the surface. The thickness of the leaching zone varies with wells; it is 18, 16, 12, and 49 m for the Ma17, Niudong9-8, Niudong9-10, and Ma19 wells, respectively.

3.3.3.3.1.4.4 Hydrolization Zone The lithology in this zone is mainly mudstone and basalt with poor integrity. Most rocks in this zone are destroyed, and core segments appear as fragments, which are decomposed into clay by weathering. Only a few oil traces were found in the Ma17 and Ma19 wells. The depth of this zone is 9, 16, 19, and 30 m in the Ma17, Niudong9-8, Niudong9-10, and Ma19 wells, respectively.

3.3.3.3.1.4.5 Final Decomposed Product Zone This zone is the top layer of the weathered crust. The completely destroyed rocks are disintegrated or decomposed into loose

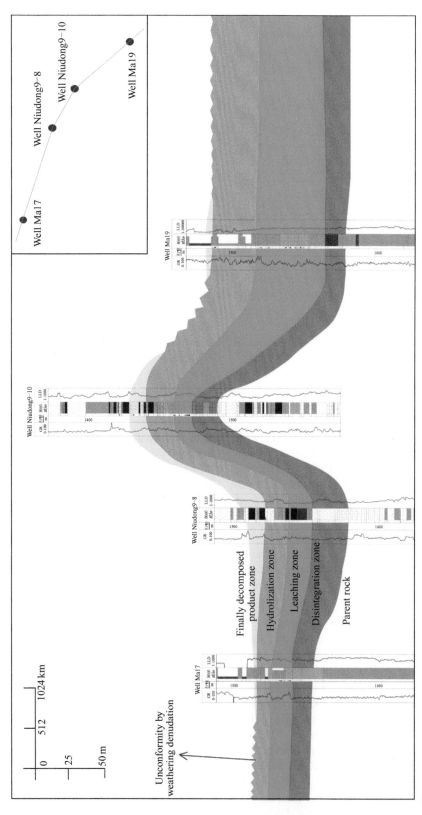

FIGURE 3.15 Vertical development model of volcanic weathered crust in Malang sag.

earthy or sandy matter in the zone, and minerals are altered into secondary minerals, such as chlorite, iddingsite, lotrite, and zeolites. Most areas underwent denudation, and their thickness varies with wells, being 1.2, 12, 10, and 4 m in the Ma17, Niudong9-8, Niudong9-10, and Ma19 wells, respectively.

3.3.3.3.2   Paleozoic Volcanic Weathered Crust Reservoir in Northwest Margin of Junggar Basin

The volcanic reservoirs of the Gu-16 well area in the northwest margin of Junggar Basin have a complex pores system which contains fractures and matrix pores. The reservoir type includes low porosity-fracture type and pore-fracture type. The reservoir mode includes directly interconnected fracture type and indirectly interconnected pore-fracture type. Pore, cavity, and fracture reservoirs are vertically distributed in the interval from 75 to 250 m from the top unconformity surface downward, and they are most favorable near the lower unconformity surface. Fractures are developed well in volcanic reservoirs. The statistics of imaging logging interpretations show that the thickness of fracture development intervals accounts for 53% of the total thickness of the volcanic rock, and field observation shows that it exceeds 53%. Based on the interpretation and comprehensive analyses of logging data, the volcanic reservoirs are vertically distributed in the middle to upper part of the volcanic rock (250 m from the unconformity surface). A wide variety of pores in Carboniferous volcanic reservoirs of northwest Junggar Basin are secondary pores related to fractures, and they vary in their degree of development. In particular, intercrystalline dissolved pores and dissolved pores are the best developed. There has many types of pore, such as gravel-edge pore, intercrystalline pore, dissolved pore, fracture pore, microfracture pore, vesicle pore, and replacement pore. And these

types of pore could be assembled to make more kinds of pore assemblages. Intercrystalline dissolved pores and dissolved pores constitute the most favorable reservoir space in the area (Figure 3.16).

3.3.3.4   Fractured Reservoirs

In fractured reservoirs, reservoir space is dominated by fractures with relatively good permeability. The formed reservoirs are characterized by high reserves and high productivity. On the genesis basis, fractures can be further divided into structural fractures (Shao-18 basalt reservoir), weathered fractures (Gao-41 basalt reservoir), and mixed fractures (Shang741 diabase reservoir).

The probability of producing fractures in Minqiao volcanic rocks varies with lithology. Autoclastic and quenching-shatter basalt is more probable than tight basalt to produce fractures, and tight basalt is more probable than vesicular and amygdaloidal basalt. In terms of the fracture distribution, autoclastic basalt occurs at the top surface of rock-flow units, quenching-shatter basalt occurs in positions where the lava enters the water body, vesicular and amygdaloidal basalt occurs at the top and bottom of rock-flow units, and tight basalt occurs in rock-flow units with great thicknesses. Therefore, fractures tend to produce in thinner rock-flow units. The high point of fault nose traps occurs on the upper sides of faults, where fractures tend to develop when other conditions are the same due to thin rock-flow units (Yang, 2005).

3.3.4   Distribution Rules of Volcanic Reservoirs

Volcanic rocks are developed in the Mesozoic-Cenozoic of eastern China, and favorable volcanic reservoirs include the

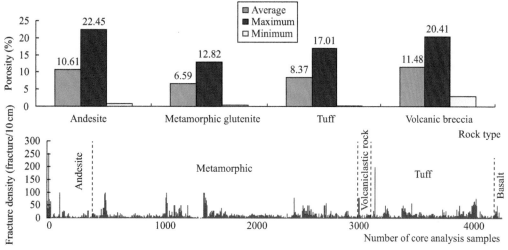

FIGURE 3.16   Characteristics of weathered-crust karst-type reservoirs in northwest Junggar Basin.

Yingcheng Formation in Songliao Basin, Suhongtu Formation of Yingen Basin, Xing'anling group volcanic reservoir in the Abei oil field of Erlian Basin, Mesozoic-Cenozoic volcanic reservoir in Bohai Bay, Jianghan, and Subei Basins, Carboniferous volcanic reservoir and Ludong-Wucaiwan Carboniferous volcanic reservoir of Junggar Basin, and Permian volcanic reservoir of Tarim, Santanghu, and Sichuan Basins (Table 3.16).

Volcanic reservoirs are characterized by wide distribution and great geologic age. It is possible to form favorable reservoirs out of any type of igneous rock or any age of volcanic rock.

Burial depth has little impact on the porosity of volcanic reservoirs. This is because volcanic rocks mainly form in the surface environment or are shallowly buried underground, and magma may form rock within a short time and narrow range as soon as it cools down. Compared with other rocks, the volcanic framework is harder and more resistant to compaction, and it resists mechanical compaction during the burial process. As a result, pores in volcanic rocks are more easily preserved than those of other rocks (Figure 3.17). Clasolite has lower porosity than volcanic rock at the same depth. For example, in the Carboniferous volcanic rock in Shixi oil field of Junggar Basin where depths exceed 3800 m, porosity ranges from 8.46% to 19.78% and averages 14.4%, showing that depth has little impact on porosity.

3.4 MAIN CONTROLLING FACTORS FOR FORMING VOLCANIC RESERVOIRS IN HYDROCARBON-BEARING BASINS

The various assemblages of reservoir space in volcanic reservoirs generally include the pore-fracture assemblage, pure fractured reservoir and matrix-dissolved pores and fractures assemblage, and phenocryst-dissolved pores and fractures assemblage. Volcanic reservoir space may experience a complex evolution at various stages, including formation, development, plugging, and reformation. Primary pores and fractures are dominated by original eruptivity (volcanic subfacies). Under the same tectonic stress, the development and preservation of structural fractures are also controlled by the original eruptivity (subfacies). Magmatic melt effusing during volcanic eruption may condense to form lava sheets covering a large area, which are unconnected and have little or no permeability even though primary vesicles are present. However, the lava sheets may possess hydrocarbon reservoir capacity after later diagenetic and structural events. As a whole, volcanism, tectonism, and fluid action are the predominant geologic processes that influence the development of reservoir space.

3.4.1 Volcanism

Volcanism not only controls the reservoir shape, scale, and mutual relationship of facies but also controls the reservoir space type and mineral composition.

3.4.1.1 Relationship Between Volcanic Rock Type and Reservoir

Within a given area, a volcanic reservoir is closely related to the volcanic rock type and lithofacies, with its physical properties mainly controlled by the volcanic rock type. Different volcanic rock types determine the mineral composition, chemical composition, and rock texture. These differences further determine the difference in volcanic reservoir type. Different types of volcanic reservoirs develop different types of reservoir systems. For example, rhyolitic volcaniclastic rock in Member 1 of the Yingcheng Formation and pellet rhyolite in Member 3 of the Yingcheng Formation are the major volcanic reservoirs in SongLiao Basin (Figure 3.18).

3.4.1.1.1 Reservoir Space of Pellet Rhyolite

The reservoir space of a pellet rhyolite mainly consists of vesicles, residual pores of filled vesicles and intra-amygdaloid pores, micropores resulting from devitrification of rhyolitic glass, pores resulting from dissolution of feldspar and carbonate minerals, and local microfractures. The major reservoir space of rhyolite is the same as pellet rhyolite except the absence of micropores resulting from the devitrification of rhyolitic glass. Vesicles, residual pores of filled vesicles, and intra-amygdaloid pores are the most common in pellet rhyolite and also the most important reservoir space. The vesicles of pellet rhyolite often occur in locally developed bands along the flow structure, forming good reservoir space. The vesicle development provides passage and space for fluid; thus vesicle development determines the secondary pore development to some extent. The quartz and albite often grow in or around vesicles, and the carbonate minerals that fill vesicles plug the reservoir space. When the margins of vesicles in the pellet rhyolite do not develop quartz and albite, large primary vesicles will form. The micropores resulting from devitrification of rhyolitic glass is a newly identified and important pore type in the pellet rhyolite. These pores are small, but they cover a large area and have good connectivity. In conjunction with the dissolution of feldspar produced from devitrification, the reservoir capacity is greatly improved.

The feldspar grains are partially dissolved along their cleavages, and complete dissolution is seldom observed. The dissolution is mainly dissolution of feldspar phenocrysts, feldspar matrix, and microcrystalline feldspar resulting from devitrification of rhyolitic glass in the pellet

TABLE 3.16 Reservoir Characteristics of Volcanic Reservoirs in China's Hydrocarbon-Bearing Basins

System	Series	Strata	Basin or sag	Lithology	Porosity (%)	Permeability ($10^{-3} \, \mu m^2$)
Neogene	Miocene	Member 1 of Yancheng group	Gaoyou sag	Grayish-black, grayish-green, and grayish-purple basalt	20	37
		Bottom of Guantao Formation	Dongying sag	Olivine basalt	25	80
			Huimin sag	Olivine basalt	25	80
Paleogene	Oligocene	Sanduo Formation	Gaoyou sag	Basalt	22	19
		Member 1 of Shahejie Formation	Dongying sag	Basalt, andesite basalt, and volcanic breccia	25.5	7.4
			Zhanhua sag	Basaltic volcanic	40-70 of vesicles, 0.03-0.1 mm	
	Eocene-Oligocene	Member 3 of Shahejie Formation	Huimin sag	Olivine basalt	10.1	13.2
			East Liaohe sag	Basalt and andesite basalt	20.3-24.9	1-16
		Member 4 of Shahejie Formation	Zhanhua sag	Basalt, andesite basalt, and volcanic breccia	25.2	18.7
		Xingouzui-Jingsha Formation	Jiangling sag	Grayish-black, grayish-green, and grayish-purple basalt	18-22.6	3.7-8.4
		Kongdian Formation	Huaibei sag	Basalt and tuff	20.8	90
Cretaceous		Yingcheng Formation	Songliao Basin	Basalt, andesite, dacite, rhyolite, tuff, and volcanic breccia	1.9-10.8	0.01-0.87
		Qingshankou Formation	Qijia-Gulong sag	Intermediate-acidic volcanic breccia and tuff	22.1	136
		Lougouqiao Formation	Jizhong depression	Volcanic breccia, tuffaceous glutenite	Oil patches or crude oil are observed in six wells	
		Suhongtu Formation	Yingen Basin	Basalt, andesite, volcanic breccia, and tuff	17.9	111
Jurassic		Xinzhuang Formation	Jizhong depression	Andesite, tuff, basalt, and breccia	Oil stains and fluorescence are observed in two wells	
		Xing'anling group	Erlian Basin	Basalt and andesite	3.57-12.7	1-214
			Haila'er Basin	Volcanic breccia, rhyolite porphyry, trachyte, tuff, andesite, andesite basalt, and basalt	13.68	6.6
Paleozoic group			Junggar Basin	Andesite, basalt, tuff, and volcanic breccia	4.15-16.8	0.03-153
Permian			Tarim Basin	Dacite, basalt, volcanic breccia, and tuff	0.8-19.4	0.01-10.5
			Santanghu Basin	Andesite and basalt	2.71-13.3	0.01-17
			Sichuan Basin	Basalt	5.9-20	

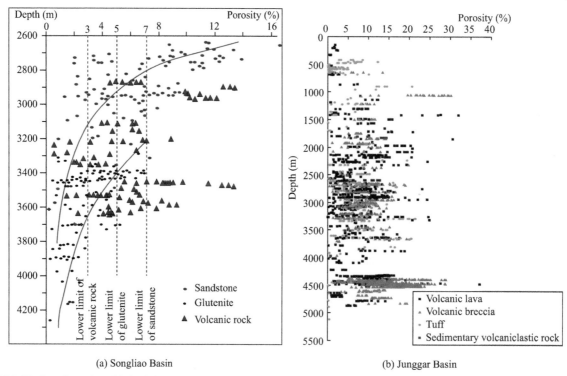

(a) Songliao Basin (b) Junggar Basin

FIGURE 3.17 Porosity varies with depth for volcanic reservoirs in Songliao and Junggar Basins.

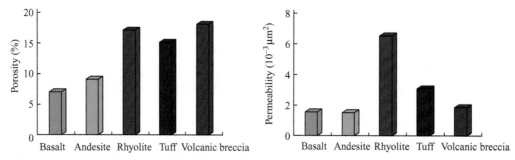

FIGURE 3.18 Physical properties of different volcanic rock types in SongLiao Basin.

rhyolite. Pores are also produced by the dissolution of carbonate minerals that replace feldspar and fill pores and local microfractures in pellet rhyolite and rhyolite.

3.4.1.1.2 Reservoir Space of Welded Tuff

The reservoir space in welded tuff mainly includes dissolved pores of volcanic ash, vesicles in the debris and plastically deformed debris, residual pores of filled vesicles and intra-amygdaloid pores, dissolved pores of feldspar and carbonate minerals, and local microfractures. The dissolved volcanic ash usually forms numerous micropores and is one of the most important types of reservoir space in welded tuff. Although these pores are small, their great number and good connectivity can result in good reservoirs. The extensive dissolution of volcanic ash can produce great cavities, forming good reservoirs.

The vesicles in the debris and plastically deformed debris, and the residual pores of filled vesicles and intra-amygdaloid pores can serve as good reservoir space. Dissolution of the feldspar in the crystal fragments and debris develops the dissolved pores that also constitute some of the most important reservoir space in welded tuff. Moreover, the pores resulting from carbonate mineral dissolution and local microfractures are also important reservoir space in welded tuff.

3.4.1.1.3 Reservoir Space of Tuff

The reservoir space of tuff mainly involves the pores resulting from dissolution of volcanic ash, feldspar, and carbonate minerals, vesicles and intra-amygdaloid pores, and local microfractures. These are the same as welded tuff except for vesicles in the plastically deformed debris, residual pores of filled vesicles, and intra-amygdaloid pores.

3.4.1.1.4 Reservoir Space of Volcanic Breccia

This reservoir space mainly involves vesicles of dissolved volcanic ash and debris, residual pores of filled vesicles and intra-amygdaloid pores, pores resulting from dissolution of feldspar and carbonate minerals, and local fissures.

3.4.1.1.5 Reservoir Space of Agglomerate

The reservoir space of agglomerate mainly depends on the composition of the rock mass and volcanic breccia. If the rock mass and volcanic breccia are pellet rhyolite, welded tuff and tuff, the reservoir space of the agglomerate is that of pellet rhyolite, welded tuff and tuff, respectively. In addition, there are local microfractures, cracks between the rock mass and breccia fragments, and dissolution of pore or fracture fillings.

For example, in the Carboniferous volcanic rock in the Wucaiwan sag of Junggar Basin, the porosity is highest in the volcaniclastic rock (1.26-30.08%, 9.84% on average), secondly in andesite (8.14%) and tuff (7.92%), and poorest in basalt (5.89%); the average permeability (2.09×10^{-3} μm^2) is highest in tuff, secondly in andesite and volcanic breccia, and poorest in basalt (0.89×10^{-3} μm^2). Because the tuff distribution is scattered with a limited thickness, in view of the porosity and permeability, the andesite and volcanic breccia are good reservoirs (Figure 3.19).

The physical properties of the volcanic reservoirs in Tahe area of Tarim Basin are dramatically different. In the volcanic rocks, the volcanic breccia develops the highest average permeability but low average porosity; in the lava, the basalt develops the highest average porosity but lowest average permeability; and the dacite develops high average porosity and average permeability. In light of the shape and genesis of reservoir space, the reservoir space of volcanic rock in the Tahe area can be classified into 2 types and 11 subtypes. According to core observations and SEM statistics, the reservoir space mainly includes vesicles (4.2% on average), intraphenocryst pores (2.1% on average), and tectonic fractures (0.5% on average). The compound reservoir space type is mainly fracture-pore space. The pore type and porosity are both different for different rocks and different for the same rock. The reservoir space of dacite is mainly intraphenocryst pores, vesicles, and intramatrix-dissolved pores and fractures; the reservoir space of basalt is vesicles, mold pores, and fractures, and vesicles are more developed than that in dacite. The primary pores of volcanic rock are mostly isolated; only when the dissolved pores and fractures connect the primary pores to produce fracture-pore compound reservoir space can the reservoir can accumulate oil and gas. The vesicles in basalt are more well developed than that of dacite, but are mostly isolated, resulting in high porosity but poor interconnectivity, hence poor physical properties. The dacite develops dissolved pores, fractures, and vesicles that are connected to form fracture-pore reservoir space; therefore, the physical properties of dacite are better than basalt.

In the Tahe area, there is an obvious paleoweathering denudation surface at both the top and base of the volcanic rock; thus the upper and lower parts of volcanic rocks are primary vesicle developed belts. Because the top of the volcanic rock suffers weathering denudation, the upper part develops secondary dissolved pores and fractures and the lower part develops fracture-pore reservoir space caused by tectonic fracture development. This creates favorable reservoirs in the upper and lower parts of volcanic rock. Taking the S87 well as an example, the porosity and permeability are high at both top and base of the dacite where it also develops fractures, and the middle is the tight zone. The upper and lower parts of the dacite interval are favorable reservoir zones.

Volcanic breccia belongs to the explosive facies and mainly develops near the crater. The impulsive force of a

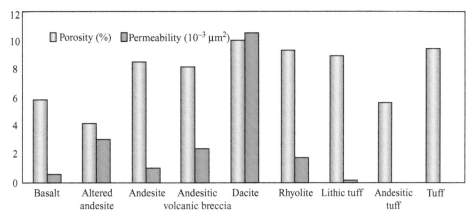

FIGURE 3.19 Physical properties of different volcanic rocks in the hinterland of Junggar.

volcanic eruption shatters the top and surrounding rocks and produces numerous fissures, cracks, and interbreccia pores. In addition, most volcanic breccias are located at structural highs or paleocorrosion highlands, subjected to extreme weathering, and prone to dissolution and leaching, resulting in the enlargement of primary pores and fractures. The fracture-pore reservoir space of volcanic breccia includes all types of interbreccia pores (fractures), fractures, and dissolution pores that are the most favorable reservoir intervals.

The fracture-pore reservoir space of tuff occurs mainly as sheetlike tectonic fractures, together with scattered intercrystalline pores and fractures and partially dissolved pores and fractures. The tuff is far away from the crater, the lava condenses quickly with a low degree of crystallization of minerals, and the cements are mainly volcanic tuffaceous or vitreous material. The tuff lies at structural lows; thus the primary pores and dissolved pores and fractures are underdeveloped, and the physical properties of the reservoir are relatively poor.

3.4.1.2 Relationship Between Volcanic Facies, Volcanic Edifice, and Reservoir Physical Properties

Volcanic facies is an important factor influencing the reservoirs. Volcanic facies is one of the major components of volcanic reservoir research, as different facies and subfacies may develop dramatically different pore types and physical properties. Different subfacies of the same facies may develop different rock textures and structures, which in turn control the combination and distribution of primary and secondary pores and fractures; the physical properties of volcanic reservoirs; and the type, characteristics, and variation of reservoir space. Therefore, understanding the volcanic facies is very important for understanding the temporal-spatial distribution of volcanic rock and reservoir performance (Wang et al., 2003a,b,c; Zhao et al., 2004; Luo et al., 1996).

3.4.1.2.1 Reservoir Characteristics of Subvolcanic Subfacies

The subvolcanic facies has a porphyritic-crystalline texture, condensation rim structure, planar flow and linear flow structure, and columnar and platy cleavage. Peripheral xenoliths are often observed in the subvolcanic subfacies. The representative reservoir space of subvolcanic subfacies involves the primary columnar, platy, and wedge cleavage fractures, tectonic fractures, and secondary pores and fractures caused by dissolution.

3.4.1.2.2 Reservoir Characteristics of Volcanic Conduit Facies

The volcanic conduit facies lies in the lower part of the volcanic edifice and develops during the whole process of the volcanic cycle, but what remains is usually the product of late activity. The volcanic conduit facies can be divided into volcanic neck and volcanic crater subfacies.

Volcanic neck subfacies: the representative lithology includes lava, brecciated lava, or tuffaceous lava, and welded breccia and welded tuff, with porphyritic, welded, brecciated or tuffaceous textures, and ring or radiated cleavage. The major pore types include interbreccia pores, primary intrabreccia pores and fractures, matrix contracting pores and fractures, all secondary pores and fractures of matrix and breccia alteration, and tectonic fractures.

Volcanic crater subfacies: this subfacies develops cryptoexplosive brecciated, autoclastic porphyritic and cataclastic textures, and cylinder-shaped, lamellar, ramiform, vein, and fracture-filling structures. The representative reservoir space involves interbreccia pores between the "*in situ* breccia"; i.e. irregular fractures cut the rock into "brecciated" shapes, and the fractures are filled by the magma with fine breccias. The unfilled parts are the *in situ* interbreccia pores. Tectonic fractures and all types of secondary pores and fractures are also observed.

The physical properties of the volcanic conduit facies are generally poor because of underdeveloped fractures, and the reservoir space mainly includes isolated vesicles and intervolcaniclastic pores. But if columnar cleavage structure is developed, it could improve reservoir space. For example, in the lava of the volcanic conduit filling facies in the Shang74-6 well in Jiyang depression, the interval from 2541.71 to 2548.27 m has measured porosity of 13.1-16.4%, the permeability is $0.10\text{-}6.95 \times 10^{-3}$ μm^2.

3.4.1.2.3 Reservoir Characteristics of Volcanic Explosive Facies

The explosive facies forms in the early and late stages of volcanism, characterized by the volcaniclastic rock created during intense volcanic eruption. The volcanic explosive facies usually lies in the central area of volcanic rock, and the impulse during the eruption will shatter the ceiling and surrounding rocks and produce numerous fractures and cracks. The volcanic breccia develops interbreccia pores and vesicles; an example is the basaltic volcanic breccia reservoir in the Jiyang depression. In addition, the volcanic explosive facies usually lies on the paleocorrosion highland and usually undergoes weathering and leaching, developing dissolved pores (cavities) and fractures that are beneficial for reservoirs. The explosive facies can be divided into fallout subfacies and hot clastic flow subfacies.

Fallout subfacies: the rocks of this subfacies usually develop agglomerate, brecciated, and tuffaceous textures, show a fining-upward sequence, and are grain-supported. This subfacies forms after compaction. The porosity changes significantly during late-stage diagenesis, and the reservoir space is similar to sedimentary rock, mainly intergranular pores, intragranular pores, and dissolved pores and fractures.

Hot clastic flow subfacies: the characteristic reservoir space is the unconsolidated layer between two condensation units, namely, between two flow units. The existence of volcanic breccia and a tuffaceous clastic component of all grain sizes results in two immiscible layers of volcanic clastic flow, forming an unconsolidated layer. Besides the unconsolidated layer, the hot clastic flow subfacies also develops various reservoir spaces, mainly including primary vesicles, intercleavage fissures, interbreccia pores, fractures, and various secondary pores and fractures.

3.4.1.2.4 Reservoir Characteristics of Volcanic Effusive Facies

The volcanic effusive facies forms in each episode of a volcanic eruption but appears mainly after a strong eruption. The lava of effusive facies develops primary vesicles, and the secondary pores result from the mineral (such as feldspar and quartz) volume shrinkage after feldspar dissolution and vitreous devitrification. Different volcanic rock types determine different mineral composition, chemical composition, rock texture, and structure. As a result, the degree of pore development is different, and the lithological difference leads to a difference in reservoir physical properties. In the Xushen gas field, the eruptive-effusion pellet rhyolite facies, explosive rhyolitic (welded) tuff facies, and rhyolitic volcanic breccia facies developed the best physical properties. The reservoir space of eruptive-effusion pellet rhyolite facies primarily includes vesicles and numerous micropores resulting from devitrification and, secondarily, feldspar dissolved pores and intergranular dissolved pores. The reservoir space of rhyolitic (welded) tuff and rhyolitic volcanic breccia mainly includes dissolved pores in feldspar and volcanic ash, as well as local fractures.

The eruptive-effusion facies can be divided into lower and upper subfacies. These two subfacies are often observed interbedded inside the depositional basin and in cross section, and the reservoir space type of each subfacies is notably different.

Lower subfacies: the rock is finely crystalline lava with strong brittleness, and primary columnar, platy, and wedge cleavages. Structural fissures are easily developed and preserved, and the reservoir space type is primarily all types of primary and secondary fractures. The fractures cause active fluid effects and also promote the development of abundant secondary pores and fractures.

Upper subfacies: the typical pore structure involves vesicles, amygdaloid pores, and lithophysa; the typical texture

is spherulitic. The reservoir space includes primary vesicles, intra-amygdaloid pores, and interlithophysa pores. The vesicles occur in bands, showing an elongation parallel to the flow direction. The vesicles may become large and dense in the upper flow unit. Among all the volcanic facies, the upper subfacies develops the best primary vesicles, which occupy more than 20% of the rock volume and are connected to each other via tectonic fractures. Because of the impact of vesicles, the tectonic fractures in the upper subfacies are mainly irregular interpore fractures and regular grouped fractures are seldom observed. According to the statistics, the upper subfacies of the eruptive-effusion facies develops the best reservoir physical properties in the Xingcheng and Shengping areas of SongLiao Basin.

There is a distinctive difference in the volcanic effusive facies in the vertical and horizontal directions. For example, the volcanic lava flow in the Erlian Basin, Inner Mongolia, can be vertically divided into four belts:

1. Volcanic cinder belt: this belt lies at the lava flow top and consists of autoclastic breccia lava, and the breccia is always vesicular and amygdaloidal lava. When the lava flows over the crater and then onto the surface, the surface layer contacts the air directly, condenses quickly, and initially indurates, but the continuous lava flow will grate this hard crust and produce a cinder belt.
2. Upper vesicular and amygdaloidal lava belt: this belt lies below the autoclastic breccia lava and contains rich vesicular and amygdaloidal structures. The elongation phenomenon is often observed, indicating the flow direction. The belt lies in the upper lava flow where the pressure is low, and the gas in the lava can easily escape to form vesicular structures.
3. Middle tight massive lava belt: this belt lies below the upper vesicular and amygdaloidal lava belt where vesicles and amygdaloids are underdeveloped because the rock is tight, and their shape is rounded if they are developed. Slow-cooling lava and high pressure mean that the gas cannot easily escape; thus there is almost no vesicular or amygdaloidal structure. Under burial conditions, the above three belts may undergo some change. Because of vesicle underdevelopment and rare fractures, the middle tight massive belt cannot develop reservoir qualities.
4. Lower vesicular and amygdaloidal lava belt: this belt is the lowest in the lava flow and develops vesicular and amygdaloidal structures caused by the lava flowing along the moist surface or water bottom.

There are fractures in the upper and lower vesicular and amygdaloidal belts, as well as the middle tight massive belt, but the fracture density in the middle belt is far less than that in the upper and lower belts. The degree of fracture development and that of direction are inconsistent in the different belts of the same lava flow layer; thus the fractures do not have a tectonic genesis, but rather are the products of lava condensation and contraction in that the fractures seem to be tensional. The fractures are often filled by calcite, chalcedony, and chlorite.

The reservoir physical properties of the eruptive-effusion facies are controlled primarily by the lithology zone and secondarily by the secondary reformation. A good reservoir must have abundant fractures. In four belts of the lava flow sequence, the interbreccia pores and cavities in the top autoclastic breccia belt are relatively large and always incompletely filled, but the reservoir usually has high porosity and permeability. In the upper and lower vesicular and amygdaloidal belts, some vesicles always remain unfilled, and the two belts develop abundant fractures, such as contraction fractures that connect partial vesicles, thus having good porosity and permeability. The reservoir space involves primary vesicles, intergranular pores, dissolved pores, tectonic and contraction fractures, and intercrystalline pores, all contributing to good reservoir physical properties. For example, in the volcanic breccia from 1975 to 1979.6 m of the Shang74-12 well in Jiyang depression, the porosity is 20.7-33.1%, and the permeability is $11.2\text{-}140 \times 10^{-3}\ \mu m^2$. The volcanic breccia in the interval from 1830 to 1838 m in the Shang74-6 well has porosity of 16.7-37.4% and permeability of $0.988\text{-}3170 \times 10^{-3}\ \mu m^2$. The eruptive-effusion facies is also a favorable reservoir belt.

3.4.1.2.5 Reservoir Characteristics of Extrusive Facies

The extrusive facies is widely distributed but difficult to identify, usually occurring in the central belt, transitional belt, and marginal belt. The identification mark of the central and transitional belts is pearlite. The three subfacies of extrusive facies play different roles in the reservoir, and the central belt is the dominant reservoir.

Central belt subfacies: the macroscopic reservoir space involves the "loose body inside dome" in the large pearlite, accumulating a series of pillow and pellet pearlites. Primary ring fractures develop inside the pellet, and loose and sandy pearlite debris develop between the pellets. The primary fractures of this subfacies are the best developed and take on a ring shape on the microscopic and macroscopic scales. On the macroscopic scale, the vitreous pearlite crumbles into volcanic vitreous balls several to tens of centimeters along the ring fissure. Because of hard deposit skeletons, as well as the pearlite in the central belt and breccia lava in the marginal belt covering and protecting the skeleton as hard crust, the reservoir space in the central belt subfacies mainly involves micropores, such as fractures, dissolved pores, and intercrystalline pores. The fractures dominate and are mainly tectonic and contraction fractures, which are beneficial for the reservoir's physical properties. This subfacies develops the most favorable reservoirs.

Transitional belt subfacies: it is always transitional or interbedded with central belt subfacies. The structure type of the transitional belt subfacies includes platy, columnar, and wedge cleavages, and the typical textures are vitreous and pearl, with less porphyritic and phenocrystal textures. The rocks of this subfacies are extremely brittle, causing tectonic fractures to form and reform easily, but still less developed than those in the lower subfacies of eruptive-effusion facies. The rock crystallization time is long enough for the development of medium porphyritic texture. The reservoir space is fractures and dissolved pores, which are beneficial for the reservoir's physical properties. This subfacies develops favorable reservoirs.

Marginal belt subfacies: it is mainly the near-crater lava that acts as the protective facies for the loose rock of central belt subfacies, preserving the central belt facies during the burial process. The marginal belt lies at the outside the extrusive dome, and its representative rock is breccia lava with deformed rhyolitic structure. The rock develops melted breccia and melted tuff textures, and deformed rhyolitic texture is often observed. The marginal belt itself cannot serve as reservoir. Because the marginal belt lies at the margin of the rock body, the lava cooling is so rapid that the resulting texture is dolerite and dolerite diabase. When the intrusion is shallow, vesicular and amygdaloidal structures often appear, and the lithology is mainly cryptocrystalline to fine diabase. The thickness of the subfacies is usually less than 20 m and shows thin interbeds on electric logs. Thin layers are usually unfavorable for the development of tectonic fractures, but some condensing contraction fractures may still develop. Thus the reservoir space includes primary vesicles and contraction fractures that form unfavorable reservoirs.

3.4.1.2.6 Reservoir Characteristic of Volcanic Sedimentary Facies

The volcanic sedimentary facies in eastern China is mainly represented by the volcanic reservoirs in the Ha'nan oil field of Erlian Basin, Qijia-Gulong sag of SongLiao Basin, and the upper Lugouqiao Formation of the Jizhong depression. The volcanic sedimentary facies generally develops at the margin of an eruptive main body, and the volcanic layer is thin and mostly tuffaceous or sedimentary tuff interbedded in the normal depositional system. Primary vesicles are underdeveloped because of thin volcanic lava, and the reservoir is generally poor because the facies usually lies in the paleocorrosion low area where secondary dissolved pores and weathering fractures are underdeveloped. However, good reservoirs may develop in the case of structural stress fractures and leaching solution. For example, in the Ha'nan oil field in Erlian Basin, the volcanic reservoir rock is basic and intermediate tuff of volcanic sedimentary facies, and abundant dissolved pores and tectonic fractures are produced by tectonic stress and leaching. Many are filled by multistage minerals, such as serpentine, chlorite, pyrophyllite, calcite, and gypsum, but the structure top and base as well as the fault zone still preserve numerous fissures, dissolution, and solution traces along the fractures and fissures. Therefore, the reservoir space comprising tectonic fissures and dissolved pores has good oil shows.

In view of the Carboniferous volcanic lithofacies drilled in the northern Xinjiang, the characteristics are generally the following: (1) there are many types of volcanic facies developed, but mostly the explosive and effusive facies; (2) the effusive lava and explosive-volcaniclastic rock alternate vertically, reflecting the Carboniferous volcanic activity featuring multistage and superimposed eruptions (Figure 3.20); and (3) in map view and controlled in two directions, to the northeast (such as Bei5-Zhang3 line in the northwest margin and east in the Junggar Basin) and the northwest (such as the distribution paralleling the Kelameili Mountain or extending to the west and in Santanhu Basin), the lithofacies is distributed in bands along the fault zone, and the crater is located at the fault intersection, belonging to centered eruptive volcanic rock along the fault.

3.4.1.3 Relationship Between Volcanic Rock Occurrence, Lithofacies, and Reservoir Space

The volcanic edifice in hydrocarbon-bearing basins in China mainly involves layered volcanoes, and to a lesser extent shield and slag volcanoes. The lithofacies composition is different for different volcanic edifice types, resulting in different reservoir physical properties.

Shield volcano: the lithology and lithofacies are relatively simple. The volcanic edifice covers a small range, usually 1-3.5 km wide and 100-250 m high. The lithology of the volcanic edifice is rhyolite and rhyolitic tuffaceous lava, the former belonging to eruptive-effusion facies and the latter belonging to the transitional type from eruptive-effusion facies to hot clastic flow subfacies of explosive facies. The volcanic edifice also develops volcaniclastic rock of the hot base surge subfacies and cryptoexplosive breccia of the volcanic conduit facies. The reservoir's physical properties are heterogeneous, vary greatly depending on lithofacies, and become the best in the vesicular rhyolite in the upper subfacies of eruptive-effusion facies. In addition, each eruption cycle top in the volcanic edifice always develops a weathering crust with a variable size that is important for this type of volcanic massif.

Layered volcano: This has a layered distribution, and lithologies change laterally but are basically continuous. The lithologies are basalt, andesitic basalt, trachyte, rhyolite, and tuff, and the lithofacies are explosive facies including fallout subfacies, hot base surge subfacies, hot clastic flow subfacies, and volcanic sedimentary facies that is sometimes interbedded with eruptive-effusion facies lava. The facies is vertically characterized by unstable development and great change, and horizontally characterized by abrupt changes, but generally showing layered characteristics. The volcanic edifice develops fractures whose logging curve is bayonet and comb-like, and the testing results show good physical properties. The major reservoirs are welded tuff of hot clastic flow subfacies and vesicular rhyolite or andesite of the upper subfacies of eruptive-effusion facies. The major reservoir space includes primary vesicles, intergranular pores, and fissures.

Slag volcano: This develops at the fault margin, and the lithology includes volcanic agglomerate, volcanic breccia, breccia tuff, tuff, andesite, and rhyolite. The lithofacies includes fallout subfacies, hot base surge subfacies, and hot clastic flow subfacies of eruptive-effusion facies and explosive facies. Fallout subfacies is well developed, and it is a mark of the near-crater facies. The eruptive-effusion facies and explosive facies alternate both vertically and horizontally. The best reservoir physical properties are developed in the explosive facies, especially where unconsolidated layers occur between condensation units, and the major reservoir space includes intergranular and intercrystalline pores (Wang et al., 2005a,b).

3.4.1.4 Volcanic Eruption Environment

In view of the volcanic rock color and vesicle development in different blocks of Junggar Basin, the volcanic rock at the base of the Wucaiwan sag is mostly strongly chloritized celadon, and the primary vesicles are extremely underdeveloped. Vesicles in the Luliang volcanic rock are developed in the hinterland of Junggar Basin, and the volcanic rock is mainly brown. This difference of "east celadon and west red" shows that the volcanic eruption environment from east to west tends to transition from underwater to above ground. The volcanic rock in the eastern Wucaiwan sag is characterized by a deep underwater eruption, and the Luliang volcanic rock in the hinterland is characterized by an onshore or shallow underwater eruption.

The eruptive environment greatly impacts the volcanic reservoir space. For example, in the Caican2 well in the Wucaiwan sag in eastern Junggar Basin, the Carboniferous is mainly interbedded volcanic rocks and sedimentary rocks with some tuffaceous breccia and tuff. The sedimentary rock contains marine fossils accumulated in epicontinental seas. When the volcano erupted in deep water, the volatile components dissolved in the lava could not escape because of great hydrostatic pressure in the deep-water environment. Thus primary vesicles are underdeveloped. In addition, water action results in obvious dissolution (chloritization) and mineral precipitation, causing the originally few primary pores to decrease even further. In contrast,

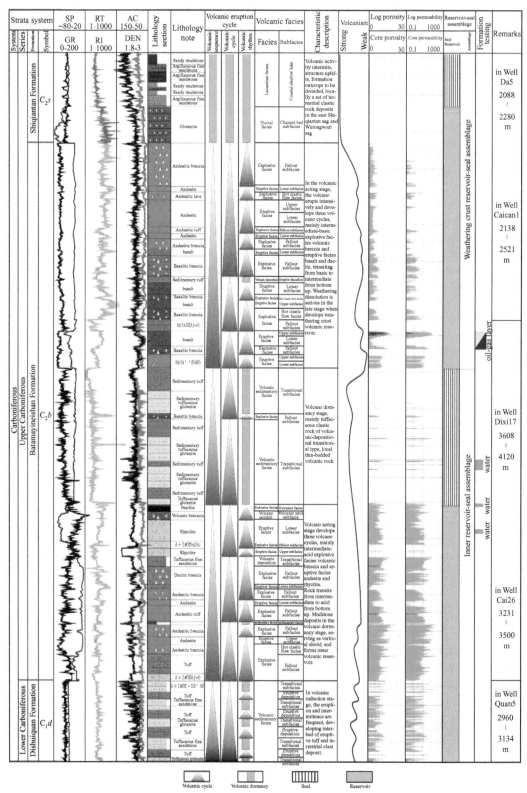

FIGURE 3.20 Carboniferous column in the northern Xinjiang.

the breccia lava in the Shixi hinterland of Junggar Basin erupted in shallow water or onshore, causing the many volatile components dissolved in the lava to escape to form vesicles. In addition, the hot magma is erupted on the water's surface and creates a quenching effect, producing numerous primary microfractures and connecting the primary vesicles, which constitute good original reservoir space. Different eruption environments in the above two regions result in great changes in reservoir physical properties (Yu et al., 2004).

3.4.2 Fluid Action

Fluid action causes mineral matter to dissolve or precipitate, creating secondary pore space as well as cement precipitation and dissolution. The fluid causes the volcanic pore texture variations, greatly improves the volcanic reservoir's physical properties, and results in diverse volcanic reservoir space types.

Volcanism, tectonism, and hydrocarbon expulsion will lead to large-scale fluid action, and the direct effect of fluid on volcanic rock is the material brought in and out, making the volcanic rock body in an open system. The fluid includes hydrothermal fluid and acidic fluid related to organic matter. The hydrothermal fluid action directly results in alteration and dissolution of the original minerals and, at the same time, new mineral generation resulting in secondary cementation and filling. The alteration and dissolution improve the volcanic rock porosity, and the cementation and filling reduce the porosity and especially permeability. Therefore, the compound effect of hydrothermal action on volcanic reservoirs varies from time to time and from place to place and is most dependent on local factors. The pure filling in the volcanic rock greatly destroys the volcanic reservoir because the authigenic minerals in vesicles can partially or completely fill pores, which greatly weaken the reservoir's physical properties. The minerals that fill fractures are even more destructive, not only by occupying additional pore space but also by greatly reducing the reservoir connectivity and permeability. SEM observations show that the pores filled by minerals are more abundant than residual pores in most samples. The mineral filling occurs in stages; i.e. the filling may be the result of a single filling or a sequence of fillings. For example, the quartz that fills vesicles of the Yingcheng Formation rhyolite at 3123.4 m in the SS1 well, SongLiao Basin, the homogenization temperature of the inclusions is 96-105 °C, which is in one temperature range, so can be regarded as the result of a single filling episode. However, the fracture filling in the Yingcheng Formation andesite at 3187.9 m in the SS2 well is two staged—the first is quartz formed at 92-99 °C, and the second is calcite formed at 115-129 °C. In summary, it is very common that the fluid action results in dissolution,

alteration, cementation, and filling, occurring from the volcanic rock burial to the present day.

The acidic fluid is a key factor for the development of secondary pores; thus the source of acidic fluid is a key issue. The fluid is mainly sourced from underground water, deeply circulating atmospheric water, and water released from kerogen maturation. The latter may blend into water that percolated to great depths and mixed with CO_2 acid and carboxy water formed during the thermal evolution of organic matter to produce compound acidic fluid. There are two major sources for H^+ in the fluid: the first source is the organic acid from the thermal evolution of organic matter and pyrolysis acid during the kerogen maturation, which is generally carboxy acid. According to much organic acid found in formation water and the tests on the dissolution intensity of different organic acids, Surdam and Crossey (1984) believed that the organic acid has strong complex capacity, and the carboxy anion will be decarboxylated in the weakly reducing environment and convert into hydrocarbon and CO_2, making the CO_2 concentration increase in the water solution. The second source is the abundant carbonic acid formed when the kerogen constantly releases CO_2 that dissolves in the pore water during the evolution of organic matter. The freshwater blends with the CO_2 and carboxy acid water into acid fluid. The acid water is expelled from the source rock under the pressure, providing a great amount of H^+ for feldspar dissolution. Moreover, the acid pore water flow dissolves the rock and produces secondary pores. The secondary pores develop after the fissures form, which provide passageways for acid water flow. Most acid water comes from the thermal evolution of kerogen, which occurs at the early stage of hydrocarbon formation. Therefore, the secondary pores develop before the formation of hydrocarbon and provide passageways and accumulation space for crude oil (Feng et al., 2008; Zhao et al., 2004).

3.4.3 Tectonism

Tectonism plays a leading role in fault and fracture development. The primary vesicles in the volcanic rock are always disconnected and impermeable; thus late tectonic movement is indispensible for the volcanic reservoir and hydrocarbon accumulation. Fractures impact the reservoir development in three aspects: (1) the fractures in the vesicular and amygdaloidal zone improve vesicle connection, and the most important thing is that the surface freshwater or subsurface water dissolves and redeposits the volcanic rock along the fractures, producing dissolved pores or even cavities on the basis of primary vesicles, residual vesicles, and matrix intercrystalline pores; (2) if there are fractures in the tight zone, a pure fractured reservoir can develop, and dissolved pores or even cavities can develop under some conditions; and (3) fractures can improve the formation water distribution and fluidity, promoting the generation and evolution of dissolution.

Tectonic movement produces many fractures for volcanic rock bodies, and multistage, multidirectional faulting produces many highly dense fractures in the volcanic rock, connecting the originally isolated pore network and greatly improving the volcanic reservoir's physical properties. This is beneficial for oil and gas migration and accumulation. The tighter the volcanic rock, the stronger its brittleness and the more easily tectonic fractures are developed and preserved. These fractures not only connect the isolated primary vesicles but also improve the volcanic reservoir space. For example, in the 3500- to 3600-m volcanic reservoir interval in the Shengshen1 well, the reservoir rock includes finely crystalline rhyolite with syngenetic breccia of the lower subfacies of the eruptive-effusion facies, where primary pores are seldom seen. However, fractures are developed because of strong brittleness, which contributes to good reservoir under late-stage tectonism. Multiple tectonic movements bring about multistage fractures, and the early fractures are often cut by late fractures. Multistage volcanic fractures provide good conditions for hydrocarbon migration and accumulation. In Santanghu Basin, Carboniferous-Permian volcanic rock developed at least two stages of tectonic fractures. Stage I fractures are large and formed early, greatly impacting the reservoir, but the fractures themselves are mostly filled. Stage II fractures are small, and the effect on the reservoir was not as beneficial as stage I fractures, but the fractures are mostly open and less filled; thus the reservoir quality was somewhat improved.

In summary, the volcanic reservoir is the core content of volcanic reservoir research and the key factor of hydrocarbon enrichment, so it is significant that we establish a thorough understanding of volcanic reservoir geology.

Chapter 4

Geology of Volcanic Reservoirs

The geologic study of volcanic reservoirs focuses on the effective configuration relationship of the reservoir, hydrocarbon-generation assemblage, and trap. This chapter introduces the hydrocarbon source of volcanic reservoirs, major play types, and reservoir types of volcanic rock. Moreover, we discuss the oil and gas enrichment law of volcanic rock.

4.1 HYDROCARBON SOURCE

4.1.1 Hydrocarbon Genesis of Volcanic Reservoirs

The understanding of volcanic rocks and hydrocarbon genesis can be traced back to the mid-eighteenth century. Before 1763, the former Soviet scholar Ломоносов observed that the genesis of hydrocarbon was related to volcanism, and he also put forth the organic origin concept that the hydrocarbon was generated from the dry distillation of coal under geothermal activity. The organic origin of hydrocarbon has been recognized by petroleum geologists both at home and abroad for a century or more. The primary organic material was deposited in a dispersed state or locally in a concentrated state (turf, coal) in marine or continental sedimentary basins and was often accompanied by other mineral matter. Under shallow burial conditions, it generated biogas. With a gradual increase in burial depth and formation temperature, the organic matter gradually transformed to oil and natural gas under a (strong) reducing environment. Oil and gas exploration practices over the past 100 years have shown that the reservoirs discovered in the world are almost all in sedimentary rocks of marine or continental basins, and the volcanic rock province is generally regarded as an unlikely to impossible zone for the discovery of hydrocarbons. However, with the deepening of oil and gas exploration, a series of reservoirs, including some major fields (e.g., Xushen gas field), have been discovered in volcanic rocks both at home and abroad. Therefore, explorationists are beginning to focus on volcanic reservoirs.

Hydrocarbon genesis theories can be traced to inorganic origin and organic origin from the 1770s to the present day. Classical inorganic generation hypotheses (e.g., carbonization theory, universe theory) and classical organic generation hypotheses (e.g., animal theory, plant theory) were developed from the eighteenth century to mid-nineteenth century, and strong controversy broke out between the two schools of thought. Long-term production practices and scientific research have confirmed that substantial hydrocarbon, especially oil, has organic origins. However, some hydrocarbon, especially some natural gas, is inorganic in origin and can form commercial gas reservoirs (Dai et al., 1995). In the same depositional basin, the organic matter evolution and the inorganic geologic action may interact and inter-couple with each other, forming hybrid origin reservoirs.

The volcanic reservoir has both organic and inorganic origins. It is generally agreed that most volcanic reservoirs have organic origins; i.e., the oil and gas are the hydrocarbon organic compounds formed after the long-term evolution of organisms in ancient oceans or lakes.

The source of organic matter can be generally divided into three types: CO_2 reduction, fermentation, and thermogenesis. The hydrocarbon in the volcanic rock is generally thermogenetic in the natural world. The formation mechanism of volcanic rock is largely different from that of sedimentary rock. Because the volcanic rock is generally not symbiotic with the source rock, the prerequisite for forming a volcanic reservoir is its association with source rock; i.e., the volcanic rock is located in, above, or below the source rock series, or there is source sag nearby. Only by being so situated can the volcanic reservoir receive migrating hydrocarbons generated in the source rock.

Volcanic rock accounts for a significant percentage of the filling in a depositional basin, and its contribution to the sediment load can reach 25% in all classes of basins. Therefore, the volcanic rock is as prone to store the hydrocarbon coming from sedimentary rock in the depositional basin as sedimentary rock is, causing the oil and gas exploration of volcanic rocks to have great prospects. Lakes also often develop in the volcanic rock formed in the onshore surficial environment, and the lake may contain hydrocarbon-rich deposits. Kirkham (1935) believed that the natural gas in the Rattlesnake Mountain gas field possibly came from the lacustrine sediments in the basalt, for the natural gas contained considerable nitrogen. In addition, some volcanism and pyrogenesis can also supply the volcanic rock with hydrocarbon and thus result in the accumulation of anorganogene oil and gas. The scale of these types of reservoirs can be appreciable, e.g., the accumulation of gas in Lake Kivu of the Congo.

The sediments rich in organic matter that are proximal to volcanic activity can become vitally important source rock. The volcanism contributes to the formation of an anaerobic environment. Before and after volcanism, accompanied by the blowout of a great deal of thermal fluid and gaseous fluid substances, the thermal fluid often contains transition metals, such as Ni, Co, Cu, Mn, Zn, Ti, and V, and substances, such as N and P. The substances in these thermal and gaseous fluids play a positive role in the growth, propagation, maturation, and conversion rate of organic matter (Table 4.1). In addition, volcanoes frequently erupt in the ocean, which can cause organisms to generate a great deal of organic acid and salt after immediate high-temperature exposure, or they may be shallowly buried and then encounter high temperatures, thus promoting hydrocarbon generation. For instance, basaltic volcanism and source rock deposition took place simultaneously in the Bohai Bay Basin, and the warm water resulting from the volcanism probably increased the productivity of organic matter.

Under the action of volcanoes and thermal fluid, the geothermal gradient in basins is apparently higher, which largely promotes the maturation of organic matter. For instance, the geothermal gradient averages 3.5 °C/100 m in the Liaohe depression, and it can reach 4.25 °C/100 m in the interval from 2200 to 3200 m in the volcanic rock distribution area. Both the volcanic rock and the intrusive rock carry high-temperature heat energy and bake the sedimentary formations containing organic matter. The source rock can reach a highly mature or overmature stage under the action of high geotemperature and thus expel a great deal of natural gas; simultaneously, the crude oil generated at an early stage can also be cracked into gas with increasing thermal evolution levels and ultimately become bitumen. In addition, the volcanic rock generally contains a higher percentage of radioelements (e.g., U, Th, and K), and the heat resulting from the long-term decay of these elements is also one of the important heat sources in the depositional basin.

The formation of inorganic origin hydrocarbon is mostly related to volcanism or magmation, but this is still a hypothesis at present. The major hypotheses are summarized below.

4.1.1.1 Carbonization Theory

This hypothesis was put forth by the well-known Russian chemist Mendelyeev. He believed that the major original substance for generating hydrocarbon was metallic carbide; therefore, his hypothesis was called the "carbonization theory." Recently, Hunt et al. (1992) expounded on the carbonization theory in more detail and believed that carbon and iron were turned into liquid at very high temperatures at the time of formation of the earth and then interacted to form ferric carbide. Because of its large relative density, the ferric carbide could not be preserved in the oxygen-rich crust but rather in the deep crust. When it met the hot water that infiltrated the deep crust along fractures, they reacted, and hydrocarbon was formed. When the generated oil and gas upwelled into the crust, it was condensed in the porous layers. When optimal conditions were reached, a reservoir was formed. Their reaction equation is expressed as follows:

$$3Fe_mC_n + 4mH_2O \rightarrow mFe_3O_4 + C_{3n}H_{8m}$$

4.1.1.2 Universe Theory

The universe theory was initially put forth by the Russian scholar Соколов in 1889. During the process that the nebula matters containing iron, silicon, calcium, hydrocarbon (primarily methane), carbon dioxide, and helium were gradually condensed and consolidated to form mantle and crust, the gases such as methane, carbon dioxide, and helium in the nebula or protoatmosphere were also "absorbed" and stored in the mantle and crust simultaneously. Subsequently, deep rupture, mid-oceanic ridge, volcanism-magmation, and earthquake effects resulted in

TABLE 4.1 Simulated Experiment Results of the Interaction Between Source Rock Organic Matter and Volcanic Minerals

Mineral type	Product type	Product amount ($cm^3 g^{-1}$)				
		300 °C	350 °C	400 °C	450 °C	500 °C
No volcanic mineral	Hydrogen gas	0.41	8.74	67.52	76.34	37.58
	Hydrocarbon	3.1	6.54	58.37	138.49	168.45
Adding volcanogenic zeolites	Hydrogen gas	0.66	9.13	68.46	79.10	90.47
	Hydrocarbon	9.83	30.21	103.67	177.41	201.95
Adding olivine	Hydrogen gas	79.82	217.66	430.55	144.78	75.43
	Hydrocarbon	41.08	45.31	148.59	268.97	309.66
Adding zeolites and olivine	Hydrogen gas	86.50	506.44	760.33	648.97	298.67
	Hydrocarbon	57.62	90.21	166.98	346.46	416.58

strong degasification and made the abiogenetic gas deep in the earth migrate to the shallow layers in different forms and scales. Some of them accumulated and formed reservoirs, but most dissipated in the lithosphere, hydrosphere, and atmosphere. Gold (1993) pointed out that substantive methane and the nonhydrocarbon resources existed deeply in the earth, and the methane existed at the time of formation of the earth. The carbon dioxide, helium, and methane in the high-temperature, warm-spring gas of Wudalianchi, Heilongjiang province, Zhegulong, Guangdong province, and the volcanic region of Tengchong county, Yunnan province, China (Dai, 1988; Dai et al., 1990, 1992) all belong to natural gas of this genesis.

4.1.1.3 Magma Theory

The former Soviet scholar Кудрявцев put forth the magma theory of hydrocarbon origin on October 3, 1949 on the same rostrum where the 60th anniversary of the publishing of universe theory was held. He believed that hydrocarbon generation was related to the synthesis of hydrocarbon that occurred during the cooling of basic magma. Because the synthesis was accomplished under high pressure, it could cause the unsaturated hydrocarbon to be polymerized into saturated hydrocarbon. After the basaltic rock and ophiolite in the oceanic plate had underthrusted to a certain depth in the mantle, they were transformed into denser eclogite due to high temperature and pressure, and some were even melted into magma. Simultaneously, in the course of transforming into magma and rock decomposition, hydrogen and carbon (and its oxide) were synthesized into methane under high temperature and pressure. Багдасарова (2000) observed and calculated that in the modern thermal fluid in places such as the Kamchatka Peninsula, substantive anatectic hydrocarbon was brought upward to the rift zone along the deep rupture in the course of volcanism and subsequent activity of thermal fluid. Гептнер (2002) found that respectable amounts of methane and heavy hydrocarbon components were contained in the modern and late Cenozoic eruptive materials in the Iceland rift zone; moreover, a certain scale of asphalt also existed to the southeast.

4.1.1.4 Theory of High-Temperature Gas Generation in Upper Mantle

Based on the experiment of man-made diamonds, a mixture of calcite, quartz, and hexahydrite was used to replace graphite and put it in a reactor. Under high pressure (6000-7000 MPa) and high temperature (1800°K), light-volatile components such as methane, ethane, propane, butane, pentane, hexane, and a few heptanes separated out in the reactor within a few minutes. As a result, he deduced that in the Gutenberg layer of the upper mantle, when the temperature exceeded 1500°K and the pressure reached 5000 MPa,

because of the participation of FeO and Fe_3O_4, H_2O and CO_2 were reduced to hydrocarbons, which were squeezed upward into the crustal sedimentary rock under strong tectonic compression. It was shown by the lithogeochemical study of the bitumen and fluid inclusions in the basement of the Непко-Ботуобин hydrocarbon province that bitumen migrated upward from deep strata under high-temperature and pressure conditions, and the oil and gas fields basically concentrated in the fault convergence region.

4.1.1.5 Metamorphic Theory

Because the oil fields in the Russian Volga-Ural oil province existed in serpentine and strongly serpentinized peridotite, ланский pointed out that the serpentinization of olivine (ultramafic rock) could generate hydrocarbon (Бека and Высоцкий, 1976). The gas that was emitted from the partly serpentinized ultramafic rock fracture in the Zambales region of the Philippines is the serpentinized product of the rock. As much as 15×10^4 tons CH_4 and 30×10^4 tons H_2 can be generated in 1 year in the course of serpentinization from the equator to the Rhine Bailey at 36° north latitude along the axial region of the mid-Atlantic ridge. As per the calculation of Сорохтин, when the ultrabasic rock is serpentinized in the ocean, 650×10^4 tons CH_4 and 800×10^4 tons H_2 can be generated each year. In addition, based on the study of Hosgrmez (2007), the Chimera gas seepage in the place where the first Olympic flame was ignited came from the strong serpentinization of ultrabasic magma. Moreover, substantive CO_2 can be generated in the course of rock metamorphism.

4.1.1.6 Radiogenic Theory

In 2001, Лесовой (2001a,b) put forth the new radiogenic theory based on the fact that the percentage of radioelements is abnormally high in many hydrocarbon-bearing rocks, and all the well-known abnormal indications of uranium and thorium, despite their formation time and geologic characteristics, both contain mineral carbon (organic carbon) and various components of high molecular hydrocarbon. The hypothesis suggests that the gamma ray generated at the time of the transmutation of radioelements can ionize the inert carbon, combine with the hydrogen atom, and form hydrocarbon compounds such as CH, CH_n, and $C_nH_{3.4n}$. Under highly ionized conditions, light hydrocarbon can synthesize different kinds of complicated high molecular hydrocarbons and even bitumen or asphalt. Furthermore, local high pressure can be generated under the action of radioactivity, which can increase the fissure volume of the recessional skin and promote the accumulation of hydrocarbon. Based on the study of Вержбицкий (2002), the high geotemperature background (70 mBT m^{-2}) in the eastern Barents Sea in the arctic zone of Russia was not created by rifting but by radioactive heat resulting from sedimentary cap rock.

Discriminating the organic versus inorganic origin of hydrocarbon is one of the critical aspects in the study of hydrocarbon genesis of the volcanic reservoir.

First, methane is the most common component in the alkane gas of both organic and inorganic origin natural gas, and the carbon isotopes of methane are one of the important parameters to distinguish the genesis of natural gas. It has different threshold values of $\delta^{13}C_1$ used for dividing inorganic origin and organic origin methane, with mainly three representative values: $>-20‰$, $>-25‰$, and $>-30‰$. Dai et al. (2008) believed that the threshold value of $\delta^{13}C_1$ could be used to divide the vast majority of organic and inorganic origin methanes, and it was relatively reasonable and practical when the value was more than $-30‰$.

Second, correlating the carbon isotope series of alkane gas is very effective to discriminate between inorganic origin and organic origin alkane gas. The so-called carbon isotope series of alkane gas refers to the ascending or descending of $\delta^{13}C$ values in turn with the ascending of the carbon number sequence of the alkane gas molecule. The ascending ($\delta^{13}C_1 < \delta^{13}C_2 < \delta^{13}C_3 < \delta^{13}C_4$) is called a positive carbon isotope series, while the descending ($\delta^{13}C_1 > \delta^{13}C_2 > \delta^{13}C_3$) is called a negative carbon isotope series. The organic origin alkane gas possesses a positive carbon isotope series, while the inorganic origin alkane gas possesses a negative carbon isotope series (Dai et al., 1989, 1992).

Third, both R/R_a and $CH_4/^3He$ parameters are important indices to discriminate between organic origin and inorganic origin alkane gas. When the R/R_a is greater than 0.1 in natural gas, it indicates that inorganic gas exists. $CH_4/^3He$ is a parameter combining inorganic and organic origins, and it has overcome the limitations of the geochemical characteristics of independent alkane gas. When $CH_4/^3He$ is 10^5-10^7 orders of magnitude, the methane has an inorganic origin; when it is 10^9-10^{12} orders of magnitude, the alkane gas has an organic origin.

4.1.2 Hydrocarbon Sources in Volcanic Reservoirs

4.1.2.1 Types of Source Rocks and Volcanic Reservoir Assemblages

The spatial configuration relationship of volcanic rock and source rock in the depositional basin can be divided into two types. Type 1 is the proximal assemblage, where the volcanic reservoir is located above, below, or between the source rocks, and horizontally, the volcanic rock is located inside or in the vicinity of the hydrocarbon-generation kitchen, for instance, the upper Jurassic-lower Cretaceous rift sequence in the deep layers of the Songliao Basin; the

Permo-Carboniferous in the eastern region of the Junggar Basin and in the Santanghu Basin; the Jurassic basalt reservoir in the Shijiutuo uplift of the Bohai Sea, where the basalt directly underlies the Paleogene-Neogene source rock series and there is Bozhong source sag in the south and Qinnan source sag in the north; and the Rehetai volcanic reservoir in eastern Liaohe, where the volcanic rock is in the Paleogene source rock series, and a Jiazhangsi source sag in the south and Huangshatuo source sag in the north. Type 2 is the distal assemblage, where the volcanic rock is located above the source rock, but is separated by many sets of strata; horizontally, the volcanic rock is located beyond the hydrocarbon-generation kitchen, and the hydrocarbon accumulation needs a connection with the help of conduction system such as a fault or unconformity. For instance, the oil and gas in the Carboniferous volcanic reservoir in the upper wall of a thrust belt in the northwest margin of Junggar Basin was migrated from the lower wall through a fault and unconformity; low-yield oil flow was obtained from the Permian volcanic rock in the Haitan1 well in the Tabei region of the Tarim Basin, where the oil came from the deep Cambrian-Ordovician, and it is the fault that connects the oil source and the volcanic reservoir. All these belong to the distal assemblage. Generally speaking, the proximal assemblage is superior to the distal assemblage.

4.1.2.2 Organic and Inorganic Sources of Volcanic Hydrocarbon

Most of the hydrocarbon in the volcanic reservoirs of China's onshore depositional basins came from organic matter in the sedimentary rock, but inorganic origin gaseous hydrocarbon was also discovered. We still lack an in-depth study on the impact of the volcanism on the formation and evolution of organic matter.

Taking the deep layer in the Songliao Basin as an example, the study on the components such as CO_2, CH_4, He, and carbon isotopes of natural gas shows that an organic origin generally predominates in the region. However, an inorganic origin also exists, as individual inorganic CO_2 content is 60%, and the He is all mantle-source inorganic origin.

(1) Basis of predominantly organic origin: (1) a 160-day study of the carbon isotopes of CO_2 and CH_4 in the gas well showed that the organic accounts for 83%; (2) 85% of natural gas was produced from the strata above the Shahezi Formation (coal measures), and a small percentage from the strata below it; (3) the gas production is in positive correlation with the thickness and area of the Shahezi Formation; and (4) the formation of substantive producing zones both in China and abroad are mostly related to an organic origin such as sedimentary rocks and coal measures.

(2) Basis of predominantly inorganic origin: (1) the mineral inclusions of mantle peridotite, basalt, and rhyolite all contain gases such as CO, CO_2, CH_4, and H_2S, and especially CO_2 and CH_4 predominate; (2) the depth of the magmatic origin is different from basicity to acidity; the basicity mostly contains CH_4, and the acidity mostly CO_2; (3) commercial gas also exists in the Huoshiling Formation below the Shahezi Formation and in the Permo-Carboniferous metamorphic rock basement; (4) inorganic origin CH_4 and CO_2 exist in all the different regions and wells, and commercial gas flow exists in some regions and wells; and (5) mantle-origin He exists in the CO_2 and CH_4 gas wells.

The Huoshiling, Shahezi, Yingcheng, and Denglouku Formations source rock series developed in the deep strata of the Songliao Basin. The organic carbon content of the source rock is 1.58-2.4%, the organic matter is type III, the R_o is 1.15-1.6%, and they are favorable gas source rocks. The Yingcheng Formation volcanic rock is the major reservoir. Apart from the discovery of 100 Bcm gaseous hydrocarbon reserves, more than 100 Bcm CO_2 reserves were also discovered. Therefore, in addition to exploring for organic natural gas, we should also focus on exploring for and studying inorganic gas.

The gas source of the Xushen gas field in the northern Songliao Basin is a mixture of coal-derived gas and inorganic alkane gas. There are two points of view regarding the priority of the two gas sources: one considers the gas source a mixture source in which coal-derived gas predominates, and the other considers the gas source a mixture source in which inorganic alkane gas predominates, or inorganic alkane gas predominates and coal-derived gas supplements. In the Xushen gas field, CO_2 predominates in the chemical constituent of natural gas in the Fangshen9 and Fangshen701 wells, and gaseous hydrocarbon predominates in the natural gas of the other wells as well as in N_2, CO_2, and few hydrogen gases and rare gases (helium and argon). Although gaseous hydrocarbon dominates the natural gas of the Xushen gas field, the carbon isotope constituent of the alkane gas presents a lightening tendency with an increase in carbon number ($\delta^{13}C_1 > -30‰$), and the carbon isotope sequence of alkane gas possesses an inorganic origin characteristic; its $\delta^{13}C_{CO_2}$ value is between $-16.5‰$ and $-5.1‰$. The variation range of R/R_a is 0.77-5.84, and the R/R_a of most samples is more than 1.0. Simultaneously, substantive source rocks developed around the gas fields, and the geochemical parameters $CO_2/^3He$ and $CH_4/^3He$ fall neither in the typical crust-source volume nor in the typical mantle-source volume (Figure 4.1). When the CO_2 in the natural gas is a simple two end-member mixture of crust source and mantle source, R/R_a and $CO_2/^3He$ should be in a linear relationship, as shown in Figure 4.1A. However, the CO_2 data points from the Xushen gas field fall below the region of the typical organic origin gas of the Ordos Basin, for the inorganic CO_2 can be reduced to CH_4 under particular conditions. Furthermore, the negative correlation of R/R_a and $CH_4/^3He$ also shows the mixture of organic and inorganic origin gases, for the $CH_4/^3He$ of organic origin gas is 109-1012, while the $CH_4/^3He$ of typical inorganic gases in the mid-oceanic ridge basalt of the eastern Pacific, hot springs gas, and volcanic exhalation is 105-107, and $R/R_a > 1.0$. The high $CH_4/^3He$ and R/R_a ratios of the Xushen gas field are possibly related to the deep activity, for the CO_2 and H_2 can be synthesized into methane through Fisher-Tropsch reactions in the course of volcanism. Therefore, the methane of the Xushen gas field mainly includes two gas sources: thermal degradation of organic matter and addition of inorganic gases (mantle-source gas and Fisher-Tropsch syntheses), while the anomalous paraffinic carbon isotopes are related to the mixture of some inorganic alkane gases. It is shown by the isotopes of gaseous hydrocarbon in the Changling fault depression in the southern Songliao Basin that the coal-related gas predominates the gaseous hydrocarbon (Figure 4.2 and Table 4.2), and simultaneously, some inorganic alkane gases were possibly intermixed.

CO_2 genesis can be divided into two types: one has an inorganic origin, including magma degasification and crustal carbon-enriched rock decomposition, $\delta^{13}C_{CO_2} > -8‰$, and the other has an organic origin; i.e., it is formed by the decomposition of organic matter, $\delta^{13}C_{CO_2} < -10‰$. The

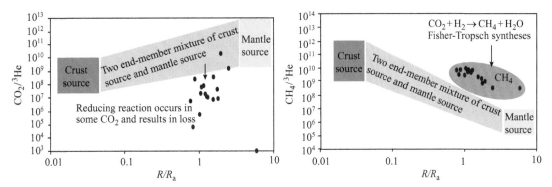

FIGURE 4.1 Relational graph of R/R_a versus $CO_2/^3He$ and R/R_a versus $CH_4/^3He$ in the Xushen gas field (Yang et al., 2009).

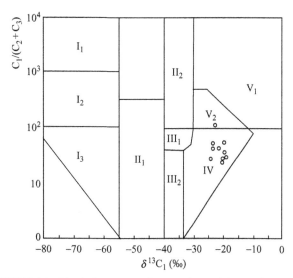

FIGURE 4.2 Discrimination chart for natural gas origin type for Changling fault depression.

$\delta^{13}C_{CO_2}$ is $-3.73‰$ to $-11.9‰$ in the southern Songliao Basin, which shows that the CO_2 is mostly of inorganic origin, and it has possibly a crustal source or mantle-derived magma genesis. The dot projection of the discrimination chart shows that the CO_2 is of inorganic origin; furthermore, it is also shown through the discrimination chart of helium isotopes that the CO_2 reservoir in the Changling fault

depression of the Songliao Basin is mantle source-magmatic origin (Figure 4.3).

The hydrocarbon accumulation mode of carbon dioxide gas in the Changling fault depression may be as follows: extensive volcanic eruption occurred in the Changling fault depression in the Yingcheng stage, which carried substantive gases, but most of them dissipated. In the Paleogene-Neogene, the Songliao Basin was tensile and its discordogenic faults became active again. The substantive gases carried by the mantle-derived magma migrated upward along the discordogenic fault and accumulated in the Yingcheng Formation volcanic rock or Denglouku Formation clastic rock. The carbon dioxide gas reservoir was formed (Figure 4.4).

The rift zone tends to be the allogeneic and homoaccumulated site of organic and inorganic hydrocarbons. The rift zone is the mantle uplift region, where the mantle asthenosphere is close to the earth's surface (30–40 km deep), the mantle fluid and the tensional faults concentrate, the crust attenuates, and the rift basin is formed. These regions are the sedimentary provinces of coal-measure organic matter and other inorganic sediments from the magmation and volcanic eruption, where the geotemperature is higher, the pressure is larger, the organic matter is apt to be cracked, and the oil and gas are apt to be formed. These regions are also the regions for blowing off, ascending, trapping, and accumulating mantle-source and crust-source inorganic gases and

TABLE 4.2 Analytic Results of Carbon Isotopes of Natural Gas in Southern Songliao Basin

Sample no.	Horizon	Carbon isotope value $\delta^{13}C_{PDB}$ (‰)					
		Carbon dioxide	Methane	Ethane	Propane	Isobutane	n-Butane
1	K_1yc	−11.6	−18.3	−25.0	−	−	−
2	K_1yc	−11.9	−22.4	−27.0	−	−	−
3	K_1yc	−7.5	−22.2	−26.9	−27.0	−	−33.7
4	K_1yc	−6.8	−23.0	−26.3	−27.3	−	−34.0
5	K_1yc	−5.26	−20.78	−20.73	−	−	−
6	K_1yc	−7.2	−26.5	−26.65	−	−	−
7	K_1yc	−6.88	−26.07	−26.99	−	−	−
8	K_1yc	−4.63	−23.17	−24.43	−	−	−
9	K_1yc	−4.83	−25.1	−23.32	−	−	−
10	K_2q^3	−4.04	−	−	−	−	−
11	K_2q^3	−4.96	−38.66	−	−	−	−
12	K_2q^4	−5.73	−43.7	−	−	−	−
13	K_2q^4	−8.44	−43.97	−	−	−	−
14	K_2q^4	−3.73	−	−	−	−	−

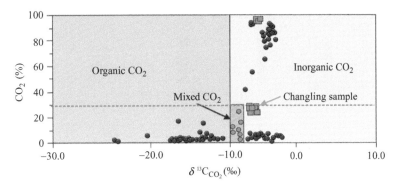

FIGURE 4.3   Genesis of CO_2 in the natural gas of the Changling fault depression, Songliao Basin. *Adapted from Mi et al. (2008).*

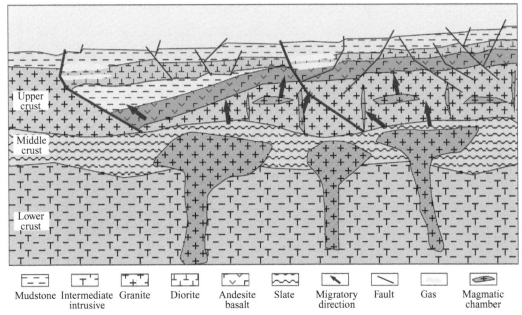

FIGURE 4.4   Hydrocarbon accumulation mode of the CO_2 reservoir in the Changling fault depression of Songliao Basin.

biological organic gas. Primary and secondary pores, cavities, and fissures develop in the volcanic rock and edifice formed by the uplift of volcanic cones and clusters of cones. They are difficult to compact and thus can more easily become favorable regions for the mixing and accumulating of organic gas, inorganic gas, mantle-source gas, and crust-source gas. The gas is allogeneic and homoaccumulated; therefore, the gas source is plentiful.

4.2   VOLCANIC RESERVOIR PLAYS

4.2.1   Volcanic Play Types

Volcanic rock cannot generate hydrocarbon by itself, but it can develop high-quality reservoirs. Therefore, a favorable source-reservoir-caprock configuration is the key to the formation of a volcanic reservoir. In terms of the vertical and horizontal configuration relationship of volcanic reservoir

and source rock, proximal and distal assemblage types are observed. The proximal assemblage means that the volcanic rock and the source rock are basically at the same layer longitudinally, and the volcanic reservoir occurs within the scope of hydrocarbon generation horizontally. The distal assemblage means that the volcanic rock and the source rock are located at different layers longitudinally, and the volcanic reservoir occurs beyond the scope of hydrocarbon generation horizontally. In general, the hydrocarbon accumulation conditions of the proximal assemblage are the most favorable.

Based on the analysis of the longitudinal source-reservoir-caprock assemblage characteristics of volcanic rocks in the major hydrocarbon-bearing basins of China, the proximal assemblage predominates the fault depressions in eastern China. For instance, the Paleogene in the Bohai Bay Basin and the deep volcanic rock in the Songliao Basin developed within the hydrocarbon-generation layers (Figure 4.5). However, both the proximal and the distal assemblages exist

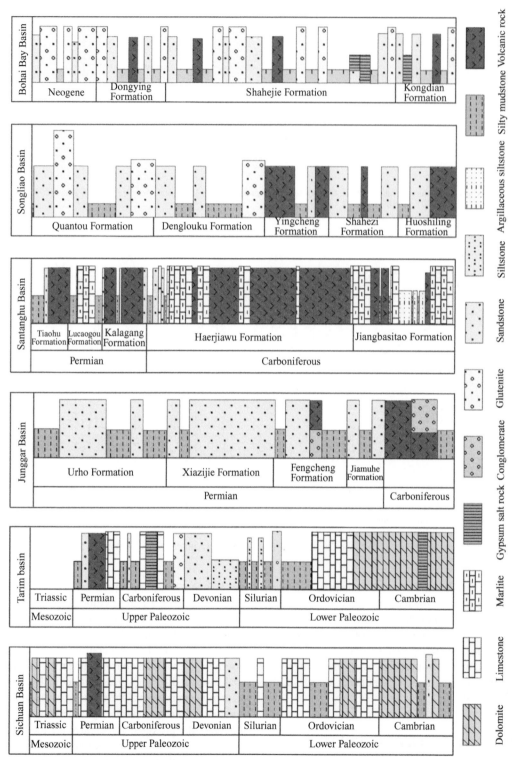

FIGURE 4.5 Longitudinal distribution of source-reservoir-caprock assemblages of volcanic rocks in major hydrocarbon-bearing basins of China.

in western China. For instance, the volcanic rocks in the Junggar Basin and the Santanghu Basin accumulated in the Carboniferous-Permian where source rock also developed, and this is a proximal assemblage type. In contrast, in the Sichuan and Tarim Basins the volcanic rocks

mainly developed in the Permian, but hydrocarbon generation mainly occurred in the lower Paleozoic Cambrian-Ordovician, and this is a distal assemblage type.

In the fault depressions of eastern China, the proximal assemblage predominates, the volcanic rock interbeds with the source rock, and mainly occurs in or in the vicinity of the source sag. Therefore, the structural-lithologic reservoir in which explosive facies predominates is formed in high positions, and the lithologic reservoir in which the effusion facies predominates is formed on the slope. However, in central and western China, proximal and distal plays develop, and the large unconformable volcanic weathered crust reservoir predominates.

4.2.2   Volcanic Plays in Eastern China Basins

In the hydrocarbon-bearing basins such as Bohai Bay, Songliao, Erlian, Hailar, Subei, and Jianghan in eastern China, volcanic reservoirs mainly developed in the fault depression period. For example, during the Paleogene in the Bohai Bay Basin and the lower Cretaceous in the Songliao Basin, the framework of the fault depression basins controlled the spatial distribution of volcanic rock, so that the volcanic rock intimately contacted the hydrocarbon-generation measures or hydrocarbon-generation center either vertically or horizontally, and formed a proximal play. For instance, the distribution of the volcanic reservoir and source rock basically overlaps in the Xujiaweizi fault depression in the deep strata of Songliao Basin, and this is a typical proximal play (Figure 4.6).

Using the fault depressions in the deep strata of Songliao Basin as an example, this book focuses on introducing its source-reservoir-caprock assemblage characteristics.

4.2.2.1   Stratigraphy

The deep strata of Songliao Basin refer to the strata below Member II of the Quantou Formation, where the burial depth of exploration targets is generally 3000-5000 m, including strata such as pre-Mesozoic and Mesozoic, middle Jurassic, and lower Cretaceous. Among these, the

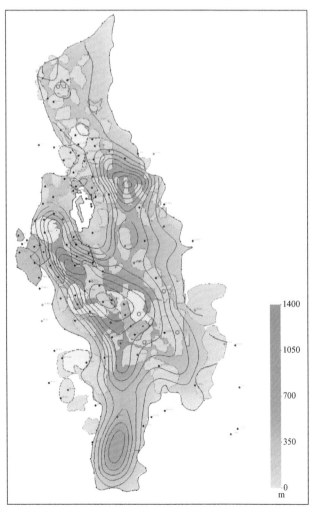

FIGURE 4.6   Superimposed distribution of dark mudstone and volcanic massif in the Xujiaweizi fault depression.

lower Cretaceous is currently the major exploration target. The lower Cretaceous is divided into the Huoshiling, Shahezi, Yingcheng, Denglouku, and Quantou Formations from bottom to top, and their stratigraphic and lithologic characteristics are described below.

The Huoshiling Formation is divided into two members from bottom to top: Member I is a clastic rock intercalated with carbargilite or coal seams, and Member II is andesite intercalated with clastic rock. As of today, the underlying stratum corresponding to the Huoshiling Formation has not yet been drilled in northern Songliao Basin.

The upper Shahezi Formation is sandy shale; bluish gray and yellowish green acidic tuff is seen locally, and glutenite increases near the margin of the fault depression. The lower Shahezi Formation is sandy shale intercalated with coal seams, and it is often the stable productive coal seam (five to six layers).

The Yingcheng Formation is divided into four members, and the lithologic characteristics are as follows from bottom to top. Member I is predominantly acidic volcanics, and the common types are rhyolite and purple and gray-white tuff. Member II is gray-black sandy shale and green-gray and motley glutenite, sometimes intercalated with several layers of coal seam. Member III is predominantly intermediate volcanics, and the common types are andesite and andesite basalt. Member IV is gray-black and purple-brown sandy shale and green-gray and gray-white glutenite.

The Denglouku Formation is mainly glutenite with no volcanic rock, and it is divided into four members. Member I is motley conglomerate, intercalated with sandstone on top. Member II is predominantly gray-black sandy shale, and gray and white thick packsands interbedded in various thicknesses. Member III is interbedded gray-white massive packsand to medium-grained sandstone and grayish-brown and gray-black sandy shale. Member IV is grayish-brown and gray-black sandy shale, and light gray green, gray-white, and purplish-gray sandstone.

The Quantou Formation is primarily interbedded gray-white and purplish-gray sandstone and dark mauve and dark brown mudstone.

4.2.2.2 Distribution of Source Rock

The source rock in which coal-bearing series predominate mainly develops in the deep strata of Songliao Basin, including the Shahezi, Yingcheng, and Denglouku Formations, among which the Shahezi Formation is the major source rock and is widely distributed in the major fault depression.

Observed from the Xujiaweizi fault depression where the exploration maturity is higher for the moment, the maximum thickness of the dark mudstone of the Shahezi Formation is 384 m, that of the Yingcheng Formation is 118 m, and that of the Huoshiling Formation is 110 m.

As a whole, the dark mudstone of the Shahezi Formation is well developed and widely distributed and was drilled into by many exploration wells.

True Shahezi Formation source rock has not yet been drilled by any exploration well in the Changling fault depression. The Xinshen1 well drilled through the Yingcheng Formation, and its lithology is a grayish-brown mudstone, but it cannot reflect the actual conditions of a source rock area. In the Dongling structure, the thickness of the shallow-lake facies dark mudstone of the Yingcheng Formation is 110-345 m. In the SN101 well, the depth of the Yingcheng Formation is 2326-2495 m, and the thickness of the mudstone is 153.6 m, among which the thickness of the dark mudstone is 110 m, accounting for 65% of the formation thickness. In the SN108 well, the depth of the Yingcheng Formation is 2484-2660 m, and the dark mudstone accounts for 56% of the formation thickness. In the SN109 well, the depth of the Yingcheng Formation is 2710-3500 m, and the dark mudstone accounts for 44% of the formation thickness. Finally, in the Tuoshen6 well, the dark mudstone accounts for 19% of the formation thickness.

The exploration maturity is relatively low in the Yingshan-Shuangcheng fault depression. It is revealed by drilling that the maximum thickness of the Shahezi Formation dark mudstone is 770 m, that of the Yingcheng Formation dark mudstone is 191 m, and that of Huoshiling Formation dark mudstone 201 m, which confirms that the source rock is extremely well developed in the fault depression. Horizontally, the distribution of source rock is controlled by the fault depression range. It is speculated based on sedimentary facies that the thickness is generally 200-500 m, controlled by boundary faults, and the maximum thickness of the source rock can be more than 1000 m in the depocenter. The thickness of the source rock can reach 1000-2000 m in the hydrocarbon-generation center in the middle of the Xujiaweizi fault depression. The Changling fault depression is divided into south and north depocenters, where the thickness of the Shahezi Formation source rock can reach 700-900 m in either one.

Based on the overall test data of the deep strata of the Songliao Basin, the organic carbon of the Shahezi Formation is 1.63-3.47%, the chloroform bitumen "A" is 0.026-0.16%, the total hydrocarbon is $220\text{-}1954 \times 10^{-6}$, and the $S_1 + S_2$ is 1.00-32.00 mg g^{-1}. The organic carbon of the Yingcheng Formation is 0.50-2.43%, the chloroform bitumen "A" is 0.02-0.176%, the total hydrocarbon is $100\text{-}1056 \times 10^{-6}$, and the $S_1 + S_2$ is 2.00-31.90 mg g^{-1}. It is shown by studies that the Shahezi Formation is the most important source rock in the deep strata of the Songliao Basin. The organic carbon of the Denglouku Formation is relatively low, 0.49-0.97%, and the chloroform bitumen "A" is 0.01-0.06%, the total hydrocarbon is $100\text{-}152 \times 10^{-6}$, the $S_1 + S_2$ is 0.24-0.76 mg g^{-1}, and it is comprehensively evaluated as a preferable source rock.

The organic matter abundance is higher in the Shahezi and Huoshiling Formations of the Xujiaweizi fault depression, high in the Yingcheng Formation, and the lowest in the Denglouku Formation. The organic matter abundance is apparently higher in the Xubei sag than in the Xu'nan sag of the Xujiaweizi fault depression. For instance, the mean value of organic carbon is 0.369% in the Denglouku Formation of Xu'nan sag, but it ascends to 0.55% in the Xubei sag; the mean value of organic carbon is 0.934% in the Huoshiling Formation of Xu'nan sag, but it ascends to 1.996% in the Xubei sag. Both the chloroform bitumen "A" and the pyrolysis $S_1 + S_2$ have the same regularity. The organic matter abundance of deep gas source rock is higher in the Yingshan-Shuangcheng fault depression, but the chloroform bitumen "A" and the hydrocarbon-generation potential $(S_1 + S_2)$ are lower, which is related to the high maturity of deep source rock and the cracking of liquid hydrocarbon.

Therefore, as a whole, the organic matter content of the Shahezi Formation source rock is at its maximum in the northern Songliao Basin, next is the Huoshiling Formation and Yingcheng Formation source rocks, and last is the Denglouku Formation source rock. Going from top to bottom, the deep source rock profiles are as follows: Denglouku Formation is a poor oil source rock; Yingcheng Formation is a moderate gas source rock and poor oil source rock; Shahezi Formation is mainly a moderate gas source rock and moderate oil source rock; and Huoshiling Formation is a moderate to poor gas source rock. As a whole, all the source rocks have gas-generating capability, and the Shahezi Formation source rock also has the capability of generating oil.

In the Dongling structure of the Changling fault depression, the organic carbon of the Yingcheng Formation source rock is 0.65-1.13%, averaging 0.89%, which belongs to type II_2, and is preferable to good gas source rock. The Yingcheng Formation of the Dongling structure was formed in a shallow-lake facies environment. It is speculated that the abundance of semi-deep lake and deep lake facies dark mudstone is higher, but the Shahezi Formation source rock has not yet been drilled by any exploration well. Based on the data of the other fault depressions in the southern Songliao Basin, we speculate that the average organic carbon content of the Shahezi Formation source rock should be around 1.31%, and the chloroform bitumen "A" should be greater than 0.05%. Observed from the sedimentary facies and the spread of strata, a higher abundance of organic carbon in the Shahezi Formation should occur in the south and north subsags; in particular, the content should be slightly higher in the south sag than in the north sag.

The analyses on deep samples taken from the northern Songliao Basin showed that the organic matter type of the Denglouku Formation source rock is III_1-III_2; that of the Yingcheng Formation source rock is II-III_1; and that of Shahezi Formation source rock is relatively diverse, including I_2-III_2, among which type I_2-II samples occur more than type III_1-III_2 samples, which reflects that this set of source rock has the best hydrocarbon-generation conditions. The maturity of Huoshiling Formation source rock is higher, and it is speculated that its organic matter is type II-III_2.

It is shown by the Tuoshen1 and Xinshen1 wells in the Changling fault depression that the organic matter type of the Denglouku Formation is III. It is believed through integrated analysis that type III organic matter predominates in the Denglouku Formation source rock, and type II organic matter occurs less often. Sample analysis has not yet been conducted on the Yingcheng or Shahezi Formations up to now, so their organic matter type is speculated as II-III_1.

Vitrinite reflectance was conducted on samples in the northern Songliao Basin, showing that the mean value of the Huoshiling Formation reached 3.1% and 2.58% in the Xujiaweizi fault depression and Yingshan fault depression, respectively, and they have all reached the overmature stage. The mean value of R_o of the Shahezi Formation has reached 2.7% in the Yingshan-Shuangcheng fault depression, and it has been in the dry gas phase for a long time. However, the thermal evolution level of the Shahezi Formation is vastly different in the Xujiaweizi fault depression—R_o ranges from 1.70% to 3.56%, and it is at a highly mature to overmature stage. The Yingcheng Formation source rock is at a highly mature to overmature stage both in the Xujiaweizi fault depression and in the Yingshan-Shuangcheng fault depression.

4.2.2.3 Reservoir

Because of its long geologic history and deep burial depth, the sandstone reservoir in the deep strata of the Songliao Basin, such as Member II and Member I of the Quantou, Denglouku, Yingcheng, Shahezi, and Huoshiling Formations, generally underwent strong and complicated diagenetic epigenesis and the physical properties of the reservoir is relatively poor. However, late tectonic movement improved the connectivity of the reservoir space through fracturing. Its porosity changed little with increasing depth. Therefore, volcanic rock is the favorable reservoir of deep natural gas.

In the deep strata of the Songliao Basin, the volcanic reservoir is characterized by multihorizon, multitype, and complicated lithologies. The volcanic rock mainly concentrates in Member I and Member III of the Yingcheng Formation and Member I of the Huoshiling Formation in the northern part of the basin, where the volcanic rock has basic, intermediate, and acidic lithologies. The accumulation capacity of the volcanic reservoir is closely related to its lithology and lithofacies.

For the moment, the major exploration targets of volcanic reservoirs in the deep strata of Songliao Basin are Member I and Member III of the Yingcheng Formation, which are widely distributed in all the fault depressions.

4.2.2.4 Cap Rock

The regional seals of the deep strata of the Songliao Basin include Member II of the Denglouku Formation and Member I and Member II of the Quantou Formation. These members have a stable distribution, a great depositional thickness of mudstone, and form good regional sealing systems.

In the depositional period of Member II of the Denglouku Formation, the basin was in a transitional period from rift to depression. Member II was a fan delta-lacustrine facies deposited under weak compensation conditions, the depositional environment was stable, and the lithology was a fine-grained mudstone that covered the whole region, forming a regional seal. The formation thickness is generally 100-200 m, the mudstone thickness is 50-100 m, the shale/formation thickness ratio is more than 50% in most of the region and can reach 90% locally, and the shaliness of the mudstone is higher. For instance, the direct cap rock is Member II of the Denglouku Formation in the Xushen1 well. The formation is 113.5 m thick and grayish-purple mudstone and silty mudstone predominate, intercalated with thin gray argillaceous siltstone, siltstone, and mudstone. The silty mudstone is 74 m thick, and the shale/formation thickness ratio is 65.2%. Three layers of perlites are more than 5 m thick, with thickness being 6.5, 7, and 9 m, respectively. This set of stratum is 80-160 m thick in the Changling fault depression, and the mudstone in it is 40-60 m thick.

In the depositional period of Member I and Member II of the Quantou Formation, it was mainly a meander stream, shore shallow-lake, and braided-stream sedimentation environment. The lithology is predominantly mudstone and thin argillaceous siltstone, and the shale/formation thickness ratio can be more than 90%. As for the mudstone of Member II of the Quantou Formation, in the Xuzhong area the maximum thickness is proximal to the Xushen2 well and is 330 m, and the minimum thickness is proximal to the Xushen9 well and is more than 100 m. In the Xuxi area, the thickness is more than 120 m; in the Xudong slope, the depositional thickness is more than 240 m; and in the whole

region, the minimum thickness is proximal to the Zhaoshen3 well in the Xu'nan area and to the Song16 well at the eastern side of Songzhan, and is generally around 90 m. The mudstone of Member I of the Quantou Formation is more than 90 m thick in the whole region. Its maximum thickness is proximal to the Fangshen7 well in the Xuxi area and to the Zhaoshen3 well in the Fengle low uplift and can reach 260 m. Its minimum thickness is proximal to the Xushen1 well in the Xuzhong area and is 60 m. The cumulative thickness of Member I and Member II of the Quantou Formation is more than 150 m in the whole region, more than 300 m in most of the region, and its maximum thickness reaches 500 m near the Xushen2 well. The single-layer thickness of mudstone of Member I and Member II of the Quantou Formation is generally more than 10 m and can reach a maximum of 37 m locally. This set of stratum is 150-260 m thick in the Changling fault depression, and its mudstone is 100-180 m thick.

4.2.2.5 Reservoir-Caprock Assemblage and Its Distribution

Four sets of source rocks and four sets of reservoirs interstratify in the Xujiaweizi fault depression and compose the favorable source-reservoir-caprock assemblage conditions (Figure 4.7). It has been confirmed by deep prospecting in the Xujiaweizi fault depression that in the major reservoirs such as Member I of the Denglouku, Yingcheng, Shahezi, and Huoshiling Formations, glutenite and volcanic reservoirs all possess preferable accumulation conditions, especially the volcanic reservoir of the rift stage, which has a porosity of 7-8%. The volcanic porosity is 11% at 3778 m in the Fangshen8 well, with a dual reservoir media of pore-fissure space. Member II of the Denglouku Formation and Member I and Member II of the Quantou Formation have a stable distribution. The mudstone deposits and the Yingcheng Formation volcanic rock and glutenite compose a "lower reservoir and upper caprock" reservoir-caprock assemblage. In addition, the explosive

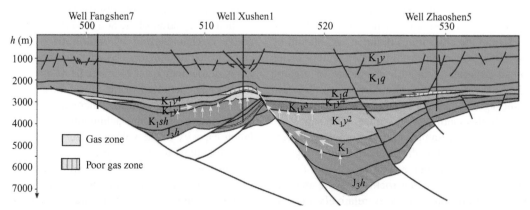

FIGURE 4.7 Source-reservoir-caprock assemblage in the Xujiaweizi fault depression in deep strata of the northern Songliao Basin (Daqing oil field).

facies volcanic breccia and rhyolite in the Yingcheng Formation and the overlying tuff constitute a "lower reservoir and upper caprock" reservoir-caprock assemblage.

The sandstone reservoir is buried relatively deeply in the Qijia-Gulong fault depression, and it is extremely tight. Three exploration wells revealed that the porosity of the sandstones of Member I and Member II of the Quantou Formation is generally 3-7% and that of the Denglouku Formation is 3-4%; the permeability is mostly less than 0.1×10^{-3} μm^2. They belong to low-porosity tight reservoirs where tight gas can be formed. As for the extensively distributed volcanic and glutenite reservoirs, the reservoir properties are less affected by the burial depth than they are by the origofacies belt and later reformation action, which creates preferable accumulation conditions.

A certain number of structural traps developed in the Qijia-Gulong fault depression, but stratigraphic overlap traps and volcanic lithologic traps are more common, and they concentrate in the periphery of the fault depression. Because they are adjacent to the source area, they possess the condition of preferentially trapping the natural gas generated in the source area. Favorable gas shows are seen in the Pushen1 well, which corroborates the exploration prospect of the Qijia-Gulong fault depression, and becomes the breakthrough direction of gas exploration at the next step. Both the tectonic movement and the volcanism are stronger in the Qijia-Gulong fault depression than in the east at the final stage of deposition; therefore, the distribution area and depositional thickness of volcanics and glutenite are larger, and the reservoir conditions are relatively superior.

The mudstone of Member I and Member II of the Quantou Formation is present over a large area in the Qijia-Gulong fault depression, its depositional thickness is great, it was buried deeply, and it has acted as a good seal. It has been confirmed by exploration in the Xujiaweizi fault depression that the source rock seals of Member II of the Denglouku Formation and Member I and Member II of the Quantou Formation are regional seals. Simultaneously, because a massive Qingshankou Formation source rock seal also exists in the region, multiple reservoir-caprock assemblages are formed, which are quite favorable to the storage of natural gas.

The hydrocarbon discovered in the southeast uplift mainly exists in the lower Cretaceous Shahezi, Yingcheng, and Quantou Formations, and natural gas is also discovered in the basement fractures. It is known from hydrocarbon-generation assessment that the Shahezi and Yingcheng Formation source rocks developed in all the fault depressions of the region, especially in Lishu, Dehui, and Wangfu fault depressions. The dark mudstone is very thick, abundant, and it has entered the hydrocarbon threshold. The mudstone is not only a source bed but also a good local caprock. In this way, a self-source self-reservoir source-reservoir-caprock assemblage is formed.

Drilling in the Changling fault depression has shown that four types of source-reservoir-caprock assemblages exist. First, taking the Shahezi Formation as the source bed, the Yingcheng Formation and Member I of the Quantou Formation as the reservoir, and Member II and Member III of the Quantou Formation as the cap rock, a lower source-upper reservoir-top caprock assemblage is formed. Second, the Shahezi Formation is not only the source bed but also the cap rock, and Huoshiling Formation is the reservoir; in such a way, an upper source-lower reservoir-upper caprock assemblage is formed. Third, the Shahezi Formation is not only the source bed but also the reservoir and cap rock, so that a self-source self-reservoir and self-caprock assemblage is formed. Fourth, taking the crust source or mantle source as the gas source, and the overlying sedimentary layer as the reservoir and cap rock, a deep source and shallow reservoir assemblage are formed. A similar play develops in the Yingtai depression where a breakthrough has been achieved.

4.2.3   Volcanic Plays in the Basins of Western China

Volcanic rock mainly occurs in the Permo-Carboniferous in western hydrocarbon-bearing basins. Their age is relatively old, their original basins have been strongly altered, and they have large play variation. For instance, the volcanic rock in the Junggar Basin mainly occurs in the Permo-Carboniferous. Longitudinally, the proximal assemblage should predominate; however, affected by late tectonic activities, the Permo-Carboniferous in the northwest margin of Junggar Basin was uplifted, weathered, and denuded, the hydrocarbon-generating capability of the strata in the thrust belt was apparently weakened, and the hydrocarbon mainly came from the Permo-Carboniferous in the footwall of the thrust belt. Therefore, horizontally, the hydrocarbon-generation scope and the distribution of volcanic reservoir disaccorded, and thus a side source play was formed.

4.2.3.1   Proximal Play, Middle and Eastern Junggar Basin

The Permo-Carboniferous in the hinterland of the Junggar Basin is preserved relatively completely. Furthermore, it has preferable hydrocarbon-generating conditions, and as a result, a proximal play is formed (Figure 4.8). The Carboniferous source rock has become a set of effective source beds in the hinterland of the Junggar Basin, which plays a crucial role in the effective hydrocarbon accumulation of Permo-Carboniferous volcanic rock.

The Carboniferous Dishuiquan Formation is an irregular interbed of dark-gray mudstone and carbargilite, intercalated with girdle and coal streaks, and an interbed of intermediate and basic lava, volcanic breccia, and volcaniclastic rock in the middle. The effective source rock is a dark-gray

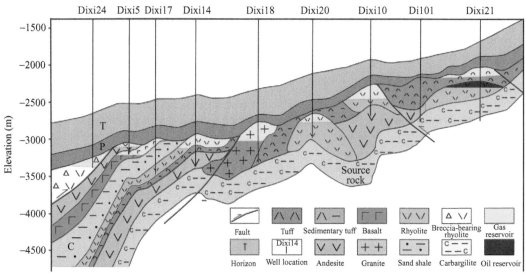

FIGURE 4.8 Assemblage relationship of Carboniferous volcanic reservoir and source rock in the Ludong region of Junggar Basin (Xinjiang oil field).

mudstone and carbargilite. It outcrops in the piedmont of the Kelameili Mountain on the surface, mainly develops in the eastern section of Luliang uplift, the Dishuiquan sag, the fault footwall of Di'nan uplift, the northern Dong-daohaizi and Wucaiwan sags in the basin, and was deposited during the short-term extensional rifting at the postcollision phase. It has been drilled into by the Lu'nan1, Sancan1, Dibei1, Dixi17, Caican1, and Caishen1 wells, and the source rock is 50-500 m thick.

The organic carbon content of the Dishuiquan Formation is 0.27-10.7%, averaging 2.19%; the chloroform bitumen "A" content is 0.0014-0.1291%, averaging 0.0471%; the total hydrocarbon content is 231.13×10^{-6} to 989.55×10^{-6}, averaging 485.49×10^{-6}; the genetic potential $S_1 + S_2$ is 0.07-2.47 mg g^{-1}, averaging 1.05 mg g^{-1}; and hydrogen index I_H is 25.0-262.5, averaging 85.56. Kerogen-type index T_i is less than -4, and it reflects the characteristics of the humic parent material type. Vitrinite reflectance R_o is 0.5-1.6%, averaging 1.35%; and T_{max} is 446-494 °C, averaging 468 °C. The Dishuiquan Formation is a moderate organic matter abundance source rock and is at a highly mature stage.

The carbon isotope of methane is very light in the natural gas of the Ludong-Wucaiwan region, but the carbon isotope of ethane, propane, and butane is all heavier. The carbon isotope ratio of methane is -34.77 to -48.4‰, that of ethane is -23.72 to -24.54‰, that of propane is -21.16 to -22.57‰, and that of butane is -21.03 to -22.33‰. The methane dominates the natural gas component, and the gas is dry gas. The natural gas is supposed to be a relatively special type, and it is possibly related to the biological remaking reservoir. The biological remaking reservoir can make the carbon isotope of methane become light and the carbon isotope of ethane and propane become heavy.

Observed from the Carboniferous maturity in the whole Ludong-Wucaiwan region, the Carboniferous Dishuiquan

Formation maturity is very high in the Ludong region and has even reached the overmature stage locally. An apparent gap can be seen between the Mesozoic and underlying Paleozoic in wells such as the Lu'nan1 in the Di'nan uplift, Dixi1, Dixi2, Dixi3, Caican1, and Caishen1. The Carboniferous source centers are located in the Dishuiquan sag and the northern Dongdaohaizi sag. The maturity of the Dishuiquan Formation source rock is high, gas generation predominates, and the major hydrocarbon generation and discharge period is Permian. Although most of the reservoirs formed at the early stage were destroyed, a few of them remain. The typical reservoirs include Carboniferous gas reservoirs of well block Dixi10 in the Di'nan uplift zone and of well block Cai25 in the Wucaiwan sag.

Due to stronger dissection at the late Hercynian, multiple source centers and hydrocarbon systems were formed in the Junggar Basin, and they combined and superimposed horizontally (Figure 4.9). When the Luliang uplift, Wucaiwan sag, and central uplift zone were in a short-term intracontinental rift zone at the postcollision phase, the Carboniferous source rock effectively developed, and natural gas was formed mainly by types II$_2$ and III kerogen of terrigenous plants. The Carboniferous volcanic rock in the Luliang uplift-Wucaiwan sag formed a self-source self-reservoir assemblage around the Dishuiquan sag, northern Dongdaohaizi sag, and Wucaiwan sag; the Carboniferous volcanic rock in the Santai and northern Santai uplift zone formed a new-source old-reservoir assemblage closely adjacent to the Permian source center in the Fukang sag; and the Carboniferous formed a self-source self-reservoir assemblage in the Jimsar sag.

Oil and gas exploration in the Ludong-Wucaiwan region confirmed that Carboniferous source rock mainly developed in the Dishuiquan sag at the eastern section of the Luliang uplift and in the northern Dongdaohaizi sag and Wucaiwan sag adjacent to it in the south. It belongs to a set of transitional facies of coaly terrigenous clastics.

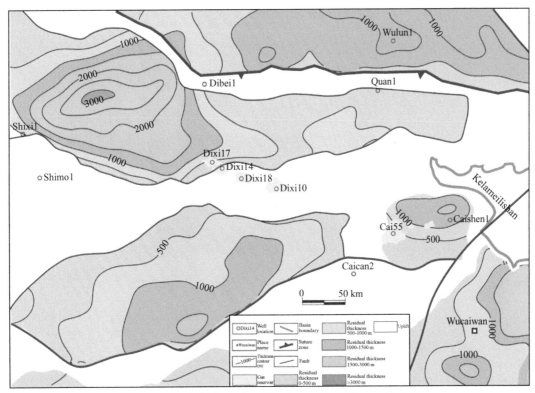

FIGURE 4.9   Distribution of source rocks and reservoirs in the Ludong-Wucaiwan region of Junggar Basin.

The volcanic reservoirs discovered in the Junggar Basin concentrate at the eastern section of the Luliang uplift and in the Wucaiwan sag, and mostly occur along the major fault, which shows that the archeovolcanic activity is closely related to the formation of faults. The developmental horizons of the volcanic reservoir consist of lower Carboniferous Baogutu Formation (C_1b), upper Carboniferous Batamayineishan Formation (C_2b), and lower Permian Jiamuhe Formation (P_1j), and the effective reservoir is mainly composed of eruptive rock. Intermediate and acidic eruptive rock and volcaniclastic rock assemblages exist in most of the Luliang uplift and Zhudong regions, where the explosive-effusion facies predominates. Based on the integrated description and assessment of the volcanic reservoir, it was determined that volcanic eruption lava and volcaniclastic rock are the major reservoirs.

Therefore, in the hinterland of the Junggar Basin, the Dishuiquan Formation is the major source bed; the Baogutu, Batamayineishan, and lower Permian Jiamuhe Formations are the reservoirs, and they form the lower source and upper reservoir proximal play.

4.2.3.2 Side Source Play, Northwest Margin of Junggar Basin

The northwest margin of the Junggar Basin is situated in a collision zone. The development and distribution of Carboniferous source rock are unclear. At present, it is believed

that the hydrocarbon mainly comes from the Permian source rock in the Mahu sag located in the footwall of the thrust belt (Figure 4.10), and the source bed is mainly consists of lower Permian Jiamuhe, Fengcheng, and middle Permian lower Urho Formations.

The Jiamuhe Formation source rock mainly occurs in the lower member and can be up to 250 m thick. The residual organic carbon content of the Jiamuhe Formation averages 0.56%, chloroform bitumen "A" content averages 0.0056%, and hydrocarbon-generation potential $S_1 + S_2$ averages 0.25 mg g^{-1}. The type of residual organic matter is predominantly III and individually II_2 and II_1; the carbon isotopes of kerogen are heavier and generally greater than -23‰. The measured R_o ranges from 1.38% to 1.9%, and it is a set of source rocks at a highly mature to overmature stage.

The Fengcheng Formation is the major source bed, occurring in the Kebai and Wuxia fault zones at the northwest margin of Junggar Basin, as well as the Mahu sag in the central depression region, and the thickness of the source rock is generally 200-300 m. The Fengcheng Formation was deposited in a marine and continental margin offshore lacustrine environment in saline waters. Its lithology is blackish gray mudstone, dolomitic mudstone, tuffaceous mudstone, tuffaceous carbonate rock, and sedimentary tuff. The residual organic carbon content averages 1.26%, chloroform bitumen "A" content averages 0.1493%, total hydrocarbon content averages 0.0820%, and hydrocarbon-generation potential $S_1 + S_2$ averages 7.30 mg g^{-1}. The type

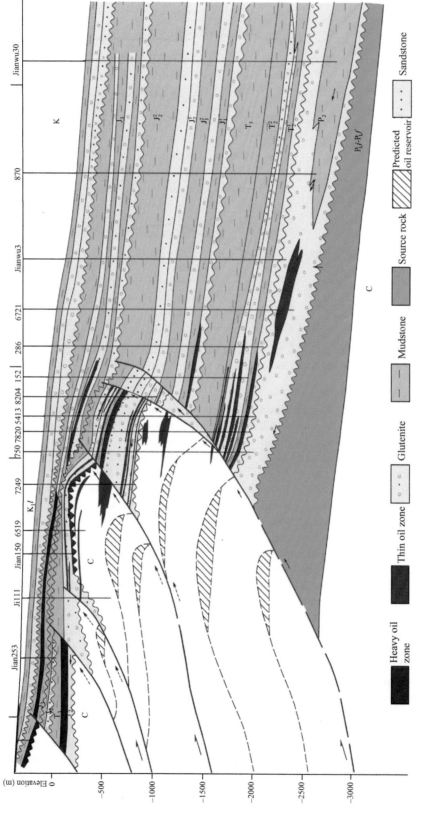

P$_1$j-P$_1$f—Jiamuhe Formation-Fengcheng Formation; P$_2$—lower Urho Formation; T$_2^1$—lower Karamay Formation; T$_2^2$—upper Karamay Formation; T$_3$—Baijiantan Formation; J$_1^1$—Badaowan Formation; J$_1^2$—San'gonghe Formation; J$_2^1$—Xishanyao Formation; J$_2^2$—Toutunhe Formation; J$_3$—Qigu Formation

FIGURE 4.10 Distal assemblage hydrocarbon accumulation mode in the thrust belt of the northwest margin of Junggar Basin.

of organic matter is mostly I-II, and R_o is 0.85-1.16%. It is at a mature to highly mature stage and is a set of preferable to good source rocks.

The lower Urho Formation is 1220 m thick in the Aican1 well located in the western slope of the Mahu sag. The dark mudstone is 178 m thick, and it belongs to the shallow-lake facies to semi-deep lake facies deposition. The organic carbon content of this set of source rocks averages 0.7-1.4%, chloroform bitumen "A" content averages 0.0088%, and the type of organic matter is predominantly III and individually II_2 and II_1. R_o averages 0.86% in the vicinity of the fault zone, 1.0% in the slope area, and as high as 1.7% in the Mabei anticline. The lower Urho Formation is at a mature to highly mature stage and is a set of poor to preferable source rocks.

The Permo-Carboniferous volcanic reservoir in the thrust belt at the northwest margin of Junggar Basin is mainly a large weathered crust, with its reservoir properties being independent of the volcanic rock type. The eruptive rock in the upper wall of the thrust belt is mostly an assemblage of basic basalt and andesite (mostly lower Carboniferous), and explosive facies predominates. In the footwall is an assemblage of intermediate and acidic eruptive rock and volcaniclastic rock, and the explosive-effusion facies predominates.

The regional seals at the northwest margin of Junggar Basin mainly consist of middle Permian lower Urho Formation and upper Triassic Baijiantan Formation. Their lithology is all lacustrine facies mudstone, they have a stable distribution, and their thickness is generally greater than 50 m. In addition, there are still some local cap rocks such as the "mud neck" at the top of upper Permian upper Ur Formation and the mudstone barrier in the middle Triassic Karamay Formation. Therefore, the Permo-Carboniferous volcanic reservoir in the fault zone at the northwest margin of Junggar Basin and the Permian source sag of Mahu form a side source play.

4.2.3.3 Proximal Play, Santanghu Basin

The lower part of the Santanghu Basin consists of lower Carboniferous Jiangbasitao Formation and upper Carboniferous Haerjiawu Formation and Kalagang Formation, where paralic facies volcanic rocks intercalated with clastic rocks were deposited. Carboniferous rocks are distributed extensively in the basin, and they are between 600 and 2000 m thick. The Kalagang Formation mainly occurs at the southwest margin of the basin, with thicknesses generally between 800 and 1000 m. It is absent in the northeastern Malang sag and the eastern Fangliang uplift. Outcrops of the Daheishan and Naomaohu show that the lower Carboniferous source rock is about 300 m thick and has good exploration potential.

The upper Carboniferous source rock occurs mostly at the top and has been drilled into in the Malang, Tiaohu,

and Hanshuiquan sags. Drilling shows that the maximum cumulative thickness in individual wells is 66 m. It is interpreted from seismic data that the source rock will be more developed in the southeast.

The lower Carboniferous source rock is mainly exposed in outcrop at the southern margin of the basin and by the Fang1 well. Oil and gas shows are seen in the cuttings and gas logging data. It is speculated that the lower Carboniferous source rock is relatively thick in the south and southeast. The thickness is estimated to be 500 m at its maximum and between 150 and 300 m on average. The lithology mainly includes black mudstone and oil shale. The organic carbon content is 1.87-8.8%, averaging 5.5%, the hydrocarbon-generation potential averages 21 mg g^{-1}, the type of organic matter is II_1, and the thermal evolution level of source rock is higher.

Both the Paleozoic volcanic rock and the Mesozoic angular unconformity occur throughout the basin. Horizontally, the Carboniferous volcanic weathered crust improved reservoirs that were superimposed and distributed successively. The weathering and leaching corrosion zone developed mainly along the upper Permian denudation line and controls the distribution of high-quality reservoir. Both the near-crater facies and the transition facies are the favorable volcanic reservoir facies belts, and the formation of strongly weathered volcanic reservoirs is the key to hydrocarbon accumulation. Four stages of volcanic rocks developed in the Kalagang Formation in the Niudong block, and an autoclastic volcanic breccia reservoir was formed at the top of each cycle in the volcanic dormant period. Weathering and leaching pores and fractures, dissolved pores, and the pore-fracture assemblage are the three effective pore types in the region.

The Carboniferous volcanic reservoir in the Santanghu Basin belongs to a proximal play (Figure 4.11). The lower Carboniferous is the potential source rock measure, both the source rock and the volcanic reservoir contact closely, and the Triassic is the high-quality regional seal. This play is distributed widely, and well drilling shows that the volcanic strata of the lower part exist in the Hanshuiquan, Tiaohu, Malang, and Naomaohu sags. The Permian is almost denuded away in the north and at both the east and west ends, but the Carboniferous occurs in the whole basin. The residual thickness is great in the central geotectogene and to the south, and the residual "subsidence" main body is in the Malang-Tiaohu sag in the middle. This shows that the lower part of the Santanghu Basin is a favorable exploration prospect. A study of the hydrocarbon accumulation mode of the lower part shows that the formation of a volcanic weathered reservoir is the key to hydrocarbon accumulation, the nose-uplift structural belt is the important tectonic setting for hydrocarbon enrichment in volcanic rock, and the fractures and microfractures control the liquid-producing capacity of the reservoir.

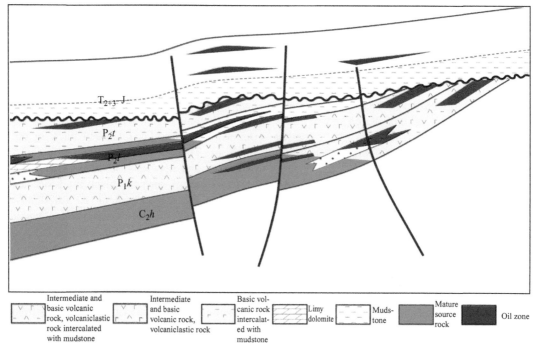

FIGURE 4.11 Permo-Carboniferous play in Santanghu Basin.

4.3 VOLCANIC RESERVOIR CHARACTERISTICS IN MAJOR REGIONS

The volcanic reservoirs discovered both at home and abroad are mostly lithologic-stratigraphic reservoirs. As a kind of special reservoir, the distribution of volcanic rock possesses a characteristic of the lithologic-stratigraphic trap itself. For instance, the centered eruption volcanic massif in the fault depression basins of eastern China is generally distributed locally, and a lithological trap is apt to be present; the volcanic reservoir extensively distributed in western China is mainly controlled by unconformity, and a stratigraphic trap is apt to be present.

Numerous volcanic reservoirs have been discovered successively in the basins such as Bohai Bay, Songliao, Junggar, and Santanghu in China. With the ceaseless discovery of volcanic reservoirs in the deep strata of the Songliao, Junggar, and Santanghu Basins, some new volcanic reservoir types have been identified. In particular, large lithologic-stratigraphic reservoirs have been discovered, such as the large lithologic gas reservoir of Xushen and the large Carboniferous stratigraphic reservoir of Kelameili in the Junggar Basin. The type and characteristics of the volcanic reservoirs are introduced briefly, basin by basin, as follows.

4.3.1 Bohai Bay Basin

Numerous medium and small volcanic reservoirs have been discovered in the Bohai Bay Basin, and most of them have been placed into production; therefore, the study level is

high. Based on the exploration and development practice of volcanic reservoirs in the Bohai Bay Basin, Jiang et al. (1999) comprehensively considered different kinds of factors related to hydrocarbon accumulation in the volcanic rock and their impact on the hydrocarbon accumulation. They used the principle of reservoir space origin type and rock type to conduct a systematic classification on these volcanic reservoirs (Table 4.3).

The reservoirs in the Bohai Bay Basin are initially divided into volcanic reservoir and volcanic-related reservoir and then subdivided as per the classification scheme. Volcanic reservoir consists of four classes (11 subclasses) of reservoirs, such as weathering and leaching, primary pore and fracture, structural fracture, and burial dissolution, and volcanic-related reservoir consists of three classes of reservoirs, such as surrounding rock contact metamorphism, volcanic rock lateral barrier, and overlap drape.

The weathering and leaching reservoir mainly develops in the Mesozoic volcanic rock. Because the volcanic rock formed before the development of the Paleogene fault depression and had been exposed to weathering and leaching by atmospheric freshwater, its original reservoir space was altered, new pore and fracture space were formed, and the reservoir properties were greatly improved. This type of reservoir is mostly located in the basement of the sags and under the regional or local unconformity in the cap rock and is represented by the Xinglongtai buried hill in the Liaohe depression and the Fenghuadian Mesozoic volcanic weathered crust in the Huanghua depression (Figure 4.12).

TABLE 4.3 Classification of Volcanic Reservoirs in Bohai Bay Basin

Reservoir type			Case
Volcanic reservoir	Weathering and leaching reservoir	Volcanic rock weathering and leaching buried hill reservoir	Xinglongtai of Liaohe (Es^{3+4})
		Volcanic rock weathering and leaching unconformity reservoir	Well block Da13 of Dapingfang in Liaohe (Ed)
	Primary pore-type reservoir	Volcaniclastic rock pore-type reservoir	Well block Shang741 of Jiyang (Es^1)
		Lava primary pore-type reservoir	Well block Bin338 of Jiyang (Es^4), Caoqiaoxi area (Es^4)
		Intrusive rock joint-type reservoir	Well block Luo151 of Jiyang (Es^3)
	Structural fracture reservoir	Volcaniclastic rock fracture reservoir	Oulituozi region of lower Liaohe (Es^3)
		Lava fracture reservoir	Well block Shao18 of Jiyang (Ek)
		Intrusive rock structural fracture reservoir	Well block Shang741 of Jiyang (Es^3)
	Burial dissolution reservoir	Volcaniclastic rock burial dissolution reservoir	Well block Bei12 × 1 of Huanghua
		Lava burial dissolution reservoir	Gaoqing of Jiyang (Ek)
		Intrusive rock burial dissolution reservoir	Well block Xia38 of Jiyang (Es^3)
Volcanic-related reservoir	Surrounding rock contact metamorphism reservoir	Slate map cracking oil reservoir	Well An40 of Zhongchakou in Jizhong (Es^4)
		Hornfels micropore oil reservoir	Well block Luo151 of Jiyang (Es^3)
	Volcanic massif lateral barrier reservoir		Well Hong5 of lower Liaohe
	Overlap-drape reservoir		Xinglongtai of lower Liaohe (Es^{3+4})

Adapted from Jiang et al. (1999).

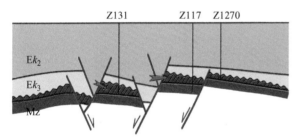

FIGURE 4.12 Fenghuadian Mesozoic volcanic reservoir in the Huanghua depression of Bohai Bay Basin (Dagang oil field).

The primary pore and fracture-type reservoir mainly developed in the volcanic or subvolcanic rock. The reservoir spaces such as pores and fractures were the primary products in the course of magmatic eruption or condensation and were weakly altered by actions such as weathering, cementation, and dissolution. The effusive facies lava often developed vesicular structures in the course of effusion. Simultaneously, it could also form a few fractures in the course of secondary action and thus enhance the connectivity between the vesicles. This has created favorable conditions for hydrocarbon migration and accumulation. This type of reservoir is represented by the Zaobei volcanic reservoir in the Huanghua depression (Figure 4.13).

The structural fracture reservoir can be developed in different kinds of volcanic rocks, and the reservoir space is mainly the structural fractures formed by tectonic compression or rifting. This type of reservoir usually has a unified water-oil-gas system, reservoir properties are good, formation pressure is high, and it is typically a prolific reservoir. The volcaniclastic fracture reservoir is represented by the volcaniclastic reservoir in the Oulituozi region of the Liaohe depression (Figure 4.14), the lava fracture reservoir is represented by the basalt reservoir in well block Shao18 in the Shaojia region of the Jiyang depression, and the intrusive rock structural fracture reservoir is represented by the Shahejie Formation Member III diabase reservoir in well block Shang741 in the Jiyang depression.

The burial dissolution reservoir is located in the tectonic cycles and magmatic cycles and is far away from the unconformity surface. The formation of this type of reservoir has a certain relationship with faults. Fractures accompanying the faulting not only increased the reservoir space but also connected the primary pores in rocks such as basalt. In addition, because the volcanic rock was located in a favorable position for burial diagenesis, dissolution easily occurred along the faults or fractures and caused the primary reservoir space lost by cementation to be recovered or increased. Because the fault action made the atmospheric freshwater become more effective, the dissolution of volcanic rock was also expedited. The reservoir spaces in this type of reservoir are pores and fractures. Pores are mainly composed of different kinds of dissolution spaces; the fractures have many origins, and the closer to the faults, the more developed are the fractures. The representative burial dissolution reservoirs include the volcaniclastic rock burial dissolution reservoir, such as the tuff reservoir in well block

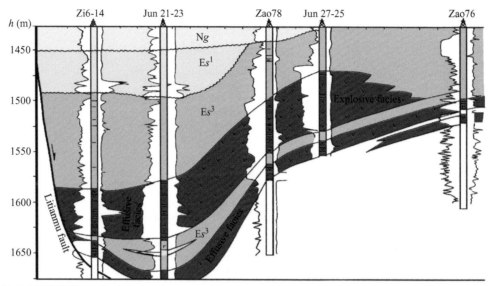

FIGURE 4.13   Zaobei Shahejie Formation Member III volcanic reservoir in the Huanghua depression of Bohai Bay Basin (Dagang oil field).

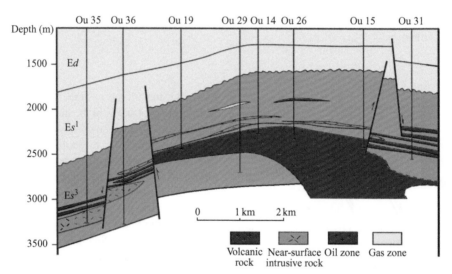

FIGURE 4.14   Oulituozi Shahejie Formation Member III fractured reservoir in Liaohe depression (Liaohe oil field).

Bei12-XI of Huanghua depression; lava burial dissolution reservoir, such as the Kongdian stage basalt reservoir in the Gaoqing region of the Jiyang depression; and the intrusive rock burial dissolution reservoir, such as the diabase reservoir in well block Xia38 of the Jiyang depression and the diabase reservoir in the Chunxi region of the Jiyang depression.

Apart from volcanic reservoirs, volcanic-related reservoirs also develop in the Bohai Bay Basin, and their main types are as follows: surrounding rock contact metamorphism reservoir, such as the slate reservoir in the An40 well in the Zhongchakou region of the Langgu sag; and the volcanic massif lateral barrier and overlap-drape reservoir, such as the Hong5 well basalt lateral barrier oil reservoir and drape gas reservoir.

4.3.2   Deep Strata of the Songliao Basin

Analyses of the typical volcanic reservoirs discovered in the Xushen gas field and Changshen gas field in the deep strata of the Songliao Basin show that they are almost all lithologic gas reservoirs. The characteristics of this kind of reservoir are described as follows by using the Xushen gas field as an example.

Xushen gas field is located in the Zhaozhou county and Anda region of Daqing city geographically, and in the Xuzhong structural belt of the Xujiaweizi fault depression in the southeast rift region of the deep structural element of northern Songliao Basin. The primary targets of the Xushen gas field are the Yingcheng Formation Member I and Member III volcanic reservoirs and Member IV conglomerate reservoir, with the volcanic reservoir predominating. The Yingcheng Formation Member I volcanic rock mainly occurs in the southern Xushen gas field, where acidic volcanic rock predominates, and the volcanic rock thickness revealed by well drilling is 77-989 m. The Yingcheng Formation Member III volcanic rock mainly occurs in the northern Xushen gas field, where intermediate and acidic eruptive rock predominates, intercalated with some glutenite and tuffaceous siltstone, and it is 200-780 m thick. The Yingcheng Formation Member IV mostly occurs in the southern Xushen gas field, where conglomerate and glutenite reservoirs predominate, and the formation thickness revealed by well drilling is 10-367 m.

The Xushen gas field consists of many lithologic gas reservoirs in blocks Xushen1, Xushen8, Xushen9, and Shengshen2-1. Many gas reservoirs are superimposed in each block, there is no unified gas-water contact, the connectivity of the gas zone is poor, the height of the gas column exceeds the structural amplitude, and macroscopically they are relatively typical lithologic gas reservoirs (Figure 4.15).

The gas reservoirs in the Xushen gas field are predominantly lithologic gas reservoirs as a whole. The volcanic lithology and reservoir performance are largely different between the well points; furthermore, the gas-bearing properties and gas-water relationships are largely different as well, and all they show the characteristics of a lithologic gas reservoir. Observed from the achievement of a volcanic rock prediction study of 3D seismic data, the volcanic rocks are distributed successively as a whole, but they are formed by the superposition and development of different volcanic edifices; the individual edifice usually forms an independent gas reservoir and has an independent gas-water contact, which also shows the characteristics of a lithologic gas reservoir.

The Xushen5 well drilled in well block Xushen1 is only 2.8 km away from the Xushen1 well. The volcanic lithology of this well is close to that of the Xushen1 well, but their reservoir performance is very different. The effective reservoir thickness of the Xushen5 well takes up a small proportion of the gross thickness of volcanic rock, while most of the intervals in the Xushen1 well are reservoirs. Tight rhyolite predominates in the whole interval of the Xushen2 well; therefore, the effective reservoir thickness is as thin as 70 m. The lithologies and physical properties change rapidly in the Yingcheng Formation Member I volcanic reservoir within the fault depression. Its heterogeneity is extreme, and the reservoir distribution is mainly controlled by the volcanic facies, with the high-quality reservoir developing in the near-crater facies belt.

A gas zone was encountered in the Xushen1, Xushen6, Xushen5, Xushen2, and Xushen4 wells, but there is apparently no effective structural trap in these wells, which shows that the accumulation of natural gas is mainly controlled by lithology. The Xushen5 well is characterized by upper gas and lower water, and the Xushen2 and Xushen4 wells in the south also have the same characteristics, but their gas-water contact is different from each other. The bottom boundary depth of the gas zone is −4025 m in the Xushen2 well and −3780 m in the Xushen4 well, and they are much deeper than the gas-water contact of −3520 m in well block Xushen5. This shows that the horizontal connectivity of the volcanic reservoirs is poor, and that different volcanic massifs can form reservoirs independently.

The fluid distribution of the Yingcheng Formation volcanic gas reservoir is characterized by apparent upper gas and lower water. Apart from producing pure gas from the volcanic reservoirs in the Xushen1, Xushen601, and Xushen603 wells located at structural highs, all other wells drilled in the Xushen gas field produce gas (or gas and water simultaneously) from the upper part and water from the lower part based on formation testing (for gas) longitudinally or integrated interpretations. Horizontally, the structural location of block Xushen1 in the southern Xushen gas field is the highest, its productivity is also the highest, and the three wells at structural highs produce pure gas. The structural location of well blocks Xushen8 and Xushen9 in the low uplift is higher, their productivity is higher, but water cut is seen in the lower part. The structural location of well blocks Xushen4 and Xushen2 in the middle of Xushen gas field is the lowest, their productivity is low, and gas and water differentiation is poor. In general, the gas column is high at structural highs where gas enriches and gas production is high, and it is low at structural lows where gas and water differentiation is poor, which shows that the structure plays a controlling role in the distribution of gas and water. The Xushen1 well has all-gas zones, but the Xushen5 well connected with it has the characteristic of upper gas and lower water; therefore, the distribution of natural gas is also controlled by structure in locally developed reservoirs. The Yingcheng Formation Member III volcanic gas reservoir in the Shengping-Wangjiatun region is also controlled by structure macroscopically (Figure 4.16), and it has a characteristic of upper gas and lower water. There is not any unified gas-water contact in the blocks in the southern Xushen gas field, the gas-bearing height exceeds the structural trap, and the lithologic trap is the main controlling factor for forming a volcanic gas reservoir.

The volcanic gas reservoir in block Shengshen2-1 in the northern Xushen gas field is clearly controlled by structure. Formation testing (for gas) was done in seven of the nine wells in the gas-bearing area, and they all have the characteristic of upper gas and lower water. The Shengshen4, Weishen3, Shengshen201, and Shengshen203 wells located at structural lows are predominantly water zones. Log interpretation and formation testing (for gas) show that the gas-water contact is almost identical in the wells, which

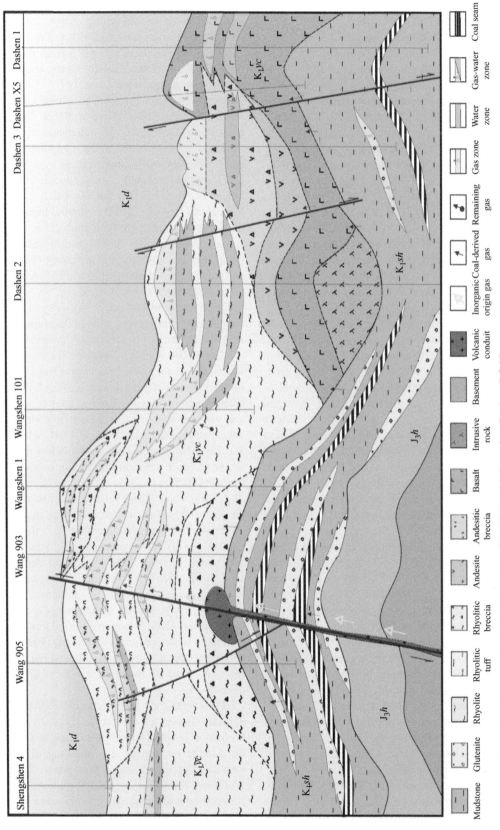

FIGURE 4.15 Distribution of gas reservoirs in the Anda region of the Xujiaweizi depression (Daqing oil field).

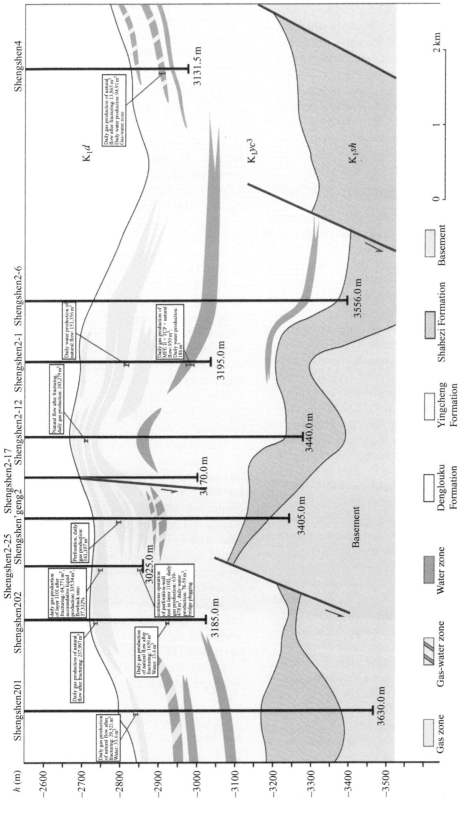

FIGURE 4.16 Distribution of gas reservoirs in the Shengping region of the Xujiaweizi depression in Songliao Basin (Daqing oil field).

indicates that structure is a major controlling factor in their gas-bearing properties. Because the volcanic reservoir facies changes rapidly laterally and the physical property difference is large, the gas-water contact is still slightly mismatched in the wells, and the water top in an individual well (Shengshen2.12 well) is higher than the gas bottom in the other wells. It is believed through integrated analysis that structural control predominates in the Yingcheng Formation Member III gas reservoir in the block Shengshen2-1 on the whole, lithology plays a certain controlling role, and the gas reservoir is a lithologic-structural gas reservoir.

4.3.3 Carboniferous in Northern Xinjiang

The Carboniferous in northern Xinjiang is an important formation for volcanic reservoir exploration in China at present, and large stratigraphic reservoirs predominate in the reservoir types discovered on the whole. Volcanic rocks can be classified into weathered crust stratigraphic reservoirs and inside lithologic reservoirs in terms of trap type; they can also be classified into layered, massive, and pectinate reservoirs in terms of reservoir shape and classified into self-source self-reservoir and new-source old-reservoir reservoirs in terms of hydrocarbon source. Based on an integrated analyses of the nine discovered typical Carboniferous reservoirs, they can be grouped into six types: gas reservoirs such as well blocks Dixi10, Dixi14, Dixi17, and Dixi18 in the Kelameili gas field are Carboniferous self-source self-reservoir volcanic weathered crust massive stratigraphic reservoirs; Carboniferous Kalagang Formation oil reservoir in the Niudong oil field is a Carboniferous self-source self-reservoir volcanic weathered crust layered stratigraphic reservoir; Carboniferous oil reservoir in the upper wall of the Ke-Bai fault zone in the northwest margin of Junggar Basin is a new-source old-reservoir volcanic weathered crust pectinate stratigraphic reservoir; Carboniferous oil reservoir in the Shixi oil field is a new-source old-reservoir volcanic weathered crust massive stratigraphic reservoir; Carboniferous Haerjiawu Formation oil reservoir in the Malang sag of Santanghu Basin is a Carboniferous self-source self-reservoir volcanic rock inside lithologic reservoir; and Carboniferous oil reservoir in the Hongche fault zone in the northwest margin of Junggar Basin is a new-source old-reservoir volcanic rock inside lithologic reservoir (Table 4.4). Different types of reservoirs are largely different in their hydrocarbon source, controlling factors, and fluid properties. Defining the characteristics and controlling factors of a reservoir can provide the basis for the subdivision and prediction of favorable blocks.

The reservoir characteristics of the northwest margin of Junggar Basin, Kelameili field and Niudong field, are described below.

4.3.3.1 Northwest Margin, Junggar Basin

A typical stratigraphic reservoir was developed in the northwest margin of Junggar Basin (Figure 4.17). Carboniferous volcanic rock formed large stratigraphic reservoirs in a weathering and leaching facies formed along the unconformity surface.

4.3.3.1.1 Geological Features of the Reservoir

The Carboniferous basement oil reservoir of the upper wall of the Ke-Bai fault zone, located at the boundary of Karamay city and the upper wall in the middle of the northwest margin fault zone of Junggar Basin, is near the Zaire Mountain folded zone in the north and superposes the overlying Triassic and Jurassic oil zones. The height difference of the surface topography is negligible, and the ground elevation is 270-350 m. The Carboniferous top structure dips northeastward at 2-7°, with a small uplift locally. The burial depth is 100-1800 m, and the oil domain is 13 km wide and 80 km long, with an area of about 1100 km². The basement of the Ke-Bai fault zone is the Neopaleozoic leveled off by long-term denudation after violent Hercynian folding and metamorphism. It lacks a complete structure suitable for storing oil and gas and is dissected into three-class terraces by two stages of northeast-southwest faults. The Carboniferous basement in the upper wall of the major fault is mainly sealed by the fault and asphalt. Commercial oil flow was obtained from the basement in the Karamay222 well initially in July 1957, and the daily oil production was 7.25 m³. It was believed at the time that the upper wall Carboniferous basement oil reservoir was a fault block reservoir controlled by lithology. In recent years, renewed exploration has been conducted following the lithologic-stratigraphic reservoir theory, and it is now believed that the upper wall Carboniferous basement is a lithologic-stratigraphic reservoir controlled by faults and microfractures, and contains oil predominantly. Furthermore, renewed understanding of the controlling factors of favorable reservoirs has enlarged the exploration scope beyond only andesite to all lithologies. It is now recognized that different types of highly weathered rocks can form favorable reservoirs, and this has caused the various lithologies in the entire upper wall of the Ke-Bai fault zone to become favorable reservoir development areas (Figure 4.18).

4.3.3.1.2 Hydrocarbon Accumulation Conditions and Main Controlling Factors

The Carboniferous and Permian in the Mahu sag have provided abundant oil source rocks, and types I, II, and III all exist. The Carboniferous source rock and lower Permian (P_1j) source rock are predominantly type III, the lower Permian (P_1f) source rock is predominantly types I and II,

TABLE 4.4 Characteristics of Carboniferous Volcanic Reservoirs in Northern Xinjiang

Reservoir classification	Self-source self-reservoir volcanic weathered crust massive stratigraphic				Self-source self-reservoir volcanic weathered crust bedded stratigraphic	New-source old-reservoir volcanic weathered crust pectinate stratigraphic	New-source old-reservoir volcanic weathered crust massive stratigraphic	Self-source self-reservoir volcanic rock inside lithologic	New-source old-reservoir volcanic rock inside lithologic
Field name	Kelameili gas field				Niudong oil field	Carboniferous oil reservoir in upper wall of Ke-Bai fault zone	Shixi oil field	Haerjiawu Formation oil reservoir	Carboniferous oil reservoir in Hongche fault zone
Reservoir name	Dixi10 gas reservoir	Dixi14 gas reservoir	Dixi17 gas reservoir	Dixi18 gas reservoir	C_2k oil reservoir	Blocks 6, 7, 9	Carboniferous oil reservoir	Ma36, Ma38	Chefeng6, Che912
Reservoir type	Fault-stratigraphic	Stratigraphic	Fault-stratigraphic	Fault-stratigraphic	Fault-stratigraphic	Stratigraphic	Structural-stratigraphic	Lithologic	Lithologic
Reservoir characteristics — Main lithology	Dacite, rhyolite	Volcanic breccia	Basalt	Granite porphyry	Basalt, andesite	Andesite, basalt, volcanic breccia, tuff	Dacite, andesite, volcanic breccia	Basalt, andesite	Basalt, volcanic breccia
Lithofacies	Explosive facies	Explosive facies	Effusive facies	Intrusive facies	Effusive facies, explosive facies	Effusive facies, explosive facies	Effusive facies, explosive facies	Effusive facies, explosive facies	Effusive facies, explosive facies
Reservoir porosity (%)	1.2-28.8	0.9-28.4	0.8-25.6	1.9-19.2	4.2-15.8	0.1-17.04	2.47-28.84	2-17.2	2.7-25.5
Average porosity of hydrocarbon zone (%)	12.2	15.3	16.15	8.63	10.6	9.929	14.2	7.8	12.94
Effective porosity cutoff (%)	7.0	6.0	6.3	5.5	6	5.5	5.6	6	6
Permeability cutoff (10^{-3} μm^2)	0.07	0.02	0.07	0.01	0.06	0.1	0.3	0.07	0.05
Reservoir type	Fracture-secondary pore								Primary pore-secondary pore-local fracture

Continued

TABLE 4.4 Characteristics of Carboniferous Volcanic Reservoirs in Northern Xinjiang—cont'd

Reservoir classification		Self-source self-reservoir volcanic weathered crust massive stratigraphic				Self-source self-reservoir volcanic weathered crust bedded stratigraphic	New-source old-reservoir volcanic weathered crust pectinate stratigraphic	New-source old-reservoir volcanic weathered crust massive stratigraphic	Self-source self-reservoir volcanic rock inside lithologic	New-source old-reservoir volcanic rock inside lithologic
Static attribute of reservoir	Burial depth of crest (m)	2995	3430	3645	3455	1200	420-1460	4260	3113-3128	1920-2430
	Elevation in the middle of hydrocarbon zone (m)	−2415	−3040	−3120	−2990	−976	−350 to −1600	−3947	−2576 to −2603	−1450
	Height of oil (gas) layer (m)	230	430	135	380	368-500	220-700	293	103	178
	Oil- (gas-) water contact (m)	−2530.15	−3250	−3185	−3176	−1226 to −1385	No obvious oil-water contact	−4093		−1765
	Pressure coefficient (100 MPa m^{-1})	1.07	1.21	1.27	1.07	0.965	1.07-1.44	1.495	0.75-1.439	1.05
	Temperature in the middle of hydrocarbon zone (°C)	89.73	114.27	115.51	114.13	52.5	26.5-53.6	120	91.6	32.8-52
	Oil-bearing (gas-bearing) area (km^2)	15.2	22.78	21.94	17.2	33.67	180.6	30.4		18.8
	Oil (gas) saturation (%)	60.1	63.3	62.7	71.3	51-67.8	53.1-67.8	57		63.6
	Average net thickness (m)	95.9	110.9			86.3	16.6-172.7	118.8		90.85
	Reserves abundance — Crude oil (10^4 tons k^{-1} m^{-2})					165.4	89.2	126.5		109.3
	Natural gas (10^8 m^3 km^{-2})	3.4	10.1	3.3	16.9					
	Water body condition	Bottom water	Bottom water	No bottom water	Bottom water	Edge and bottom water	Bottom water	Bottom water		Bottom water

Fluid characteristics	Fluid property		Natural gas	Natural gas + gas condensate			Crude oil				
Fluid characteristics	Natural gas	Relative density	0.633	0.648	0.636	0.644	0.97–1.19	0.64–0.83	0.7305		0.7548
		CH$_4$ content (%)	87.5	85.19	86.73	83.6	58.43–67.32	78.25–87.75	78.5		95
		CO$_2$ content (%)	0.258	0.144	1.0	0.109		0.04–0.22	0.156		0.1
		N$_2$ content (%)	6.26	5.81	4.08	4.67			6.3		3.38
	Volume factor (dimensionless)		0.00372	0.00336	0.00326	0.00359	1.09–1.178	1.06–1.15	1.8266		1.1
	Crude oil	Surface crude oil density (g cm^{-3})	0.771	0.774	0.773	0.766	0.8536	0.7308–0.9309	0.809	0.8516–0.8534	0.772–0.9054
		Crude oil viscosity at 50 °C (mPa s)	1.0	1.06	1.18	0.95	26.7	11.95–5600.35	3.06	11.34	73.4–168.64
		GOR (%)					44	30–90	329	5.6–950	30.78–60
		Wax content (%)	1.66	1.61	3.22	1.18	12.1	1.44–3.96	8.29		3.0
		Freezing point (°C)	−13.2	−3.3	1.3	−9.7	7.5	5.0 to −23.89	7.3		−17.42
		Initial boiling point (°C)	111.5	93	89.2	76.1	56	105	95		250.88
	Formation water	Total salinity (mg L^{-1})	20,404–226,06	8776–137,43	11,569–18,016	No water sample	3000–7500	22,322–28,655	14,774–26,273		41,102
		Water type	CaCl$_2$	CaCl$_2$	CaCl$_2$	CaCl$_2$	CaCl$_2$	NaHCO$_3$	CaCl$_2$		CaCl$_2$
Oil/gas well productivity	Crude oil (tons d^{-1})		5.17	6.41	19.56	26.93	88.98	145.86	66.8	36.5	20.61
	Natural gas (10^4 m^3 d^{-1})		20.24	9.14	25.17	25.01			0.8571		0.091
	Representative well		Dixi10	Dixi14	Dixi17	Dixi18	Niudong9-9	Gu3	Shixi1	Ma36	Chefeng3
Hydrocarbon source	Horizon of source		C$_2$b, C$_1$d	C$_2$b, C$_1$d	C$_2$b, C$_1$d	C$_2$b, C$_1$d	C$_2$h, C$_1$j, C$_2$b	P$_1$f	P$_2$w	C$_2$h	P$_1$f, P$_2$w
	Type of source rock		III				II				
Main controlling factors of reservoir formation	Position of source rock		Below gas reservoir				Below oil reservoir	Flank of oil reservoir	Flank of oil reservoir	Between oil zone	Flank of oil reservoir
	Reservoir-caprock assemblage		C/P				C/P	C/T	C/T	C/C	C/C
	Conduction system		Fault-unconformity				Fault-unconformity	Fault-unconformity	Fault-unconformity	Fault-directly	Fault-discontinuous surface
	Tectonic setting		Monocline-nose-like structure				Faulted anticline	Monocline	Anticline	Tectonic setting	Tectonic setting

C$_1$j, lower Carboniferous Jiangbasitao Formation; C$_1$d, lower Carboniferous Dishuiquan Formation; C$_2$b, upper Carboniferous Batamayineishan Formation; C$_2$h, upper Carboniferous Haerjiawu Formation; P$_2$w, middle Permian lower Urho Formation; P$_1$f, lower Permian Fengcheng Formation.

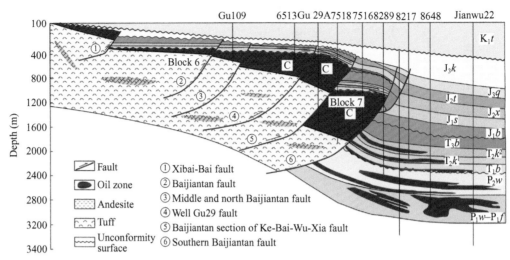

FIGURE 4.17   Profile of the reservoir at the northwest margin of Junggar Basin.

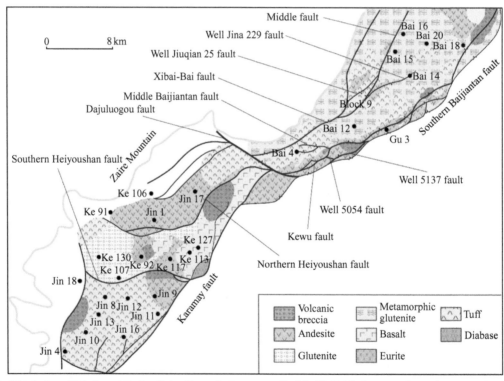

FIGURE 4.18   Distribution of lithology at the top Carboniferous in the upper wall of the Ke-Bai fault zone at the northwest margin of Junggar Basin.

the upper Permian (P_2w) source rock is predominantly types II and III, and the upper wall Carboniferous possesses favorable hydrocarbon source conditions.

The upper wall basement of the Ke-Bai fault zone is mainly a set of intermediate and basic intrusions of dark gray, sandy slate-siliconized mudstone, metamorphic conglomerate, and sandstone, as well as intermediate and acidic eruptive rock. The lithology is tight and fragile, the fold is complex, the dip angle is very large, and the stratigraphic

dip is 55-80°. Because of violent compression, the nappe of the upper wall Carboniferous overlying the Permian is about 15-20 km wide, and the reservoir distribution scope is large. Long-term denudation and weathering took place before the deposition of Mesozoic, and irregular reticular fractures developed in all lithologies. Generally, there are 5-150 fractures developed in a 10-cm long core. The pores and fractures constitute the major accumulation and permeation space. The pore types consist of phenocryst dissolved pores, interparticle dissolved pores, vesicles, intragranular dissolved pores, and matrix dissolved pores. The effective porosity is 5-23%, and the permeability is $0.5\text{-}2000 \times 10^{-3}\ \mu m^2$.

The upper wall of the Ke-Bai fault zone underwent violent Hercynian folding and metamorphism and formed three-class terraces by two faults. From bottom to top, the footwall basement of the major fault is class I terrace, the basement fault block between the two faults is class II terrace, and the upper wall above the second fault is class III terrace. The oil-bearing properties of class II terrace are the best. The oil source is mainly provided by the first fault, a semi-sealing trap is formed by the second fault, and the basement oil reservoir is intact because of the protection of the Baijiantan Formation mudstone. The hydrocarbon is kept in a relatively stable and balanced state in the fault block. The basal surface is the upper boundary of the reservoir, but the lower boundary is not obvious and varies from 300 to 1200 m. Oil is saturated above the boundary, but the oil saturation decreases gradually to zero below the boundary, and the middle consists of an anhydrous layer and apparent barrier.

4.3.3.1.3 Hydrocarbon Accumulation Mechanisms

The oil-generation threshold depth of the source rock in the Mahu sag is about 2460 m. The Permian source rock had reached the mature stage at the end of the Triassic and started to expel and migrate hydrocarbons. The northwest slope zone facing the source sag was at the highest position of the uplift at that time and became the favorable region for hydrocarbon accumulation. Initially, the hydrocarbons migrated parallel along the Permian and started to migrate along the basal surface and Triassic unconformity surface (including the upper part of the basement and basal conglomerate of the Karamay Formation) after it was overridden by the Triassic. After the migration pathway was cut off by the major fault at the northwest margin, the oil source in the footwall contacted the basement directly in the upper wall, and then the hydrocarbon passed through the fault plane and continued to migrate into the fractured basement, and formed the basement oil reservoir.

The fault activity in the oil province was ceaseless at the time of Triassic deposition. The fault throw was still minor at the final stage of the Triassic when the reservoir was initially formed, and the oil source in the footwall infused hydrocarbon only into the upper part of the basement in the upper wall. In time, the fault throw was enlarged, and the hydrocarbon was infused gradually into the middle and lower part, and at present the infusion continues. As a result, the larger the fault throw, the longer the oil-bearing interval of the base rock is. The distance from the basal surface of the oil-bearing interval to the basal surface of the footwall that acts as the hydrocarbon migration pathway corresponds to the height of the basal fault throw. The hydrocarbon saturation decreases gradually from top to bottom—saturated reservoir is located in the upper part and unsaturated reservoir is located in the lower part or only oil and gas shows can be seen.

The hydrocarbon migrated into the base rock through a complex fracture network upward into the basal surface, then passed through the unconformity surface, and entered the overlying Triassic Karamay Formation pluvial fan glutenite. At that time, diagenesis such as compaction and cementation of the glutenite was much weaker than that seen in the Karamay Formation conglomerate at present, and the reservoir properties were better. After the Karamay Formation reservoir was saturated with hydrocarbon, it was sealed by the overlying Triassic Baijiantan Formation tight mudstone. At this point, the hydrocarbon could only continue to migrate along the updip direction of the basement massif and Karamay Formation conglomerate until it reached the Laoshan outcrop. Apart from resulting in oil saturation in a large area of the base rock, a large stratigraphic unconformity also developed in the diachronous porous formations (J, K, and N) overriding the base rock. Together with the basement oil reservoir, they occur in an area starting from Chepaizi in the west, ending in Xiazijie in the east, and more than 200 km long and several kilometers wide along the basin margin.

4.3.3.1.4 Distribution Characteristics of Reservoirs

Fractures linked different pore types in the basement of the upper wall of the major fault into a networked body; the hydrocarbon was controlled by the shape of the basement massif, the lithology, and the fractures; and a fractured irregular massive oil reservoir was formed. Because there is no active bottom water, a unified and flat oil-water contact is not present. Generally, the oil-bearing properties are good in the upper part and become gradually poorer downward until oil ceases to exist. Although the productivity of each interval is different because of differences in physical properties and lithologies, they are connected to each other, and the pressure system of each interval is uniform. The 250-km long major fault from Chepaizi to Xiazijie surrounded the Mahu source sag in an arc and intercepted the extensive hydrocarbon migration from the sag to the northwest slope zone. The hydrocarbon was continually infused into the basement massif of the upper wall via the fault plane since the end of Triassic. Because there is no barrier between the base rock and the overlying Triassic Karamay Formation conglomerate, some hydrocarbon passed through the unconformity surface and entered the Karamay Formation and thus caused the base rock and Karamay Formation of the upper wall to contain oil within an area of $1550\ km^2$ from Chepaizi to Xiazijie. The oil saturation and the thickness of the reservoir change with different regional sealing conditions, oil supply conditions, reservoir properties, and late preserving conditions. The thickness of the oil-bearing interval of base rock in the upper wall of the major fault at the northwest margin is 100-1400 m on the whole. The thickness of the oil zone is 200-500 m, the oil-bearing area is approximately 1100 km², and the reserves in the basement oil reservoir have not yet been completely assessed.

The Karamay basement oil reservoir is a fractured heterogeneous massive oil reservoir. It is close to the major fault, its thickness is great, fractures are well developed, the crude oil properties are good, and oil production is high in the zone of the upper wall basement oil reservoir close to the major fault. When the major fault separates and merges again along the strike, slim front fault blocks occur. Fractures develop in these fault blocks, and the fault creates good sealing conditions. The oil zone thickness is relatively large, the crude oil properties are good, and they are usually prolific zones.

The upper wall of the Carboniferous base rock of the major fault underwent violent compression, folding, and faulting. Its dip angle is very large and cleavage fractures are well developed. Long-term weathering further enlarged the secondary pores. Its porosity is less than 6%, but fracture pores account for one-third of the total porosity, which have given the oil well higher output and recoverable reserves. However, a common characteristic of the fractured oil zone is that it has marked heterogeneity, and the large stratigraphic dip aggravates the variance of lithology horizontally. The variance in lithology affects the reservoir properties; in addition, it is closely related to the development of fractures. Generally, fractures develop more in the brittle rock than in the ductile rock, which results in high heterogeneity of the oil zone and great differences in individual well production.

The reservoir belongs to the elastic-solution gas drive type. The bottom water is inactive and there is no definite oil-water contact, but the solution gas volume is larger than that of an ordinary basement oil reservoir. The gas/oil ratio (GOR) is 40-152 $m^3 m^{-3}$, oil saturation is high, saturation pressure is close to the formation pressure, and the elastic energy is low. Poor sealing conditions result in rapid production decline, while good sealing conditions contribute to good production.

4.3.3.2   Kelameili Gas Field

The Kelameili gas field of Ludong in the Junggar Basin is a typical volcanic weathered crust massive stratigraphic gas reservoir. It is located in the Di'nan uplift at the eastern section of the Luliang uplift in the hinterland of Junggar Basin. It was discovered in 2005 by the Dixi10 well that a high-yield commercial gas flow was obtained from the Carboniferous. The Kelameili gas field comprises four gas reservoirs in well blocks Dixi10, Dixi14, Dixi17, and Dixi18 (Figure 4.19), which are weathered crust massive stratigraphic gas reservoirs, and the reservoir controlling parameters of each gas reservoir are different.

The Carboniferous strata in the Di'nan uplift zone become gradually older from west to east, and their top is an angular unconformity contact with the Permian (Figure 4.20). The top structure shape of the Carboniferous is a large nose-like structure dipping westward and cut by boundary faults at both the south and north sides. A series of approximately east-west- and northwest-trending faults developed in the Di'nan uplift. The northern Dishuiquan fault at the north side is one of the large faults. It is a southeastward-dipping reverse fault, and the disconnected horizon is Carboniferous-lower Jurassic. The Carboniferous volcanic lithology reflects that the volcanism had an apparent regularity horizontally: intermediate and basic lava developed in the west, intermediate and acidic lava developed in the east, intermediate to acidic intrusive rock developed along the fault in the south, and volcaniclastic rock to sedimentary tuff were deposited discontinuously among these areas. The Carboniferous volcanic massif underwent long-term weathering and denudation, and the volcanic reservoir was altered significantly. Therefore, the reservoir is complicated and diversified, and the entire suite of shallow intrusive rock, lava, and volcaniclastic rock breccia can form effective reservoirs. The primary pore types of the Carboniferous volcanic reservoir are mainly vesicles, intragranular pores, and intergranular pores formed at a solidifying diagenetic stage of the rock. The secondary pores are mainly dissolved pores formed by hypergenesis after diagenesis of the volcanic rock. Structural fractures developed universally, then the dissolved fractures and condensing contraction fractures developed in the lava. The pore types are diverse, and the reservoir is predominantly secondary dissolved pores and fractures, and it has the dual-medium characteristics of pore fracture.

4.3.3.2.1   Characteristics of the Gas Reservoir

The Kelameili gas field is located in the local nose-like anticlinal structure in the Di'nan uplift zone at the eastern section of the Luliang uplift. The developmental horizons of source rock are the lower Carboniferous Dishuiquan Formation and upper Carboniferous Batamayineishan Formation lacustrine swamp coal and dark mudstone intervals. The organic carbon content of the Dishuiquan Formation coal is 5.7-28.6%, averaging 13.3%, and that of the dark mudstone is 1.15%; the organic carbon content of the Batamayineishan Formation coal is 6.82-40.24%, averaging 23.6%, and that of the dark mudstone is greater than 1.54%. The cap rocks of the gas reservoir constitute the weathered clay layer at the top of the Batamayineishan Formation and the upper Permian upper Urho Formation dark-gray mudstone and silty mudstone.

The developmental horizon of the reservoir is the upper Carboniferous Batamayineishan Formation volcanic rock. Its lithology is mainly basalt, andesite, rhyolitic ignimbrite,

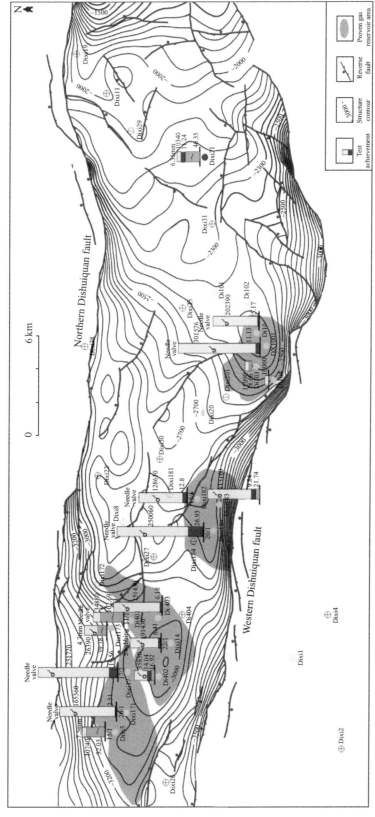

FIGURE 4.19 Horizontal distribution of the Kelameili gas field in Junggar Basin.

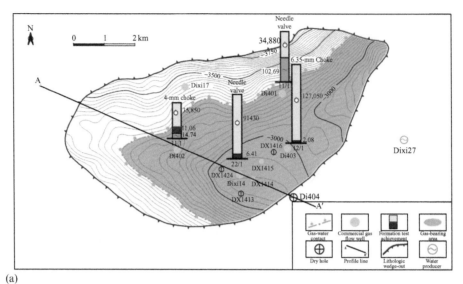

(a)   Gas-bearing area of the Carboniferous in well block Dixi14 of Junggar Basin

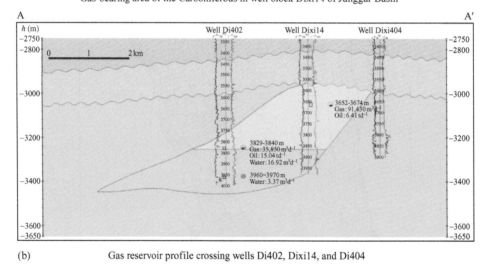

(b)   Gas reservoir profile crossing wells Di402, Dixi14, and Di404

FIGURE 4.20   Composite map of the gas reservoir in well block Dixi14 of the Kelameili gas field in Junggar Basin.

crystal tuff, volcanic breccia, and shallow intrusive granite, and its lithofacies are mainly explosive facies and effusive facies. The reservoir porosity is 0.9-28.4%, averaging 14.85%, and the permeability is $0.02\text{-}123 \times 10^{-3}$ μm^2, averaging 0.618×10^{-3} μm^2. The variation and heterogeneity in physical properties are considerable, and the lithology, lithofacies, and physical properties are interrelated. The permeability of rhyolite and ignimbrite are generally greater than 1×10^{-3} μm^2 and that of agglomerate and tuff with no fractures is generally less than 1×10^{-3} μm^2.

The burial depth of the top of the gas reservoir is 2995-3645 m, and the height of gas zone is 130-430 m. Its formation pressure coefficient is 1.07-1.27 MPa/100 m, the temperature in the middle of the gas reservoir is 89.73-115.51 °C, the effective thickness of the reservoir is 59.3-236.4 m, its average gas saturation is 60.1-71.3%, and the OOIP is $3.3\text{-}16.9 \times 10^8$ m^3 km^{-2}, with bottom water, belonging to a massive oil reservoir.

The Carboniferous gas reservoir in well block Dixi17 is a fault or stratigraphic condensate gas reservoir. The constant production per kilometer of well depth is 2.2×10^4 m^3 km^{-1} d^{-1}, recoverable reserves are 3.3×10^8 m^3 km^{-2}, and it belongs to a low-yield, medium-abundance, and deep medium-sized gas reservoir. The Carboniferous gas reservoir in well block Dixi14 is a condensate gas reservoir with bottom water. The constant production per kilometer of well depth is 1.6×10^4 m^3 km^{-1} d^{-1}, recoverable reserves are 10.1×10^8 m^3 km^{-2}, and it belongs to a low-yield, high-abundance, and deep medium-sized gas reservoir. The Carboniferous gas reservoir in well block Dixi18 is a stratigraphic or lithologic condensate gas reservoir with bottom water. The constant production per kilometer of well depth is $3.3 \times 10^4 \, m^3 \, km^{-1} \, d^{-1}$, recoverable reserves are $16.9 \times 10^8 \, m^3 \, km^{-2}$, and it belongs to a medium-production, high-abundance, and deep medium-sized gas reservoir. The Carboniferous gas reservoir in well block Dixi10 is a stratigraphic or lithologic condensate gas reservoir with bottom water. The constant production per kilometer of well depth is $2.6 \times 10^4 \, m^3 \, km^{-1} \, d^{-1}$, recoverable reserves are $3.4 \times 10^8 \, m^3 \, km^{-2}$, and it belongs to a low-yield, medium-abundance, medium-deep medium-sized gas reservoir.

4.3.3.2.2 Characteristics of the Natural Gas Component

Gaseous hydrocarbon predominates in a natural gas reservoir, and the total hydrocarbon content is 77.5-97.79%, averaging 93.53%. The methane content is 83.6-87.5%, averaging 84.80%; the C_2^+ content is 3.05-13.02%, averaging 6.47%; and the aridity coefficient is 0.88-0.98, averaging 0.94. The nonhydrocarbon gas content is very low, and it is mainly nitrogen at 4.08-21.72% and averaging 6.16%; the carbon dioxide content is low, 0-2.05%, and averaging 0.30%; and the oxygen content is 0.055%. The relative density of natural gas is 0.633-0.664, the density of the gas condensate is $0.774 \, g \, cm^{-3}$, the crude oil viscosity is 0.95-1.18 MPa s at 50°C, the wax content is 1.18-3.22%, the freezing point is 1.3 to -13.2 °C, and the initial boiling point is 76.1-115 °C. The type of formation water in the gas reservoir is $CaCl_2$, with a total salinity of 11,569.61-22,606.33 mg L^{-1} and a chloride ion content of 6796.19-16,022.47 mg L^{-1}.

4.3.3.2.3 Main Controlling Factors of the Gas Reservoir

Many gas reservoirs are distributed in the Di'nan uplift from west to east, and it is a natural gas-enriched area on the whole. Well block Dixi17 is located at the westernmost end, and the lithology of its major pay zone is basalt. This basalt is truncated and pinches out eastward and in the updip direction is dark mudstone between upper and lower volcanic sequences, which forms a barrier together with the overlying Permian mudstone, creating an unconformity barrier gas reservoir. Well block Dixi14 is located in the middle part, and the reservoir lithology is a complicated effusive facies basic basalt, acidic rhyolite, and explosive facies tuffaceous breccia. The mudstone interval forms a barrier in the updip direction, creating a complex volcanic cone lithologic gas reservoir. The lithology in the Dixi18 well is shallow intrusive granite, which is sheltered by a fault at the south side. Well block Dixi10 is located at the easternmost end, where the facies changes in the fallout facies tuff, and it is also sheltered by a fault at the south side. The gas reservoirs in well blocks Dixi17, Dixi14, Dixi18, and Dixi10 all have their own independent gas-water contacts.

4.3.3.2.4 Hydrocarbon Accumulation Processes

The natural gas in the Ludong-Wucaiwan region is mainly characterized by kerogen pyrolysis gas, but it does not mean that the kerogen pyrolysis gas in this region directly comes from by Carboniferous strata. The impact of the hydrocarbon accumulation process on the natural gas components and carbon isotopes is more remarkable. There is an apparent difference among the gas maturity levels reflected by different parameters of the Carboniferous natural gas in the region. For instance, the aridity coefficient of Carboniferous natural gas is 0.88-0.96, and this reflects its high to overmature nature. Based on the dissection of gas reservoirs in the Ludong-Wucaiwan region and the analyses on gas generation and accumulation characteristics during tectonic evolution and natural gas accumulation, the Carboniferous reservoirs in the region appear to have mainly experienced the hydrocarbon accumulation processes of the late Hercynian, Indosinian, and middle Yanshanian (Figure 4.21).

In the late Hercynian, even though the Carboniferous source rock became mature earlier and entered the mature stage in the Permian in the Wucaiwan region, the source rock only entered the low-maturity stage in the western Di'nan uplift and was not yet mature in the other areas. The inclusion of calcite veins in the amygdaloids in the Batamayineishan Formation sandstone presents a group directional distribution at 3232 m in the Cai25 well. The homogenization temperature of salt water is 86.2-88.5° C, which reflects hydrocarbon charging at that stage in the Wucaiwan region.

In the late Indosinian, because of deposition of the Triassic and lower Jurassic, the Carboniferous source rock entered the final stage of maturity or the initial stage of high maturity in the Wucaiwan region and entered the mature stage at the western section and low-maturity stage at the eastern section of the Di'nan uplift. The homogenization temperature of inclusions formed by the secondary enlargement of quartz in the lower Permian sandstone is 96.6-105.6 °C at 2990 m in the Cai25 well in the Wucaiwan region. The hydrocarbon inclusions associated with the Carboniferous basalt calcite vein are predominantly in a gaseous state at 3477.1 m in the Dixi17 well at the western section of the Di'nan uplift, and the homogenization temperature of salt water inclusions is 98.9-117.6 °C, which reflects hydrocarbon charging during the late Indosinian.

Violent tectonic activity during the early Yanshannian resulted in violent uplift and faulting of strata, which caused the gas that had accumulated in the late Indosinian Carboniferous reservoir to be dissipated almost completely; i.e., the hydrocarbon that was generated by Carboniferous source rock before R_o reached 0.8-1.2%. Simultaneously, a secondary gas reservoir might have been formed in the Jurassic reservoir. The hydrocarbon destroyed in the

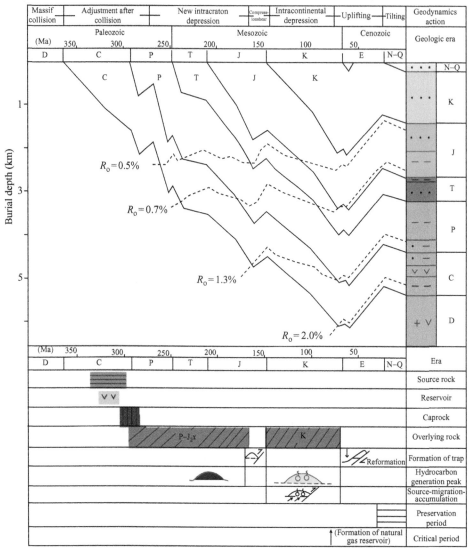

FIGURE 4.21 Hydrocarbon accumulation events of the Carboniferous hydrocarbon system in the Ludong-Wucaiwan region of Junggar Basin.

western Di'nan uplift was the product generated by Carboniferous source rock approximately before the R_o was 0.8-1.0%, while that destroyed in the Wucaiwan region was the product generated by Carboniferous source rock approximately before the R_o was 1.2%. By this time, there were only a few inclusions in the lower Jurassic sandstone; for instance, the temperature of a quartz inclusion is measured as 73.6 °C in well block Dixi17, which reflects hydrocarbon charging at a late stage.

The middle Yanshannian is a critical period for hydrocarbon accumulation of Carboniferous source rock in the Ludong-Wucaiwan region. Massive deposition of Cretaceous strata decided the ultimate maturity level of the Carboniferous source rock in the region. The Carboniferous source rock entered the highly mature wet gas phase in the Wucaiwan region, highly mature gas condensate to wet gas phase at the western section and mature stage at

the eastern section of the Di'nan uplift, and low-maturity stage at the Dibei uplift. The hydrocarbon accumulated during this epoch was natural gas generated by Carboniferous source rock after the R_o had reached 0.8-1.2%, which results in the difference between the gas maturity reflected by gas parameters and the actual value. The homogenization temperature of the Batamayineishan Formation tuff quartz fissures and secondary enlargement side inclusions is 133.9-139.6 °C in the Caican2 well, and that of salt water inclusions in the late calcite vein in the basalt sample at 3637.8 m, Dixi17 well, occurs between 140 and 150 °C, which are consistent with the hydrocarbon (mainly natural gas) charging during the middle Yanshannian.

Faulting activity has been very weak from the late Yanshannian to the present day, which is favorable for the preservation of primary gas pools accumulated under the regional cap rock of upper Permian Urho Formation mudstone at an early stage. During the late Yanshannian, natural gas pools were adjusted locally and a secondary natural gas pool was formed in the Jurassic and Cretaceous, which resulted in the dissipation and accumulation of natural gas from the Carboniferous to Jurassic and Cretaceous.

The formation of a natural gas pool is characterized by "early accumulation and late preservation" in the Ludong-Wucaiwan region; although as a whole, this region had experienced multiple stages (late Hercynian, late Indosinian, and middle Yanshannian) of hydrocarbon charging and reservoir formation, the middle Yanshannian should be the critical period for forming natural gas pools in the region. The conditions for forming a primary natural gas pool originating from the Carboniferous humic source rock are available under the Permian Urho Formation regional seal. For instance, the gas in the Carboniferous gas reservoir of the Dixi10 well mainly originated from the Carboniferous overmature humic natural gas, and the $\delta^{13}C_1$ and $\delta^{13}C_2$ values of natural gas are -29.1 to -29.5‰ and -26.6 to -26.7‰, respectively. The gas in the Wucaiwan Carboniferous gas reservoir mainly originated from the Carboniferous overmature natural gas, and the $\delta^{13}C_1$ and $\delta^{13}C_2$ values of natural gas are -29.5 to -31.0‰ and -24.2 to -26.8‰, respectively. Violent compression and shear tectonic activity in the late Hercynian became relatively weak in the Indosinian, and faulting activity was violent in the early Yanshannian but became relatively weak in the late Yanshannian-Himalayan, both of which were favorable for preserving natural gas pools formed at an early stage.

The hydrocarbon generated in Carboniferous source rock migrated along the fault longitudinally and along the weathered rock mass horizontally, accumulated, and formed a reservoir in the weathered volcanic rock mass. In such a way, a self-source self-reservoir volcanic weathered crust stratigraphic reservoir was formed, in which the fault and unconformity are the major conduction systems. Because the individual Carboniferous volcanic edifice is small in northern Xinjiang, the volcanic rock and sedimentary rock are interbedded and the weathered volcanic rock mass at the top of the weathered crust interstratifies with the sedimentary rock in the highly oblique strata. The reservoir scale depends on the thickness of the volcanic layer and the stratigraphic dip. The favorable reservoir formed by long-term weathering and leaching of volcanic rock is the major location of hydrocarbon accumulation, which is one of the key elements of reservoir formation. An effective cap rock is the key to the preservation of hydrocarbon, and an area with a positive structural setting is a favorable location for hydrocarbon accumulation. A proximal hydrocarbon accumulation is one where hydrocarbons are captured in the effective trap near the source or hydrocarbon migration pathway, thus forming a reservoir.

When there is an insufficient hydrocarbon source or poor preservation conditions higher in the structure, a reservoir is not always formed in the trap at structural highs. Based on the hydrocarbon accumulation mode, we suggest searching for effective traps proximal to the source rock rather than traps at structural highs that are far from the source rock during oil and gas exploration.

4.3.3.2.5   Distribution Rules of the Gas Reservoir

Horizontally, the reservoir boundary is controlled by a favorable weathered volcanic rock mass, a lateral barrier condition in the updip direction, and hydrocarbon enrichment. Because the individual Carboniferous volcanic edifices are small in northern Xinjiang, the weathered crust stratigraphic reservoir formed is also small; however, a large gas reservoir group can be composed of many weathered volcanic rock mass gas reservoirs, and in such a way, a large gas field with reserves of 1000×10^8 m^3 is formed. Large weathered volcanic reservoirs can also be formed in the developmental area of large volcanic edifices.

The reservoir occurs along the top of the Carboniferous longitudinally and is sheltered by impermeable rock laterally and from a weathered clay layer and overlying sedimentary mudstone at the top. The reservoirs are not connected in the strata with a large stratigraphic dip, and the thickness of the gas reservoir is controlled by the thickness of the weathered rock mass; however, gas reservoirs are connected in the fault developed area that have experienced long-term weathering and leaching, and the hydrocarbon column is generally 100-350 m. Because the source sag mainly originates from the downdip direction of the tilted stratum, this kind of gas reservoir is possibly formed in the volcanic weathered crust stratigraphic trap at the updip position along the hydrocarbon source direction; however, being controlled by the hydrocarbon conduction system, the trap at the structural high is not always suffused, or a reservoir cannot be formed. The favorable volcanic reservoir is not developed in a downdip position, and the thickness of the hydrocarbon zone is generally small; the optimal weathered volcanic reservoir mainly occurs on the paleotectonic slope.

4.3.3.3   Niudong Kalagang Formation Weathered Crust Reservoir

The Niudong oil field is located at the No. 2 Niudong structure of the Niudong nose-like structural belt in the northern Malang sag of Santanghu Basin. The Ma17 well was drilled in the No. 2 Niudong structure in 2006, formation testing was conducted in the Carboniferous Kalagang Formation, and 28.5 m^3 daily oil production and 3732 m^3 daily gas production were obtained, thus defining the Carboniferous volcanic reservoir of Niudong oil field (Figure 4.22). The Niudong oil field is a typical volcanic weathered crust layered stratigraphic reservoir.

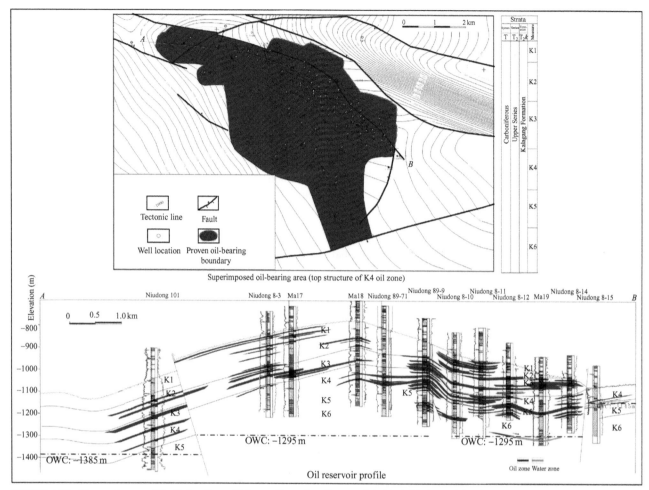

FIGURE 4.22 Composite map of the Carboniferous Kalagang Formation oil reservoir in block Niudong of the Malang Sag, Santanghu Basin (Tuha oil field).

4.3.3.3.1 Geological Features of the Reservoir

The source rock of the volcanic weathered crust reservoir in Niudong oil field is composed of three sets of lacustrine swamp coal-measure argillaceous source rocks, such as the lower Carboniferous Jiangbasitao Formation (C_1j), upper Carboniferous Batamayineishan Formation (C_2b) and Haerjiawu Formation (C_2h). The Santanghu Basin was in a back-arc basin environment in the Carboniferous, and massive paralic facies were deposited. Carboniferous source rocks developed in all sags, thickened from the north to the south, and mostly developed in the Malang and Nao-maohu sags. The upper Carboniferous source rock is mainly composed of oil shale, dark-gray mudstone, marlite, black carbargilite, and coal seams; its thickness is 19-66 m; and the maximum thickness is speculated to be 100-300 m. The lower Carboniferous source rock is mainly composed of marine dark mudstone and oil shale, it is very thick (the thickness of the effective source rock in Daheishan is 688 m), and it occurs all over the periphery of Santanghu Basin. The major argillaceous source bed occurs in the upper Carboniferous Haerjiawu Formation. It is a set of continental eruptive rock and volcaniclastic rock deposits, and dark mudstone, carbargilite, and oil shale develop locally. It is 100-300 m thick and constitutes the major source rock of the lower part of the Santanghu Basin. The organic carbon content is between 1.87% and 8.8% and averages 5.5%, the average hydrocarbon-generation potential is 21 mg g^{-1}, the type of organic matter is II_1, and it is considered preferable to good source rock.

The reservoir is the upper Carboniferous Kalagang Formation weathered volcanic rock mass. Four stages of volcanic rocks developed in the Kalagang Formation; four sets of volcanic rocks present mounded reflection characteristic on the seismic profile, with clear texture. Thin volcaniclastic rock developed in the eruption interval is generally sedimentary tuff or tuffaceous siltstone and packsand, with a thickness of 2-20 m. Generally, a rhythm is formed from the lower lava to the upper volcanic breccia and tuff, and each period or stage can be composed of numerous rhythms. The sedimentary tuff or tuffaceous sedimentary rock that represents the dormancy of a volcano is the marker between periods or stages. The lithology of the reservoir is basalt, andesite, rhyolitic ignimbrite, tuff, and volcanic breccia, and the lithofacies is mainly explosive facies and effusive facies. The cap rocks of the oil reservoir are the weathered clay layer at the top of the upper Carboniferous Kalagang Formation and the middle Permian Lucaogou Formation dark-gray mudstone and silty mudstone.

Reservoir porosity is 4.2-15.8%, averaging 10.6%; permeability is 0.01-1.2×10^{-3} μm^2, averaging 0.46×10^{-3} μm^2. The physical properties change markedly with each reservoir. The permeability of basalt and andesite is generally greater than 0.5×10^{-3} μm^2; the permeability of rhyolite, tuff, and volcanic breccia where fractures are not developed is generally less than 0.5×10^{-3} μm^2.

The burial depth at the top of the oil reservoirs is from 1200 to 1785 m, the height of the oil zones is from 368 to 500 m, and the oil-water contact is -1226 to -1385 m. The reservoir pressure is 14.96-16.17 MPa, and the pressure gradient is 0.82-0.94 MPa/100 m. The reservoir temperature is 52.5-57.5 °C, and the temperature gradient is 2.62 °C/100 m. The average net thickness of the reservoir is 86.3 m, and the average oil saturation is 57.8%. Both edge and bottom water exist in these volcanic weathered crust layered reservoirs.

4.3.3.3.2 Fluid Characteristics

The surface crude oil density is 0.8405-0.8692 g cm^{-3}, averaging 0.8536 g cm^{-3}. The freezing point is 4-13 °C, averaging 7.5 °C; the initial boiling point is 51-72 °C, averaging 56 °C. The GOR is 22-65%, averaging 44%; wax content is 7.2-15.5%, averaging 12.1%; and the crude oil viscosity at 50 °C is 6.33-87.05 mPa s, averaging 26.7 mPa s. The alkane content in the family component of crude oil is 58.68%, the aromatic hydrocarbon content is 17.3%, the colloid + asphaltene content is 2.68%, and the nonhydrocarbon content is 11.56%. The relative density of dissolved gas in the reservoir is 0.97-1.19, the methane content in the component is 58.43-67.32%, the ethane content is 11.02-12.13%, the propane content is 8.42-9.26%, the butane content is 3.60-4.76%, the average content above pentane is less than 2.5%, and it is characterized by a high percentage of intermediate components. The formation water is $CaCl_2$ type, and the total salinity is 3500-7500 mg L^{-1}. The formation water salinity increases gradually northeastward and decreases gradually southwestward on the whole, and it is 3000-3500 mg L^{-1} in the main position of well block Ma17.

4.3.3.3.3 Controlling Factors of the Oil Reservoir

The formation of the volcanic reservoir in the Niudong oil field of Santanghu Basin is mainly controlled by unconformity-related weathering and leaching, lithology, and tectonic setting. The nose-uplift structural belt is an important tectonic setting for hydrocarbon enrichment in volcanic rock. The Niudong nose-uplift zone has been situated at the higher structural location for a long time. It not only controls the direction of hydrocarbon migration and accumulation but also controls the distribution of oil and gas. For example, in the No. 2 Niudong structure, the Ma801 and Ma23 wells were drilled in the lower structural location, and they predominantly produced water, with a few oil blooms in the formation test.

Although the majority of the reservoir space had been filled by minerals such as zeolites, calcite, and chlorite at an early diagenetic stage, in the course of dissolution, the calcite in the lava amygdaloids, intergranular pores, and granular edge fissures was extremely prone to be dissolved. The zeolites that were difficult to dissolve in other areas also had their own features, being characterized by high calcium and low silicon. Under the action of weathering and leaching, mandelstone where amygdaloid is completely or partly dissolved, and autoclastic brecciform lava and volcanic breccia that are dissolved along the breccia edge fissure are formed. The autoclastic brecciform lava has the best reservoir performance and oil-bearing properties, next is the amygdaloidal basalt and andesite, and last is the volcanic breccia.

The Kalagang Formation volcanic reservoir is a complex lithologic-stratigraphic reservoir controlled by a nose-uplift tectonic setting at the front of the Tiaoshan uplift, lithofacies, lithology, long-term weathering and leaching, and a fracture network (Figure 4.23).

4.3.3.3.4 Hydrocarbon Accumulation Processes

Systematic detection of the fluid inclusions in the phenocrysts, amygdaloids, vugs, and mineral veins in the Carboniferous Kalagang Formation volcanic rock in the Malang sag of Santanghu Basin shows that there are a great deal of hydrocarbon inclusions. Observing the hydrocarbon inclusions with fluorescence, the inclusions give off saffron, yellow, and blue-white fluorescence in the zeolites and calcite veins, which indicates that three stages of hydrocarbon-charging processes may exist, i.e., one stage of low mature oil charging, one stage of mature oil charging, and one stage of highly mature oil charging. Oil inclusions giving off yellow fluorescence can be seen in the Kalagang Formation zeolites and calcite veins, as well as a few feldspar veins at 1532.0-1550.1 m in the Ma17 well, which indicates that there is at least one stage of mature oil charging. Oil inclusions giving off a yellow fluorescence can also be seen in the zeolites and calcite veins at 2661.95 m in the Ma17 well, which indicates that there is also at least one stage of mature oil charging in this interval. Oil inclusions giving off saffron, yellow, and blue-white fluorescence can be seen in the Kalagang Formation zeolites and calcite veins at 1521.1-2332.8 m in the Ma19

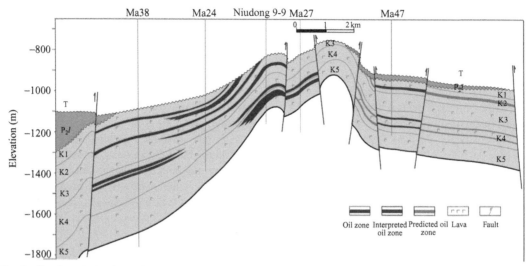

FIGURE 4.23 Profile of the Carboniferous Kalagang Formation oil reservoir in block Niudong-Mabei of the Santanghu Basin.

well. The measured temperature of the oil inclusions is 86.8C, 108.8, 125.5, and 160.9 °C, and the temperature of mature stage 106.9-130.8 °C predominates. This indicates that multiple stages of reservoir formation exists in the Carboniferous volcanic reservoir of Santanghu Basin—it is characterized by late mature oil charging, and the middle and late Yanshannian were the critical periods for the forming the reservoir. From the above, the hydrocarbons of the Niudong entered the low mature stage in the Indosinian to early Yanshannian, and reservoirs started to form; the hydrocarbon entered the mature stage in the middle and late Yanshannian, migrated and accumulated heavily, and formed reservoirs.

4.3.3.3.5 Distribution Rules of the Oil Reservoir

Horizontally, the reservoir boundary is controlled by a favorable weathered volcanic rock mass, lateral barrier conditions, and hydrocarbon fullness. A large oil reservoir can be formed in the successive distribution area of favorable volcanic lithologies of multiple volcanic edifices, and its OOIP can be more than 0.1 billion tons. In the developmental area of medium and small volcanic edifices, if the favorable volcanic lithology does not occur successively, multiple disconnected reservoirs can be formed. This is because the weathering extent of different volcanic lithologies is different, and a prolific well, stripper well, and dry hole can coexist laterally.

The reservoir occurs along the top of the Carboniferous longitudinally, is sheltered by an impermeable lithology or fault laterally, and sheltered by a weathered clay layer and overlying sedimentary mudstone at the top. The connectivity among reservoirs is controlled by the connectivity of the favorable volcanic lithologies, which is in turn controlled by the volcanic edifice. The reservoir thickness is controlled by the thickness of the weathered rock mass and is generally

100-200 m. Because the source sag primarily originates from the downdip direction of a tilted stratum, this type of reservoir is possibly formed in all the effective volcanic weathered crust stratigraphic traps located in the updip position of the hydrocarbon source direction. Hydrocarbon accumulation is more favorable at a faulting point because of the reconstruction of structural fractures, forming prolific reservoirs. The optimal weathered volcanic reservoir mainly occurs in the highest position of a paleostructure. In the multiple sets of volcanic rock and sedimentary rock formations formed in the same time slice of the Carboniferous, the weathering and leaching that occurs during the dormant period of a volcanic eruption can create a weathered crust stratigraphic reservoir composed of multiple layers.

4.3.3.4 Niudong Haerjiawu Formation Inside Oil Reservoir

The Carboniferous Haerjiawu Formation in the Niudong oil field is a typical volcanic interlayer inside a lithologic oil reservoir. An "inside lithologic reservoir" refers to hydrocarbons that accumulated in volcanic rock under an unconformity, and the reservoir occurs interlayered with a lithologic reservoir. Oil production of 36.5 m³ d⁻¹ was obtained by natural flow with a 6-mm choke in the Ma36 well in 2008, and thus this reservoir was discovered. To the present day, commercial oil flow has been obtained from wells Ma36, Ma19, Niudong103, Ma38, Tiao16, Ma29, and Ma20, and the 3P reserves (Proved, Probable, and Possible Reserves) exceed 5000×10^4 tons, which show that the Carboniferous inside volcanic rock is a favorable exploration prospect in the Santanghu Basin or even in the northern Xinjiang.

The interstratified deposition of volcanic rocks and transitional facies sedimentary rocks developed in the Carboniferous in northern Xinjiang. At the time of deposition, an effective reservoir-caprock assemblage developed in the fault depression, the source rock that was deposited within the sedimentary sequence provided a hydrocarbon source, and the volcanic reservoir provided reservoir space. In this way, a Carboniferous self-source self-reservoir inside hydrocarbon system was formed.

4.3.3.4.1 Oil Reservoir Characteristics

The source rock of this self-source self-reservoir mainly develops inside the Carboniferous fault depression, and the distribution of effective source rock controls the horizontal distribution scope of development of the inside reservoir. The fault development inside the fault depression is the primary area where volcanic rock develops, with the fault connecting the source rock and the volcanic reservoir, which is favorable for the longitudinal migration of oil and gas. The inside-type volcanic reservoir is controlled by the volcanic edifice, lithofacies, and lithology. The explosive facies and effusive facies are the favorable lithofacies, and basalt is the most favorable volcanic lithology. The lava generally has a three-layer structure: vesicles and contraction fractures develop at the top and bottom, but in the middle, vesicles do not develop and the crystallization and cyclicity are more apparent. This reflects the original physical properties of the volcanic reservoir and the potential for constructive dissolution at a later stage. Simultaneously, in the dormant period of the volcanic eruption, because the volcanic rock is generally situated at a relatively high position, surface exposure causes weathering and leaching, which is favorable for the formation of reservoirs. During thermal evolution of the source rock and generation of oil and gas, the generated organic acid migrates upward along the fault under pressure, which also has a reconstructive role on the volcanic rock and is favorable for the formation of reservoirs. Furthermore, the sandstone

present inside the Carboniferous source rock can also form a favorable reservoir (Figure 4.24).

The lithology of the Haerjiawu Formation reservoir is basalt, andesite, dacite, and the same kind of autoclastic volcanic breccia, showing that a relatively complete volcanic edifice has been preserved in the formation. The reservoir space is predominantly structural fractures, amygdaloid contraction fractures, amygdaloid dissolved pores, phenocryst contraction fractures, and matrix dissolved pores. Most cores are broken, and multiple stages of fractures are developed, which were cemented by calcite at an early stage, and opened or semifilled at a later stage, with oil stains on the fracture surface. The block core porosity is 2-17.2%, averaging 7.8%, and the permeability is low. The inside-type volcanic reservoir developed below the mudstone cap rock or in the dormant period of volcanic eruption. Covered by the mudstone, fluids such as compaction water and organic acid generated at the early stage of mudstone formation or in the course of hydrocarbon generation concentrated, transported, and dissolved the surface of the volcanic rock, forming a secondary pore reservoir. The physical properties of the Haerjiawu Formation (C_2h) inside-type volcanic reservoir are generally poorer than that of the Kalagang Formation (C_2k) weathered crust volcanic reservoir. The formation pressure gradient fluctuates greatly and is 0.755-1.439 MPa/100 m, the formation temperature gradient is 2.942 °C/100 m, and it belongs to a low geotemperature area.

4.3.3.4.2 Fluid Properties

The crude oil density is 0.8516-0.8534 g cm^{-3}, saturated hydrocarbon content is low at 61.71-66.85%, crude oil viscosity at 50.0 °C is 11.34 mPa s, and the dissolved gas content changes greatly and is 5.6-950 m^3 m^{-3}.

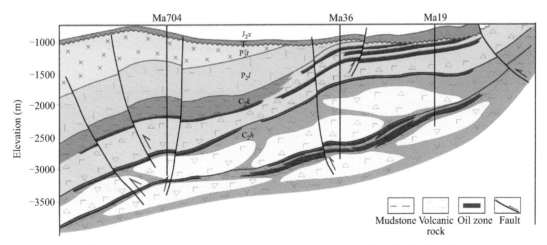

FIGURE 4.24 Profile of the Carboniferous Haerjiawu Formation oil reservoir in well block Ma19-Ma36-Ma704 of block Niudong in the Malang sag of Santanghu Basin.

4.3.3.4.3 Controlling Factors of the Oil Reservoir

The Haerjiawu Formation is mainly an interlayered lithologic oil reservoir, with the reservoir and source rock developed in an interstratified manner. The lithology is mudstone, tuff, and lava interbeds, which experienced leaching in the dormant period of the volcanic eruption, and this leaching and late hydrocarbon generation were the keys to forming this reservoir type. Apart from lithofacies, the high-quality reservoir in the Ma36 well is also related to protection of fluid compartment, hydrocarbon generation and discharge due to thermal maturation of source rock, dissolution by acidic water, and development of fractures. The inside-type oil reservoir develops during the dormant period of a volcanic eruption cycle, and clastic rock and volcanic rock form an effective reservoir-caprock assemblage. The present oil shows or the oil zone rechecked out are all located at the top or bottom surface of the volcanic massif. The degree of fracture development and the lithofacies are the critical elements for hydrocarbon accumulation. The paleoslope geomorphology setting controls the distribution of favorable facies belts.

4.3.3.4.4 Hydrocarbon Accumulation Processes

The Haerjiawu Formation volcanic rock and source rock developed in interbeds in the Santanghu Basin. At the time of volcanic eruption, the bottom of the volcanic massif condensed at an early stage, and thus microfractures were formed. In the dormant period of the volcanic eruption, the rock experienced short-term weathering and leaching, and a favorable reservoir was formed at the top. Both the top and bottom of the volcanic massif were dissolved by the organic acid expelled from the upper and lower source rocks, and the reservoir properties were further improved. At the place where faulting developed, the fault promotes reservoir alteration and oil and gas migration and can control the enrichment of oil and gas.

The inside-type volcanic reservoir of the Haerjiawu Formation developed during the dormant period of volcanic eruption, and the clastic rock and volcanic rock alternated and composed a favorable (source) reservoir-caprock assemblage. The hydrocarbon zone and hydrocarbon show interval occur at the top and bottom of a violently erupted volcanic edifice. It is a self-source self-reservoir assemblage, mainly interlayer volcanic (inside lithologic) reservoir, and forms an inside assemblage mode in a volcanic reservoir (Figure 4.25). The hydrocarbon generated in the source rock migrated vertically along the fault and formed the inside lithologic reservoir in the effective reservoir-caprock assemblage developed between the source rock and the volcanic rock.

4.3.3.4.5 Distribution Rules of the Oil Reservoir

The volcanic rock and source rock became interbedded, the volcanic rock experienced short-term weathering and leaching during the dormant period of volcanic eruption, and formed a preferable reservoir at the top and bottom surface. The organic acid resulting from thermal evolution of the source rock would dissolve the volcanic rock and increase the reservoir space. Both volcanic rock and source rock constituted an effective reservoir-caprock assemblage, where the hydrocarbon generated in the source rock entered the volcanic reservoir directly or along the fault, accumulating and forming a reservoir. The scope of an inside lithologic oil reservoir is mainly controlled by the size of the volcanic edifice. The favorable lithofacies and lithozones of each volcanic edifice can all form reservoirs, and each reservoir has an independent oil-water and pressure system.

Apart from being controlled by the distribution of lithofacies and lithology, the reservoir is also controlled by the tectonic setting horizontally, and a reservoir is more easily formed on a positive structure at a high position. Horizontally, this type of reservoir is in the Carboniferous. Only

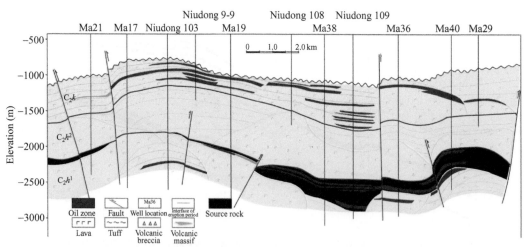

FIGURE 4.25 Profile of the Carboniferous inside reservoir in Niudong region of Santanghu Basin.

when controlled by the horizontal lithologic association can the favorable volcanic lithology form a favorable reservoir. Faulting is an important controlling factor to the formation of reservoir, as it plays a transportation role in hydrocarbon migration and accumulation. Simultaneously, the late closed fault plays a sealing role to oil and gas. Therefore, the oil-water contact position of this type of reservoir is different in different fault blocks.

4.4   DISTRIBUTION RULES OF VOLCANIC RESERVOIRS

The volcanic reservoir is a kind of unconventional reservoir in hydrocarbon-bearing basins. As an unconventional hydrocarbon-bearing geologic body, the formation of a volcanic reservoir has its own unique requirements. In order to further discuss the mechanisms of formation of volcanic reservoirs, studies on the reservoir formation conditions and distribution rules have been conducted on volcanic reservoirs both at home and abroad.

4.4.1   Rift Structure Environment and Large Volcanic Oil Provinces

In the tectonic setting of a continental plate, the continental rift, continental margin arc, island arc, passive continental margin, and mid-oceanic ridge are the structural environments where volcanic rock usually concentrate. As a result, the regional structural belt most favorable for the formation of volcanic reservoirs is the probable place where large oil provinces will be found. In recent years, a series of important discoveries related to volcanic rock exploration in China have been achieved in the rift structural environment in the east and island arc environment in the west. The large volcanic oil provinces in eastern China and northern Xinjiang are beginning to be understood, and this is closely related to the developmental characteristics of volcanic rocks in China. The volcanic rocks of China mainly developed in the Neopaleozoic, Mesozoic, and Cenozoic. The development of Neopaleozoic volcanic rock is mainly related to the Tianshan-Xingmeng trough and Paleotethys Ocean, and it occurs in regions such as northern Xinjiang, Tarim, Sichuan, and Tibet. For the moment, the oil and gas exploration of Permo-Carboniferous volcanic rock has made a significant breakthrough in the Junggar Basin of the northern Xinjiang region and the Santanghu Basin, which further indicate that the formation of large oil provinces in northern Xinjiang is related to the formation of the Tianshan-Xingmeng trough and the development of relevant volcanic rocks and source rocks. The volcanic rocks in the northern Xinjiang region mainly developed in the rift inside the massif and the island arc at the margin of the massif, and its

development was usually associated with the marine and transitional depositional environments. The marine and transitional source rocks were deposited, creating favorable hydrocarbon accumulation conditions. The Mesozoic and Cenozoic volcanic rocks developed widely in the eastern continental rift zone, and the basins such as Bohai Bay and Songliao are parts of this rift system. In the basins, the volcanic rocks developed in the dormant period or the corresponding period of continental lake basin deposition and composed favorable plays. To the present day, a large gas province has been discovered in the deep strata of the Songliao Basin, and a volcanic oil field has been discovered in the Bohai Bay Basin.

Analyzed from the structural environment, the Permian volcanic rocks in the Sichuan Basin were formed in the rift environment and related to the elongational structure environment of the Tethys Ocean. Furthermore, the Emei basalt represented a fissured eruption that was distributed relatively widely, and the underlying multihorizon source rock and the volcanic rock composed a proximal or distal association. High-yield gas flow has been discovered, and the potential for further exploration is very promising. The Permian volcanic rocks in the Tarim Basin are widely distributed, and their development is related to the subduction of the Tethys Ocean to the Tarim plate in the Paleotethys stage, belonging to the back-arc rift. The Permian volcanic rock and the underlying Cambrian-Ordovician source rock compose a distal association. A favorable oil and gas show has been seen, and it is a region worthy of further exploration.

4.4.2   Source Center Area and Giant Oil and Gas Fields

4.4.2.1   Source Center Area Favors Giant Oil Fields

The volcanic rock system cannot generate oil by itself and generally does not have the hydrocarbon accumulation conditions, and its pairing with an effective source rock is the key to the formation of a reservoir. When the volcanic rock is adjacent to or located in the major source rock measures, it is favorable for the formation of a volcanic reservoir. At present, two kinds of source center areas—oil generation and gas generation—have been discovered.

In the rift basin, the expansion stage is not only the time when the water area is extensive and the source rock is deposited, but also the time when volcanic activity is the most violent. Therefore, the hydrocarbon accumulation conditions of volcanic rock in rift basins are favorable. The Paleogene-Neogene Shahejie Formation Member III depositional stage in the Bohai Bay Basin is not only the maximal expansion stage but also the maximal lake transgression stage of the Paleogene-Neogene lake basin. In that period, the volcanic activity was frequent. Therefore, the

Shahejie Formation Member III volcanic rock has the hydrocarbon accumulation conditions of accessible and abundant oil sources. In the Liaohe depression, the Paleogene-Neogene consists of multiple sets of source rocks, such as the Shahejie Formation Member IV, Member III, Member I, and Dongying Formation. The Shahejie Formation Member III source rock is the most widely distributed, and the maximum thickness can be more than 2000 m. The maximum thickness of the Shahejie Formation Member IV source rock can reach 700 m in the western sag and the Damintun sag. The Shahejie Formation Member I source rock is less well developed than the Shahejie Formation Member III source rock, with a maximum thickness of 400-600 m. The area of Dongying Formation source rock is smaller, and it is thinner as well. The organic matter abundance of these source rocks is high, and they have an immature to overmature evolution degree. The volcanic rock in the Liaohe depression mainly develops in the lower Member III of the Shahejie Formation, which was covered by mudstone during the high position depositional stage; therefore, the hydrocarbon source is quite abundant.

A series of coal-bearing deposits predominate the source rocks in the rift basin of the deep strata of the Songliao Basin. These are the lower Cretaceous Shahezi Formation lacustrine mudstone and coal seam, and then the Denglouku Formation Member II gray-black mudstone. The maximum thickness of the Shahezi Formation dark mudstone is 384 m, and it is well juxtaposed with the Yingcheng Formation and Huoshiling Formation volcanic reservoirs developed above and below it.

Carboniferous marine and transitional source rocks developed in the Junggar Basin, and with the upper Carboniferous volcanic rock, they form a favorable assemblage. Thicker source rocks occur at the northwest margin and in the Luliang and Zhundong zone, and the maximum thickness can reach 1000-2200 m. The kerogen type in Ludong is mainly type III, the organic matter abundance is higher, and the maturity is higher. Therefore, it has entered the peak time of gas generation, providing the Carboniferous volcanic rock with an abundant gas source and creating a gas generation center area.

Lower Carboniferous marine source rock and upper Carboniferous transitional source rock developed in the Santanghu Basin, and they provided the Permo-Carboniferous volcanic reservoir with an oil source. The lithology of the lower Carboniferous source rock is a thick marine dark mudstone and oil shale. The thickness of the effective source rock in Daheishan is 688 m, it outcrops all over the periphery of the Santanghu Basin, and it has shown great exploration potential. The lithology of the upper Carboniferous source rock is mainly oil shale, dark-gray mudstone, marlite, black carbargilite, and coal seams. The thickness is 9-66 m, with a maximum thickness speculated to be 100-300 m. The quality of the source rock is favorable to good oil generation. The evolution degree of the source rock is high, and the hydrocarbons were generated at an early stage and over a long duration. It constitutes the oil source center area and controls the distribution of giant oil fields.

4.4.2.2 Proximal Assemblage Is Favorable; Distal Assemblage Needs Connecting by Fault or Unconformity

The relationship between volcanic rock and source rock can be divided into two classes: proximal assemblage and distal assemblage.

In the proximal assemblage, the source rock is located above, below, or to the side of the volcanic reservoir, the volcanic reservoir occurs in or in the vicinity of the source sag, and the hydrocarbon generated in the source rock has the maximum opportunity to contact the reservoir. The deep strata of the Songliao Basin and the Paleogene of Bohai Bay Basin in eastern China, as well as the Permo-Carboniferous inside Junggar Basin and the Permo-Carboniferous of Santanghu Basin in western China, all belong to the proximal assemblage. For instance, source rocks develop in both the upper and lower Carboniferous in the Santanghu Basin, five major reservoir-caprock assemblages develop inside the Carboniferous, volcanic rocks develop in or in the vicinity of source rock, and the hydrocarbon accumulation conditions are quite superior. Two volcanic cycles develop in the Carboniferous of Junggar Basin, sedimentary rocks including source rock were deposited between the volcanic cycles, the source and the reservoir are immediate neighbors or in the same sedimentary formation, and good plays are formed.

In the distal assemblage, the volcanic reservoir is separated from the source rock by multiple sets of strata or the volcanic rock is far away from the source sag center, and only with the help of oil source faulting or unconformity can the oil and gas migrate to the volcanic rock. For instance, the oil and gas shows or oil flows discovered in the Permian volcanic rock of Tarim Basin come from the underlying Cambrian-Ordovician source rock; the natural gas in the Permian volcanic gas reservoir of Sichuan Basin comes from the underlying source rocks at different horizons.

Generally speaking, the proximal assemblage is the most favorable for the enrichment of oil and gas.

4.4.3 High-Quality Reservoirs Controlled by Volcanic Explosive Facies, Effusive Facies, and Weathering Dissolution

4.4.3.1 Favorable Reservoirs a Prerequisite

Exploration practices have proven that both the explosive facies and effusive facies volcanic rocks can form high-quality reservoirs with favorable reservoir properties after undergoing unconformity dissolution and rifting. A high-quality reservoir is an important condition for the enrichment and high production of a volcanic rock. The volcanic reservoir is characterized by dual-medium fracture-pore space, e.g., vesicles, joint fractures, structural fractures, and dissolved vugs. The reservoir is characterized by a thick pay zone and high production and can form a giant field.

The formation of a volcanic reservoir is controlled by its volcanic lithofacies, lithology, and secondary actions such as late dissolution and faulting.

The volcanic lithology and facies belt can control the development and distribution of the volcanic reservoir and are one of the major reasons for hydrocarbon enrichment and high production of volcanic rocks. Good reservoirs can be developed in many types of rocks such as tuff, trachyte, basalt, diabase, and volcaniclastic rock. In terms of lithofacies, the genesis and abundance of volcanic reservoir space in different lithofacies are also different. Primary fractures and pores such as vesicles, intercrystalline pores, interbreccia pores, contraction fractures, and explosive fractures develop in the course of volcanic eruption. Taking the Paleogene-Neogene volcanic rock in the Jiyang depression as an example, the intrusive facies is mainly composed of hypabyssal and ultra-hypabyssal diabase and basalt, and its reservoir space is characterized by condensing contraction fractures, structural fractures, intercrystalline pores, and dissolved pores developed along the fracture; in addition, abundant vesicles are also developed in the ultra-hypabyssal basalt. The effusive facies volcanic rock is characterized by vesicles, intercrystalline pores, condensing contraction fractures, structural fractures, and dissolved pores. The explosive facies volcanic rock is characterized by intergranular pores, diagenetic contraction fractures, vesicles, and tectonic microfractures.

Secondary reservoir space refers to different kinds of pores and fractures formed by exogenic forces such as hydrothermal alteration, dissolution, tectonic stress, and weathering after the volcanic rock had been consolidated. Faulting and erosion surfaces are the important preconditions for the formation of secondary pores. No matter what kind of space it is, communication via different sizes of fractures and fissures is necessary for forming effective reservoirs, and a network of pores and fractures is the prerequisite for forming high-quality volcanic reservoirs. Frequent tectonic activity is an important mechanism that forms fissures and facilitates hydrocarbon migration and accumulation. The development of structural fractures is controlled by faulting and local structural changes.

Weathering and leaching can effectively reform the reservoir, and the thickness of the dissolved pore development zone can reach 100-1000 m.

4.4.3.2 Distribution of Reservoirs in Eastern China

The volcanic rocks in the fault depression basins of eastern China distribute in a linear arrangement along the faults. The explosive facies reservoir develops in the high position, and the effusive facies reservoir tends to develop on the slope. The reservoir is kept in the original position, late weathering and leaching is relatively weak, and a large-area primary reservoir is generally formed.

Volcanic rocks are widely developed in the Mesozoic and Cenozoic rift basins of eastern China, where primary reservoirs occur. They are favorably juxtaposed with the source rocks and are the key targets of oil and gas exploration.

The development of volcanic rock in the rift basin is associated with the large-scale pulling apart and rifting of the continental plate, as well as uplifting of the mantle. Generally, at the early stage of each cycle, the volcanic intensity is great and the volcanic rock distributes widely. The effusive facies basalt develops, as well as subvolcanic facies diabase and explosive facies volcanic breccia and tuff. The area and thickness are large, and fissured eruption predominates. At the later stage, the magmation weakens, a centered eruption predominates, and the volcanic rock distribution is relatively limited with minor thickness. Minor subvolcanic facies develops and then effusive facies; volcaniclastic rock is rarely seen. Generally, the age of the volcanic rock is old at the edge of rift and becomes younger toward the center, and this is related to the evolution characteristics of faulting.

In the Paleogene magmation cycle of the Bohai Bay Basin, the depositional period from the Kongdian Formation to Member IV of the Shahejie Formation was a period when the tectonic faulting was the most violent, and the violent action of major faults resulted in the volcanic rock depositing a widely distributed fissured eruption. At the late stage of the cycle, the faulting activity changed from being widespread to occurring locally, and the magmation gradually evolved to centered eruptions and intrusions.

In the rift basin, the development of a volcanic reservoir is controlled by faults. The faulting formed by early tectonic activities provided channels for magmation, and the tectonic activities developed at the late stage resulted in the development of structural fractures in early-formed volcanic rock. For instance, the three strike-slip NNW discordogenic fault zones, such as Shenyang-Weifang, Huanghua-Dezhou-Dongming, and Baxian-Shulu-Handan, basically control the distribution of volcanic rocks in the Bohai Bay Basin. Within the depression, the major fault associated with the strike-slip fault zone controls the distribution pattern of volcanic rock. The activity of the strike-slip fault zone controls the major fault associated with it, and the active mode and level of these major faults decide the distribution of volcanic rocks inside the region. The existence of these major faults, especially the existence of large shearing-tensional-sliding faults, makes the magma ascending along the strike-slip fault zone erupt rapidly or intrude into the veneer of crust. The place where faults intersect is the most violent eruption center of volcanic activity.

Therefore, volcanic rock mostly distributes in a linear fashion along class II faults derived from class I faults. In the Dongying sag, the scale of Jinjiazhuang rock sheet and Shicun-Caoqiao rock sheet distributed along the Shicun fault is the maximum; its area is 160 km^2, and the maximal thickness is 134-160 m. The occurrence shape is predominantly effusive facies lava flow and sheet, and explosive and extrusive facies are very rare; the eruptive environment is predominantly continental facies, and there is also sub-aquatic eruption. The volcanic rock sheet distributes mostly in an equiaxial shape and isolated state, and it is speculated that a centered eruption predominates. A small number of them are controlled by faults and arranged in a linear shape, and the eruption type is predominantly the quiet overflowing Hawaii type, with large thickness, wide area, and gentle occurrence.

In the fault depression of the deep strata of Songliao Basin, the volcanic massif occurs mainly along the fault, the explosive facies tends to develop near the fracture, and the large-area effusion facies occurs in the slope.

In the fault depression basins of eastern China, the fault depression boundaries not only control the development of the basins but also control the development of volcanic rocks and the spread of their facies belts (Figure 4.26). Explosive facies occurs in the vicinity of the major fault, and effusive facies occurs near the crater and distributes extensively in the synclinal slope and depressed area. Generally speaking, the reservoir near the fault is liable to be altered by forming fractures, thus improving the reservoir. Regarding the effusive facies reservoir, which is related to effusion and diagenesis, it is also controlled by secondary actions such as dissolution. Because the position near the fault is usually the position where the structure is relatively

higher, the lithologic-structural reservoir predominates, and the lithologic reservoir is mainly formed in the slope.

4.4.3.3 Distribution of Reservoirs in Western China

The volcanic rocks in the superposed basins of western China mostly experienced multiple tectonic movements, and large-scale weathering and leaching type reservoirs occurred along the unconformity surface.

The volcanic rocks in the basins of middle and western China are closely related to the formation and closing of the Paleoasian Ocean and Paleotethys Ocean, as well as orogeny. For instance, the volcanic rocks in the basins such as Junggar, Santanghu, and Tuha developed on the evolution background of the Tianshan-Xingmeng trough and were uplifted, weathered, and denuded because of a cessation of deposition in the Neopaleozoic, forming a regional unconformity. The elongation of the Paleotethys Ocean and its subduction to Eurasia in the Neopaleozoic resulted in back-arc elongation and formed the Permian volcanic rocks of the Sichuan and Tarim Basins. In terms of the tectonic setting, the volcanic rocks in the basin mainly resulted from cratonic or tensional activity in the continental segment, while the volcanic rocks at the basin margin and in the orogenic belt were formed from oceanic and island arc settings.

The cumulative thickness of Carboniferous volcanic rocks in the Santanghu Basin can exceed 7000 m at their maximum, and their lithology is primarily basalt, and secondarily andesite. The volcanic rock mostly occurs in the vicinity of faults, as the volcanic activity was controlled by the discordogenic fault, and a multicrater fissured eruption predominated. The Kalagang Formation volcanic rock is predominantly the effusive facies, and explosive facies volcanic breccia is seen in some areas. The thickness variation is significant laterally, as the volcanic rocks spread in bead-like shapes along the discordogenic fault. The Kalagang Formation volcanic rocks can be divided into

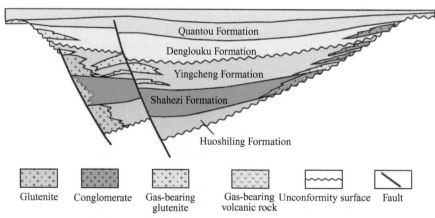

FIGURE 4.26 Development mode of volcanic rock in the fault depression basin.

four sets, and there is an unconformity surface between each set of volcanic rocks. Analysis shows that the physical properties of strongly weathered volcanic reservoirs are the best, and secondly the weakly weathered volcanic reservoirs. The matrix porosity of volcanic rock that was not exposed to weathering is universally very low, being 3-8%, the permeability is less than $0.05 \times 10^{-3} \ \mu m^2$, and the oil-bearing properties of the reservoir are also poor. The physical properties of volcanic reservoirs that were exposed to weathering and leaching are obviously increased, with the porosity being between 11% and 16% and a maximum of 25%, and the reservoir permeability is also improved significantly. Dissolved pores are the main effective reservoir space. The explosive facies volcanic breccia and the upper subfacies leached and denuded interval of the effusive facies has the best physical properties: porosity is 14-16%, and 25% maximum. The Kalagang Formation volcanic rock underwent two stages of weathering and leaching. The first stage occurred after the third stage of volcanic eruption; in that period, the dormant duration of the volcano was the longest, the rock had experienced long-term weathering and leaching in which the physical weathering predominated, and the chemical weathering played a subsidiary role. Reservoir fractures developed, and the reservoir performance is the best. This is the major oil-bearing interval in this region. The second stage of weathering and leaching occurred in the period when the Carboniferous volcanic rock transitioned to the Permian sedimentary rock. The Kalagang Formation volcanic rock experienced extensive and prolonged chemical leaching for a second time, and the fourth set of high-quality reservoir inside the volcanic massif was formed.

At the northwest margin of Junggar Basin, after experiencing strong weathering and leaching, different lithologic Carboniferous volcanic rocks all formed good reservoirs. The reservoir mostly develops within a range of 600 m below the weathering surface but can reach 1000 m below the weathering surface.

In the Ludong region of Junggar Basin, the volcanic rock is predominantly intermediate lava. There is also basic and acidic lava, and volcaniclastic rock also occurs. They are controlled by faulting, and the volcanic massif occurs in bead-like shapes along the fault.

4.4.4   Regional Seal Controls Large Stratigraphic Reservoirs

The source rock related to volcanism is the organic-rich mudstone. If the mudstone overlies the volcanic rock, it not only is in favor of oil and gas migration but also provides good sealability, and this is extremely favorable to hydrocarbon accumulation. The volcanic rock itself also has alternating tight and porous/fractured intervals, and

the tight interval itself can act as a favorable cap rock. For instance, in the Ludong region of Junggar Basin, the Permian mudstone overlying the Carboniferous is a regional seal, while the mudstone and tuff inside the Carboniferous are local barriers. The regional Triassic cap rock at the northwest margin has provided favorable conditions for the formation of large Permo-Carboniferous volcanic stratigraphic reservoirs.

4.4.5   Oil and Gas Enrichment at Structural Highs

The tectonic setting of volcanic rock development plays a critical role in the formation and enrichment of a volcanic reservoir. The structural high is not only the major target area of hydrocarbon migration but also the place where favorable reservoirs and different types of traps develop.

Eruptive material usually forms paleotopographic pyramidal uplifts, and there is usually an inherited structure above it, which is favorable for hydrocarbon migration and accumulation. The basalt oil reservoir in the Shijiutuo uplift of the Bohai Sea has been an inherited uplift on the buried hill since the end of the Mesozoic. The Neogene "green tuff" oil reservoir in the eastern Xinxi gas field and Jingcheng oil and gas field in the Xinxi region of Japan is formed by the inherited anticline of a volcanic paleogeographic pyramidal uplift and the oil and gas entrapped in it. Both the coastal plain oil field and the Cretaceous basalt oil reservoir in the State of Texas, United States, are formed by the inherited domes of volcanic cone lava and the oil and gas entrapped in them.

The volcanic rock structural high is the position where the tectonic stress is stronger, and violent tectonism can improve the reservoir performance of volcanic rock. The tectonism induces a great deal of structural fractures and enhances the permeability, and the structural fractures generated by the tectonism can promote the activity of subsurface fluid and thus form secondary reservoir space. When the structural fracture system coincides with the other oil-generation conditions, a reservoir can be formed.

In the fault depression basin, the volcanic rock located at inherited structural highs shows explosive facies developing near the crater, and structural fractures are likely to be developed near the fault. Therefore, the reservoir properties are generally good, creating a favorable position for hydrocarbon accumulation. In the slope, the effusive facies volcanic rock develops. In this position, it is closer to the oil source, and a lithologic reservoir is usually formed.

Weathering dissolution is usually stronger in the relative structural high in the intracraton or intracontinental depression basin of middle and western China, forming widely distributed dissolved reservoirs, thus forming a large uncompartmentalized oil and gas field. For instance, in the paleonose uplift belt of Junggar and Santanghu Basins, the weathering dissolution is strong and faults developed on a large scale, creating fractures and forming good reservoirs. These reservoirs may capture hydrocarbons over the long term, which should make them a major target when searching for volcanic reservoirs. Large stratigraphic reservoirs related to volcanic rock weathering and leaching have been discovered at the northwest margin of Junggar Basin and in the Santanghu Basin.

4.4.6 Lithologic-Stratigraphic-Type Volcanic Reservoirs

The volcanic reservoir has high heterogeneity generally; therefore, the hydrocarbon accumulation is usually controlled by dual factors such as structure and lithology, but overall, it is controlled by lithology. The hydrocarbon-bearing area of a volcanic reservoir is generally small, but its reserves abundance is high, and therefore it can possibly form a giant field.

Widespread lithologic reservoirs are generally formed in the fault depression basins of eastern China. Because the prototype basin suffered from weak reformation at a late stage, the original lithofacies determined the reservoir performance and then controlled hydrocarbon accumulation. The volcanic rocks in the fault depression basin present a zonal distribution along the fault. The explosive facies reservoir develops in the relative structural high around the fault zone, and a fault block structural reservoir and structural-lithologic reservoir are mainly formed. While effusive facies occurs across a wide area on the slope, a lithologic reservoir universally develops, and multiple measures are superposed and joined together.

Large unconformable stratigraphic reservoirs are mainly formed in the middle and western regions of China. For instance, in the basins such as Junggar and Santanghu, the regional tectonic setting where the volcanic rock is formed is the elongation action inside the craton or in the intracontinental depression basin. Most of these basins experienced multicycle evolution, Paleozoic volcanic rocks were mainly formed, and they experienced multiple stages of tectonic movement. Controlled by an unconformity surface, the weathering and leaching reservoirs can be developed on a large scale, distribute in layers, and form a large uncompartmentalized stratigraphic reservoir. The large Carboniferous volcanic stratigraphic reservoir at the northwest margin of Junggar Basin is an example of a large uncompartmentalized stratigraphic reservoir.

Chapter 5

Areas of Exploration in Volcanic Reservoirs

Volcanic reservoirs have become one of the major exploration areas in China. Since 2002, major breakthroughs and significant advances in exploration in volcanic rock have been made in both eastern and western China. Giant gas provinces with reserves scales of $3000\text{-}5000 \times 10^8\,\text{m}^3$ have been preliminarily discovered in deep layers of the Songliao Basin and at Ludong-Wucaiwan in the Junggar Basin. Exploration has also shown a reserves scale of 100 million tons in the Santanghu Basin. These indicate that volcanic rocks, as an exploration area with special lithology reservoirs, have become significant to practically increase and replace hydrocarbon reserves in China and promise a broad exploration potential.

5.1 RESOURCE POTENTIAL IN VOLCANIC ROCKS

China has a wide distribution of volcanic rocks. Among the major hydrocarbon-bearing basins already discovered, the distribution area of payable volcanic rocks features three sets of strata worthy of exploration: the Carboniferous-Permian System, Jurassic-Cretaceous System, and Paleogene System. These strata have reached $36 \times 10^4\,\text{km}^2$, including $5.0 \times 10^4\,\text{km}^2$ in Songliao Basin and $2.0 \times 10^4\,\text{km}^2$ in Bohai Bay Basin in eastern China; $6.0 \times 10^4\,\text{km}^2$ in Junggar Basin, $1.0 \times 10^4\,\text{km}^2$ in Santanghu Basin, $2.0 \times 10^4\,\text{km}^2$ in Tuha Basin, $13 \times 10^4\,\text{km}^2$ in Tarim Basin, and $7.0 \times 10^4\,\text{km}^2$ in the Sichuan-Tibet region in western China.

Across China today, volcanic reservoirs have been discovered in a number of basins, including Songliao, Bohai Bay, Junggar, Santanghu, Hailar, and Erlian. However, the overall degree of exploration is very low. Based on preliminary predictions, the total oil resources in volcanic rocks are $19\text{-}26 \times 10^8$ tons. The natural gas resources are $4.2 \times 10^{12}\,\text{m}^3$, the proven rate of oil resources is 19-25%, the proven rate of natural gas resources is 2%, the total hydrocarbon equivalent is $52\text{-}59 \times 10^8$ tons, and the proven rate of resources is 6-7%. These numbers indicate that volcanic rocks have abundant resources and massive exploration potential and are an important reserves replacement area for current hydrocarbon exploration efforts.

For future volcanic rock exploration activities in two major volcanic gas provinces—the deep layers in Songliao Basin and the Carboniferous System in Junggar Basin—we should fully utilize new technologies to study secondary volcanic reservoirs such as dissolution type and fractured type, as well as payable reservoir bodies such as volcaniclastic rock type (explosive facies) and lava type (eruptive-effusion facies). Areas each with a reserves scale of 100 million tons should be established in the Santanghu and Bohai Bay Basins by reinforcing volcanic rock exploration therein. New breakthroughs should be made by actively exploring new volcanic rock areas, including the Carboniferous-Permian System in Tuha Basin, peripheral Carboniferous Basins in northern Xinjiang, Sichuan Basin, and Permian System in Tarim Basin and Ordos Basin.

In the Songliao Basin, the hydrocarbon gas reserves scale should be expanded by intensifying exploration efforts in structural high locations of the volcanic explosive facies, breaking through volcanic eruptive-effusion facies in structurally low locations, and concentrating on fault depressions, such as Xujiaweizi, Changling, and Yingtai. More reserves replacement areas should be discovered by actively exploring new fault depressions.

In the Junggar Basin, more reserves replacement areas should be discovered by actively exploring new fault depressions (sag) through strengthening the comprehensive evaluation of hydrocarbon accumulation conditions, reservoir distribution prediction, and determining targets in the Upper Carboniferous fault depression (sag), and expanding the reserves scale in areas of Junggar Basin. These areas would include Ludong-Wucaiwan, the northwest margin, Beisantai, and Chepaizi and should focus on weathered crust volcanic reservoirs.

In the Santanghu Basin, large-scale reserves areas should be established by focusing on the northeastern nose-uplift belt of the Malang depression based on integrated research, overall deployment, and step-by-step implementation. More reserves replacement areas should be discovered by reinforcing problem-tackling efforts and actively exploring other western depressions such as Tiaohu and Hanshuiquan, and the south margin piedmont thrust belt.

In the Bohai Bay Basin, new large-scale reserves should be established by a renewed commitment to understanding volcanic reservoirs, focusing on volcanic rocks in the Shahejie Formation of the hydrocarbon-rich depression, and primarily studying the northern part of the Huanghua

Volcanic Reservoirs in Petroleum Exploration. http://dx.doi.org/10.1016/B978-0-12-397163-0.00005-1

depression and the eastern depression of the Liaohe River. More reserves replacement areas should be discovered by actively breaking through new plays.

In the Tuha, Sichuan, Tarim, and Ordos Basins, as well as peripheral basins in northern Xinjiang, a comprehensive search for payable basins and layer series favorable for volcanic hydrocarbon accumulation must be made through integrated research and a renewed commitment to understanding volcanic rocks. Breakthroughs should be accelerated by focusing on overall research of the Carboniferous-Permian System in Tuha Basin; new breakthroughs should also be made by intensifying the research effort and preparing for distal volcanic hydrocarbon accumulation in other basins such as the Sichuan, Tarim, and Ordos.

5.2 VOLCANIC HYDROCARBON REGIONS IN EASTERN CHINA

These mainly include volcanic rocks developed in hydrocarbon-bearing basins such as deep layers in the Songliao (J-K$_1$), Bohai Bay (J-E), Erlian (K$_1$), and Hailar (K$_1$) Basins. The horizons of volcanic rocks include Mesozoic-Cenozoic strata and comprise volcanic rocks developed in intracontinental rift valleys within tensional environments. The distribution of volcanic rocks is related to major faults in fault depressions. The hydrocarbon accumulation assemblage in reservoirs is controlled by the development of the fault basin. The formation environments of volcanic rocks and source rocks are identical, their distribution basically overlaps, and they form self-generation-self-storage, proximal hydrocarbon accumulation assemblages. Because all epigenetic basin evolutions are different, the deep volcanic rocks in Songliao Basin mainly feature gas reservoirs, while oil reservoirs are dominant in other basins, including Bohai Bay, Erlian, and Hailar.

5.2.1 Songliao Basin

Deep layers in the Songliao Basin feature a total of 35 fault depressions, which mainly align in the northeastern direction. The central paleohigh has divided deep fault depressions into two fault depression belts, namely, the western belt and eastern belt, forming a northeast-trending structural framework of "one uplift and two depressions" (Figure 5.1). The eastern fault depression belt includes fault depressions such as Xujiaweizi, Yingshan, Wangfu, Dehui, and Lishu, where uplifting, denudation, and transformation occurred during later stages. The most strongly transformed area is located at the southeastern uplifted region, where only strata below the Qingshankou Formation are present today. The western fault depression belt consists of major fault depressions such as Changjiaweizi, Gulong, Changling, and Yingtai, which had subsided continuously during later stages and are overlain with extremely thick deposits of the depression period.

Deep layers in the Songliao Basin have two sets of regional mudstone cap rocks, namely, Member 2 of the Denglouku Formation and Members 1 and 2 of the Quantou Formation. As controlled by these cap rocks, three sets of source-reservoir-caprock assemblages have formed vertically. The first assemblage features Shahezi Formation-Yingcheng Formation as the gas source and matrix weathered crust, Huoshiling Formation and Shahezi Formation as the reservoir, and internal mudstone of the Shahezi Formation as the cap rock. The second assemblage features Shahezi Formation as the source rock, the volcanic rock and glutenite of the Yingcheng Formation as the reservoir, and the lake-facies mudstone of Member 2 of Denglouku Formation as the cap rock and is the major deep hydrocarbon-bearing assemblage. The third gas-bearing assemblage features Shahezi Formation as the source rock, the fluvial-facies sandstone of Members 3 and 4 of the Denglouku Formation as the reservoir, and the shore-shallow-lake-facies mudstone of Members 1 and 2 of the Quantou Formation as the cap rock.

The major deep source rock in Songliao Basin is the coal-bearing Shahezi Formation. Its total distribution area is approximately 3.0×10^4 km^2. The source rock area of the Shahezi Formation covers five major fault depressions, i.e., the Xujiaweizi, Changling, Gulong-Changjiaweizi, Yingshan, and Lindian depressions, and is approximately 2.5×10^4 km^2, accounting for 83% of the total area.

In 2002, a major breakthrough was achieved in the Xushen 1 well, succeeding in the endeavor to explore for major gas fields in deep volcanic rocks. Through reinforced exploration and prospecting efforts thereafter, a major discovery was achieved in venture well Changshen 1 in the south, creating a flurry of deep exploration activities across the south and north boundaries. So far, several hundred billion square meters of natural gas have been proven in deep layers of the Songliao Basin, and continuous breakthroughs are being made in new depressions.

Deep layers in the Songliao Basin feature abundant remaining resources and have a good basis for natural gas development. Based on preliminary predictions, the area covered by volcanic rocks is 11,300 km^2, promising a massive exploration potential (Figure 5.2). Future exploration activities should focus on hydrocarbon gases, where exploration of the volcanic explosive facies in structurally high locations should be intensified, and more exploration breakthroughs should be made in the volcanic eruptive-effusion facies occurring in structurally low locations. The reserves scale should be expanded based on fault depressions such as Xujiaweizi, Changling, and Yingtai; more reserves replacement areas should be discovered, and new breakthroughs should be made by actively exploring new fault depressions such as Yingshan and Gulong.

The area of the Xujiaweizi fault depression is 5350 km^2, including the Anda subsag, Xuzhong structural belt, and Xuxi, Xudong, and Zhaozhou subsags. In 2002, a major breakthrough was achieved in the Xushen 1 well, and

FIGURE 5.1   Distribution of fault depressions across the Songliao Basin.

thereafter exploration was accelerated. In 2005, $1019 \times 10^8 \, \text{m}^3$ of natural gas was proven in the Xushen gas field. In 2006, the preliminary prospecting and evaluation at the Xudong slope and Anda-Fengle area was further intensified, where many wells obtained commercial gas flows. Since 2007, new advances have been made and the proven reserves at Xujiaweizi as a whole will exceed several hundred billion square meters.

The area of the Changling fault depression is 7044 km^2. Integrated exploration and development efforts have been intensified since 2005 when a major breakthrough was

achieved in the venture exploratory well Changshen 1. As of 2007, the newly added 3P reserves (Proved, Probable, and Possible Reserves) in the Changshen gas field had exceeded 100 Bm^2. The Changling fault depression is, based on basal structural shapes and through-going characteristics and the interpretation of regional seismic facies and sedimentary facies, divided into five secondary structural units. These are the western slope belt, eastern slope belt, northern fault sag, southern fault sag, and central uplift zone. Located at the center of Changling fault depression, the central uplift zone is an uplift developed over the long

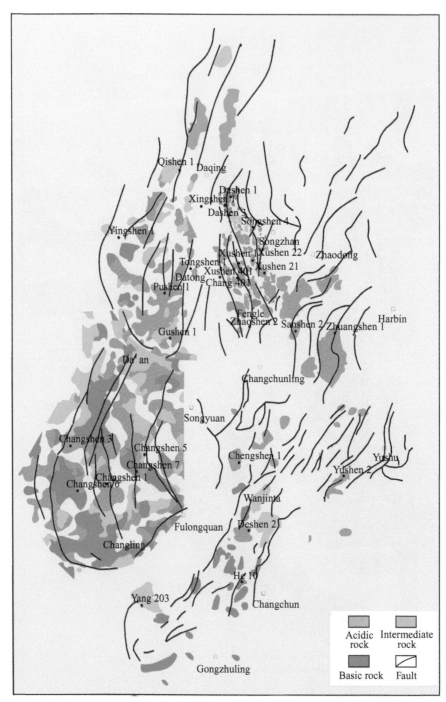

FIGURE 5.2   Distribution of volcanic rocks and faults in the Songliao Basin.

term and is the destination of long-term migration of hydro-carbons in the southern and northern secondary depressions. Volcanic activities were frequent during the fault depression period, and volcanic massifs such as Haerjin, Shenbei, and Qianshenzijing were developed. Epigenetic faulting activities in the fault sag and its eastern areas are not significant, which are mainly the development regions of hydrocarbon gases.

5.2.2   Bohai Bay Basin

Bohai Bay Basin is an intracontinental rift valley basin formed by subduction of the Pacific plate along the north-northeast direction. The major oil source beds feature the development of a large amount of volcanic rocks, such as Members 1 and 3 of the Shahejie Formation. The volcanic rocks and relevant magmatism are important to generating hydrocarbons, developing reservoir pores, and trapping the accumulated hydrocarbons, leading to the formation of a huge number of reservoirs related to volcanic rocks. For example, the Oulituozi oil field in the middle member of the eastern depression of Liaohe Basin features trachyte and intrusive rocks as its main reservoirs.

Within the 30 years from the first discovery of volcanic oil reservoirs to the present day, 30 volcanic oil reservoirs have been discovered in Bohai Bay Basin, promising reserves of nearly 100 million tons. Volcanic rocks are mainly distributed in the Mesozoic Erathem and Shahejie Formation, with a distribution area of approximately 11,000 km^2. The main volcanic reservoirs include (1) volcanic oil reservoirs in the eastern depression of Liaohe Basin, including Huangshatuo, Oulituozi, Rehetai, and Member 3 of the Shahejie Formation at Qinglongtai; (2) Mesozoic volcanic oil reservoirs in the western depression of Liaohe Basin, including Dawa, Niuxintuo, Huanxiling, and Shuguang; (3) the Gangxi buried hill and Fenghuadian Mesozoic volcanic oil reservoirs at the Qikou depression; (4) the volcanic oil reservoirs in Zaoyuan, Wangguantun (Mesozoic), Kongdian Formation, and Member 3 of the Shahejie Formation in the Cangdong-Nanpi depression; (5) the volcanic oil reservoir in Member 3 of the Shahejie Formation at Caojiawu in Langgu depression; and (6) the volcanic oil reservoirs in the Luojia and Shahejie Formations of Shang 741 in the Shengli oil field.

Four stages of volcanic rocks are mainly developed in Bohai Bay Basin: Kongdian Formation-Member 4 of Shahejie Formation, Member 3-Member 2 of Shahejie Formation, Member 1 of Shahejie Formation-Dongying Formation, and Guantao Formation. Mesozoic volcanic rocks are also present in the basin. As the basin evolved, the development of volcanic rocks migrated toward the center of the basin. Volcanic rocks are mainly distributed along three strike-slip fault zones, i.e., the Shenyang-Weifang (Tanlu) fault zone, Huanghua-Dezhou-Dongming fault zone, and Baxian-Shulu-Handan fault zone. Their lithology includes basalt, diabase, trachyte, and andesite.

The Mesozoic and Cenozoic volcanic oil reservoirs in Bohai Bay Basin will become one of the major targets of hydrocarbon exploration. The magmatic activity edifices in Mesozoic and Cenozoic rift valley environments across eastern China are mainly mid- and small-sized volcanic cones or intrusive massifs, where the area generally does not exceed 20 km^2. Among the Mesozoic and Cenozoic volcanic oil reservoirs of commercial recovery value, the oil reservoir with the largest proven reserves is the Huangshatuo trachyte oil reservoir, which has proven reserves of 1640×10^4 tons and where the scale of proven reserves in monolithic massifs is approximately $300\text{-}600 \times 10^4$ tons. In general, it features oil resources generated in mid- and small-sized lithosomes, where their scale is comparable to those of lithosome oil reservoirs in eastern

hydrocarbon-generation depressions. Oil reservoirs of such scale have economic recovery value in onshore old oil regions.

The next exploration step regarding volcanic rocks in Bohai Bay Basin should be to establish large-scale reserves by renewing our commitment to understanding volcanic reservoirs, focusing on volcanic rocks in the Shahejie Formation of the hydrocarbon-enriched depressions, the northern part of Huanghua depression, and the eastern depression of Liaohe River. Additional reserves replacement areas will be added by actively discovering new plays. The next-step key exploration areas include the Mesozoic Erathem and Member 3 of the Shahejie Formation at the middle and southern sections of the depression zone in the eastern Liaohe Basin, Members 3 and 4 of the Shahejie Formation along the northern section of the depression zone in the western Liaohe Basin, Member 3 of the Shahejie Formation at Jiannan-Xingang of the Huanghua depression, Member 3 of the Shahejie Formation-Kongdian Formation along the Wangguantun-Zaoyuan structural belt, Member 1 of the Shahejie Formation along the Beibao-Nanbao #5 structural belt, and Member 3 of the Shahejie Formation along the Liuquan-Caojiawu structural belt in the Jizhong depression.

5.3 VOLCANIC HYDROCARBON REGIONS IN NORTHERN XINJIANG

Regions in northern Xinjiang where volcanic rocks are distributed mainly include Carboniferous-Permian Systems in basins such as Junggar, Santanghu, and Tuha, which were mainly formed during the active stage of the Xingmeng trough. Due to intensive reconstruction of primary basins during later stages, the hydrocarbon accumulation assemblage changed dramatically, which includes both proximal assemblages, such as the Ludong-Wucaiwan region in Junggar Basin and the Malang depression in Santanghu Basin, and distal hydrocarbon accumulation assemblages, such as the northwest margin of Junggar Basin. Hydrocarbon distribution is mainly controlled by unconformities and faults.

5.3.1 Junggar Basin

As a sedimentary basin developed over the Junggar massif, Junggar Basin features a two-layer structure, comprising Precambrian crystalline basement and Caledonian-Hercynian fold basement, being a typical median-massif composite superimposed basin. Volcanic rocks in this basin were mainly developed during the basin-forming period of the sedimentary basin. They were formed via large-scale eruptions along faults during the tectonic relaxation period after the stage of collision and fold orogenesis around the Zhayier, Kelameili, and Bogda-Yilinheibiergen faulted depression trough surrounding the Junggar terrane. These volcanic rocks mostly feature calcalkalic bimodal characteristics of an intracontinental collision and intracontinental rift valley and are distributed along fundamental fractures. Volcanic rocks were mainly formed in two structural environments, namely, collision island arc and intracontinental rift valley during the collision relaxation period.

Deposited on the Hercynian fold basement, Junggar Basin mainly features the development of an intracontinental rift valley volcanic rock series of a collision relaxation period during the Upper Carboniferous epoch (Batamayineishan Formation C_2b). Volcanic rocks are mainly distributed along inherited thrust belts and uplift zones. Volcanic reservoirs are distributed closely around Upper Carboniferous (and Lower Carboniferous) and Permian hydrocarbon-generation centers, forming self-generation-self-storage and Cenozoic-generation Mesozoic-storage hydrocarbon accumulation assemblages. The volcanic exploration activities in Junggar Basin have also undergone the three major phases of accidental discoveries, key breakthroughs, and overall discoveries. As exploration technologies continuously advance, new breakthroughs will continue to be achieved in exploration efforts in the area of basinal Carboniferous volcanic rocks.

Since 2005, it has been fully recognized that Carboniferous volcanic rocks in Junggar Basin have natural gas exploration potential. Aiming at the key plays of natural gas exploration in the Ludong-Wucaiwan region, the following work has been conducted effectively: (1) the primary basin at the foothill belt of Kelameili has been restored and the overall exploration potential of the Carboniferous System has been evaluated; (2) the challenges in protection technologies for drilling in volcanic gas-bearing zones have been tackled; (3) seismic problems have been addressed, the 2D seismic problem-tackling framework survey grid has been established, and the vertical development sequence and superimposition structure of volcanic rocks as well as the lateral variations of volcanic massifs and lithofacies have been determined; and (4) lateral prediction of volcanic reservoirs has been conducted by comprehensive geophysical prospecting interpretations. At the same time, important breakthroughs have been achieved successively in the Dixi 14, Dixi 17, Dixi 18, Dixi 21, and Cai 54 wells in the Ludong-Wucaiwan region by focusing on large-scale stratigraphic reservoirs as main targets, demonstrating that China's sixth major onshore gas area has preliminarily reached a large scale. Meanwhile, important advances have also been made in oil exploration at the fault zone along the northwest margin and in volcanic rocks at Beisantai.

5.3.1.1 Self-Generation–Self-Storage Hydrocarbon Accumulation Assemblages

Multiple hydrocarbon-generation centers and multiple hydrocarbon-bearing systems have been formed and are compositely superimposed on the plane because of the quite strong separability of the Junggar Basin during the late Hercynian. The effective hydrocarbon accumulation in Upper Carboniferous volcanic rocks depends on the contribution of developed Upper Carboniferous source rocks and their adjacent Permian source rocks. The northwest margin is located at the collision island arc belt, where the potential in Carboniferous source rocks has not been proven so far, and it is believed that the hydrocarbon supply mainly came from Permian source rocks. Upper Carboniferous source rocks were developed effectively in intracontinental rift zones after the collision period, including Luliang uplift, Wucaiwan depression, and the central uplift zone, which mainly feature Type II_2 and Type III kerogen as their gas source (Figure 5.3). Upper Carboniferous source rocks have become a set of effective source beds in Junggar Basin. Therefore, profitable plays have developed: the Carboniferous-Permian volcanic rocks in the fault zone along the northwest margin have formed a Cenozoic-generation Mesozoic-storage hydrocarbon accumulation assemblage around the Mahu Permian hydrocarbon-generation depression; the Carboniferous volcanic rocks in the Luliang uplift-Wucaiwan depression have formed a self-generation-self-storage hydrocarbon accumulation assemblage around the Dishuiquan depression, Dongdaohaizi northern depression, and Wucaiwan depression; the Carboniferous volcanic rocks in the Santai and Beisantai uplift zone have formed a Cenozoic-generation Mesozoic-storage hydrocarbon accumulation assemblage closely around the Permian hydrocarbon-generation center in Fukang depression; and the Carboniferous volcanic rocks in the Jimusaer depression have formed a self-generation-self-storage hydrocarbon accumulation assemblage in the depression.

Hydrocarbon exploration in the Ludong-Wucaiwan region has shown that another set of source rocks in the Lower Carboniferous Dishuiquan Formation (C_1d) was deposited in the Dishuiquan depression in the Ludong region along the eastern section of the Luliang uplift, and at the adjacent Dongdaohaizi northern depression and Wucaiwan depression to the south, which is described as a set of coal-bearing terrigenous clastic layer series of transitional facies. The Dishuiquan Formation features a set of dark gray mudstone, irregular interbedded carbonaceous mudstone, and thin intercalated coal layers and coal seams, and in the center, it features interbedded intermediate-basic volcanic lava, volcanic breccia, and volcaniclastic rock. It outcrops along the foothill belt at Kelameili. The basin mainly features deposits of short-term tensional rift valleys and faulted depressions after the collision period in the Dishuiquan depression along the eastern section of the Luliang uplift as well as in the Dongdaohaizi northern and Wucaiwan depressions along the fault footwall of the Di'nan uplift. Wells Lunan 1, Sancan 1, Dibei 1, Dixi 17, Caican 1, and Caishen 1 have drilled through them, exposing source rocks between 50 and 500 m.

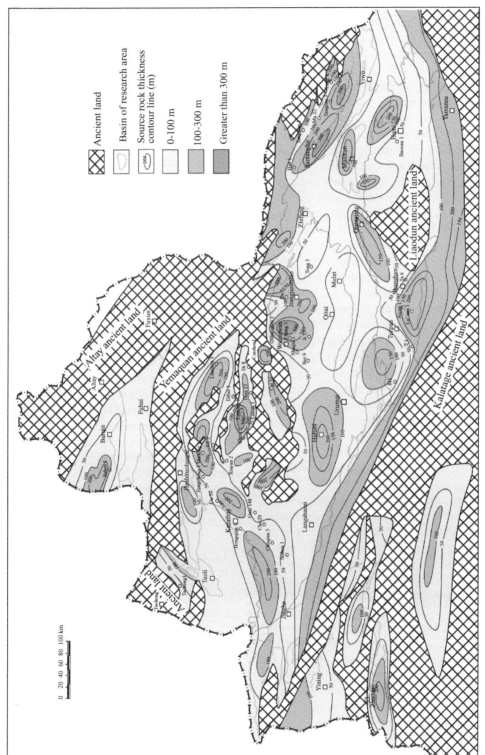

FIGURE 5.3 Isopach map of the Upper Carboniferous argillaceous source rock in northern Xinjiang.

5.3.1.2 Major Carboniferous oil and Gas Fields; Low Exploration Effort to Date

Carboniferous volcanic rocks in Junggar Basin are widely distributed and developed along fundamental fracture zones. They are mainly distributed along fault zones at the northwest margin, piedmont forelands at Kelameili, and the Zhundongzhangbei uplift zone and central uplift zone. The distribution area of volcanic rocks is up to 22,500 km², and the area of profitable exploration plays in volcanic rocks is 8300 km². Volcanic rocks along the northwest margin were mainly formed in structural environments of active continental margin island arc belts and feature the development of continental margin island-arc basic-intermediate acidic volcanic rock assemblage sequences. The volcanic rocks in the Luliang-Wucaiwan region and the central uplift zone were mainly formed in tensional rift valley structural environments after the collision period and feature intracontinental rift-valley intermediate-basic intermediate-acidic volcanic rock assemblage sequences. Today, proven oil reserves are mainly concentrated at the upper wall of the northwest margin fault zone and Shixi uplift, while proven natural gas reserves are mainly concentrated at the lower wall of the northwest margin fault zone and in the Ludong-Wucaiwan region. Massive hydrocarbon exploration potential is present because the overall exploration effort to date has been very low.

The remaining oil resources in volcanic rocks across Junggar Basin are mainly distributed along the northwest margin fault zone and Zhundongzhangbei fault fold zone and, mostly, fall within the profitable exploration category in mature regions. Recently, good progress has been made in the Carboniferous System at the Ke-Bai fault zone, Carboniferous System at the Hong-Che fault zone, and at the Beisantai uplift.

The Hong-Che fault zone is located in the southern section of the northwest margin fault zone and close to the Shawan depression and has abundant hydrocarbons. New exploration progress is being made continuously in the Neogene, Paleogene, Cretaceous, Jurassic, and Carboniferous Systems, demonstrating this zone's composite oil-bearing characteristics in a multiple layer series. The Carboniferous System is the important layer series to explore because its hydrocarbon is distributed in a south-north direction along major faults and it features a huge number of oil reservoirs with syngenesis of oil and gas. Offering a large reserve of drilling and exploration targets, the Carboniferous System in this zone is another key play and target layer series for profitable exploration in the fault zone along the northwest margin.

By strengthening our comprehensive evaluation of weathered crust reservoirs and their hydrocarbon accumulation conditions, reservoir distribution predictions, and target determinations in the Upper Carboniferous fault depressions (sags), and through expanding the reserves scale in the Ludong-Wucaiwan region, more natural gas reserves replacement areas should be discovered. The Carboniferous volcanic rocks developed in the Ludong-Wucaiwan region in the middle of the basin are mainly distributed along fundamental fractures and have formed three profitable volcanic development zones in a nearly east-west direction: Dibei uplift, Di'nan uplift, and Wucaiwan depression, where the total distribution area of volcanic rocks is at 2800 km². At the Kelameili gas field, the present proven GOIP has exceeded 100 Bm². Moreover, further exploration and breakthroughs are needed in the Zhundong region and depressions such as Sannan, Yingxi, Ulungur, and Mahu.

5.3.2 Santanghu Basin

Santanghu Basin abuts Junggar Basin to the west, Tuha Basin to the south, and Mongolia to the northeast. Moving in a northwest-southeast direction, it has an area of 2.3×10^4 km² and a scale of discovered reserves of more than 3×10^8 tons.

Santanghu Basin had frequent tectonic movements and development of unconformity surfaces because of its special structural location; namely, it is located at the southwest margin of the Siberian Plate and closely abuts the Kelameili-Maiqinwula stylolite. As affected by the bidirectional subduction of the Ebnur Hu-Shaquanzi and Kelameili-Dahei Shan suture zones, the main body of the Neopaleozoic Basin at Santanghu is characterized by the development and evolution of back-arc basins and intercontinental rift valleys. Since its initial formation, the basin has undergone a total of six tectonic movements. The late Hercynian, late Indosinian, and late Yanshanian activities were the three largest scale tectonic movements, having affected the entire Santanghu region. Both controlled the sedimentary framework and structural formation in the basin and created conditions for hydrocarbon migration and accumulation.

5.3.2.1 Well-Developed Carboniferous Source Rocks

Drilling and outcrops in Santanghu Basin have proven that the Carboniferous System developed two sets of source rocks, a Lower Carboniferous series and an Upper Carboniferous series, with a cumulative thickness between 350 and 1200 m and a wide range of distribution. Featuring dark-colored mudstone and gray tuffaceous mudstone, the source rocks are comprehensively evaluated as fairly good to good oil source rocks.

The Lower Carboniferous source rocks feature dark-colored mudstone layers and oil shales of marine facies, great thicknesses, and points of outcrop spreading all over the perimeter of the basin, demonstrating a quite high exploration potential. In the Dahei Mountain region, for example, the effective source rocks are mainly tuffaceous mudstone, mudstone, and graphitized oil shale, with a total thickness of up to 688 m. Becoming increasingly thicker from north to south, the source rocks are the best developed in the Malang and Naomaohu depressions. Source rocks as thick as 48 m had been penetrated in the upper part of the Jiangbasitao Formation in the Fang 1 well, while major source rock intervals in the middle section have not yet been drilled.

Upper Carboniferous source rocks mainly comprise oil shale, dark gray mudstone, marl, black carbonaceous mudstone, and coal beds. The source rock thickness is from 9 to 66 m as revealed by a total of 12 wells, and the speculated maximum thickness is 100-300 m. The entirety of source rocks becomes increasingly thicker from northeast to southwest, and the thickness of oil shale in the main depression areas is greater than 30 m.

5.3.2.2 Wide Distribution of Carboniferous Weathered Crust Volcanic Reservoirs

Santanghu Basin is a sedimentary basin developed over the Devonian-Lower Carboniferous fold basement, where a set of Upper Carboniferous-Lower Permian clastic rock and volcanic rock of continental-oceanic alternative facies were deposited. The Carboniferous System is widely distributed in the basin and has been penetrated in three exploratory wells and on the profiles of eight field outcrops. From bottom to top, the Carboniferous is divided into the Jiangbasitao, Batamayineishan, Haerjiawu, and Kalagang Formations. The Upper Carboniferous Dahei Mountain group features marine facies primarily, is distributed in the southeast area of the basin, and comprises a total of 1100 m of clastic and carbonate rock (limestone). The Carboniferous Kalagang Formation (C_2k) features a set of extremely thick intermediate-acidic volcanic rocks, the top of which is intercalated with a small amount of normal clastic rocks and where phytolite is produced. As inferred from seismic data, the Upper Carboniferous-Lower Permian series is widely distributed in the basin, and the Carboniferous features great thicknesses; the cumulative thickness of volcanic rocks is 4000-6000 m, and the maximum thickness can exceed 7000 m.

The volcanic rocks in Santanghu Basin have been significantly weathered and denuded, providing conditions for the formation of weathered and leached reservoir bodies. Since the late Hercynian Period, the northeast area of the Malang depression was continuously elevated, and volcanic rocks were exposed in outcrop long term. Significant reservoir and permeation spaces such as dissolved pores, recrystallized pores, and fractures were formed by weathering, denudation, and leaching. Thus, a weathered and leached dissolution zone distributed along stratigraphic unconformity surfaces and northeast fault zones formed, and the zone is developed in the south-north direction mainly along the Upper Permian denudation line. This weathered and denuded zone entirely controlled a huge area at Niudong and Mazhong, allowing the reservoir capacity of volcanic rocks at the eastern Malang depression to be dramatically improved. Stratigraphic unconformity set the foundation for the formation of volcanic reservoirs by controlling both the range of the weathering and leaching zone and the distribution of high-quality volcanic reservoirs.

5.3.2.3 Good Hydrocarbon Accumulation Assemblages

Santanghu Basin features the development of proximal hydrocarbon accumulation assemblages between volcanic rock and sedimentary rock. Three types of reservoirs—volcanic rock, limy dolomite, and clastic rock—are developed in the Carboniferous-Permian volcanic rock series. Composite hydrocarbon-bearing areas in the volcanic rock series have been formed within a profitable structural setting by communicating oil sources through faults. The two kinds of hydrocarbon accumulation assemblages that may be developed are discussed below.

5.3.2.3.1 Volcanic Reservoirs (Carboniferous System-Lower Permian Series)

Volcanic rocks with relatively favorable reservoir conditions in a volcanic rock series such as vesicular basalt and volcanic breccia can develop if fractures are present or weathered crust conditions are available. The favorable reservoir bodies can include a volcanic rock weathered crust reservoir, which may primarily feature stratigraphic unconformity trap reservoirs within structural settings because of the interbedded distribution of vesicular volcanic rocks, and a fractured reservoir in volcanic rock, which is mainly distributed beneath Upper Permian-Triassic regional mudstone cap rock and may also exist at the bottom of the Lower Permian series.

5.3.2.3.2 Possible Reservoirs with Conventional Reservoir Beds (Carboniferous-Permian Systems)

The Upper Carboniferous volcanic-sedimentary facies belt of continental facies or transitional facies may accumulate hydrocarbons where conventional reservoir bodies of the clastic rock sedimentary system are developed in the volcanic layer series. Secondary reservoirs may also be formed in the Mesozoic-Cenozoic continental-facies clastic rock layer series overlying the Carboniferous-Permian volcanic rock layer series.

The hydrocarbon accumulation models in different regions are different, as they are affected by stratigraphic characteristics such as stratigraphic distribution, sedimentation, and structure. At the southwestern Malang depression and the southeastern Tiaohu depression, the development of a Carboniferous-Permian sequence is relatively complete and mainly features composite trap reservoirs of the tectonic-igneous rock (fracture-vesicle) type. At the same time, the Upper Permian continental-facies clastic rock sedimentary system is fairly highly developed, which may be a profitable region within the conventional clastic rock reservoirs. The Carboniferous-Permian System at the northeastern Malang depression has been denuded to different extents, mainly featuring igneous rock-weathered crust composite reservoirs.

5.3.2.4 Rich Oil and Gas Resources

In 2006, in the Tuha oil field we primarily conducted (1) high-accuracy gravity-magnetic geophysical prospecting reprocessing; (2) seismic problem tackling in deep layers of the foothill belt, and determination of deep structural sequences and volcanic massifs; (3) comprehensive interpretation of geophysical prospecting data, and prediction of volcanic lithofacies and reservoir development zones; and (4) research and overall evaluations on the relationship between deep and shallow hydrocarbon accumulation assemblages. In 2006, high-yield hydrocarbon flows were obtained from the Upper Carboniferous Kalagang Formation (C_2k) of the Ma 17 well, as guided by the exploration effort of driving shallow explorations with deep outcomes and focusing on the area of deep lower-assemblage volcanic rocks. Since 2007, commercial oil flows have been obtained successively in multiple wells from the lower assemblage at eastern Malang, thus expanding the development of the Tiaohu depression, where the Niudong nose-uplift belt offered a total reserves scale of up to 2×10^8 tons. This demonstrates a good exploration prospect for lower-assemblage volcanic rocks in the Santanghu Basin.

During the Carboniferous Period, the Santanghu Basin was a monolithic back-arc basin, where extremely thick deposition of continental-oceanic alternative facies occurred. Carboniferous source rocks are developed in individual depressions, which offer abundant potential resources. Geologic conditions favorable to hydrocarbon accumulation similar to those at the Malang depression are present, making it a profitable exploration area.

Major breakthroughs have been made in hydrocarbon exploration of the Carboniferous System at the Niudong region in Santanghu Basin of the Xinjiang region, demonstrating superior hydrocarbon geologic conditions and great exploration potential in this region. In recent years, key discoveries have been made during the exploration of lower-assemblage Carboniferous reservoirs in the Niudong region in the Malang depression, further expanding its resources scale.

The next step in the exploration of Carboniferous volcanic rocks of the Santanghu Basin should be to identify large-scale reserves regions by focusing on the northeast nose-uplift belt of the Malang depression on the basis of integrated research and overall deployment. At the same time, more reserves replacement areas should be discovered by reinforcing problem-tackling efforts and actively exploring in the western depression and along the south margin piedmont thrust belt.

The northeast nose-uplift belt of the Malang depression offers a profitable exploration area of 1020 km². The Carboniferous Kalagang Formation mainly features the development of four stages of volcanic rocks. Volcanic massifs are mostly distributed near faults, and the volcanic rock series features broad distribution and great exploration potential. The breakthrough in the volcanic rock weathered crust reservoir at the top of the Carboniferous Kalagang Formation indicates that the enrichment of hydrocarbon in this area is closely related to the quality of volcanic reservoirs and that hydrocarbon mainly occurs in volcanic dissolved pores interconnected by fractures. Vertically, good reservoirs are distributed near unconformity surfaces and interfaces between volcanic rocks and normal clastic rocks. In map view, good reservoirs are distributed near overlying stratigraphic denudation zones or major faults. Analysis based on this indicates that the Chahaquan nose uplift and Niudong-Mazhong nose uplift are the major regions for expanded exploration of the uncompartmentalized oil field and offer great exploration potential.

Located at the main area of middle Permian-Carboniferous source rock sedimentation, the south margin piedmont thrust belt has an exploration area of 620 km² and good oil source conditions. Located at the middle Permian Lucaogou Formation and in the Carboniferous hydrocarbon-generation center, the Baiyishan thrust belt along the southwest margin of the Malang-Tiaohu depression comprises two structural belts, namely, Shibandun and Heidun. The area of the Shibandun structural belt at the Tiaohu depression is 460 km², where volcanic fractured reservoirs have been discovered in its Lower Permian series. The area of the Heidun structural belt at the Malang depression is 160 km², where fractured reservoirs have been discovered in its middle Permian series, but its Carboniferous System was not drilled through. Therefore, this region is a profitable play for executing preliminary prospecting in the lower-assemblage layer series and for evaluating oil reservoirs of the Lucaogou Formation.

5.3.3 Tuha Basin

The area of Tuha Basin is 5.3×10^4 km². Commercial hydrocarbon flows have been obtained in four sets of layer series, including the Triassic, Jurassic, Cretaceous, and Paleogene-Neogene Systems. Exploration and research activities as a whole in the Carboniferous System-Lower Permian series are just beginning, and the level of understanding is low. Therefore, the exploration potential is high.

5.3.3.1   Volcanic Rock Distributed in the Southern Basin

The change of Carboniferous-Lower Permian lithology and lithofacies in Tuha and its nearby basins is very abrupt. In general, belts of volcanic activity are present from south to north, including the south margin of Tuha Basin, Bogda. Its Carboniferous System has multiple sets of carbonate rock assemblages. Its Permian System features clastic rock primarily, which is intercalated with thin layers of tuffaceous mudstone and locally with thick layers of volcanic rock.

The Upper Carboniferous-Lower Permian series is widely distributed in the southern Tuha Basin. It is thick in the south and thin in the north, typically in the range of 1300-3100 m. Only 16 exploratory wells in this region have encountered this layer series, with a maximum thickness of more than 1200 m. In outcrop areas, however, Carboniferous System-Lower Permian series are widely developed, with great thicknesses ranging between 2000 and 2400 m, and the maximum thickness exceeds 1000 m. The formation in the lower assemblage at the south foreland features complex lithologies composed of interbedded clastic rock, carbonate rock, and volcanic rock.

The volcanic rocks in Tuha Basin are mainly distributed at the southern Tuha Basin, with an area of nearly $2 \times 10^4 \, km^2$. Two foreland basins—Turpan and Hami—are developed along the southern part of the basin, ranging from the Toksun depression at the west to Liaodun uplift at the east. Having an area of $1 \times 10^4 \, km^2$, the foreland basin in Turpan includes the Jueluotake thrust belt, central depression zone, and Huoyanshan low uplift, where the overall exploration effort is very low in the Carboniferous System-Lower Permian series. The Hami foreland basin includes the Xiao-huangshan thrust belt, Huoshizhen depression, and Erguquan low uplift, with an area of $8000 \, km^2$. The compression occurred nearly in a south-north direction during late Hercynian Period and helped form the foreland basin in south Kumul. By the time of the middle to late Triassic epoch, the southern foreland gradually died out. During the Jurassic Period, the northern foreland dominated the development of the basin.

5.3.3.2   Favorable Assemblage of Source Rocks and Volcanic Rocks

In the Tuha Basin, the source rock of the Upper Carboniferous-Lower Permian series, where it is mainly distributed along both flanks of the central uplift zone and is as thick as 400-800 m, forming two hydrocarbon-generation centers. Three sets of source rocks of the lower assemblage in the basin widely crop out around the perimeter of the basin. Lower Permian and Upper Carboniferous top source rocks have been encountered downhole at Tainan. Oil source rocks are also developed at the Lower Carboniferous Yamansu Formation where outcrops emerge. The

Upper Carboniferous source rock may be the major source rock of the basin. Centered on Toksun, Tanan, Taibei, and Sanbao, it is typically 400-600 m thick, with a maximum thickness of 700-800 m. The maximum thickness of the source rock as revealed in the Lunan 1 well is 714 m. Its lithology mainly features mudstone, silty mudstone, and tuffaceous mudstone, where only a small amount of limestone can become source rock. The source rock is comprehensively evaluated as fairly good to good, promising a high resource potential.

The configuration between volcanic reservoirs and source rocks is good. As inferred based on drilling, outcrops, and geotectonic settings, the Upper Carboniferous deposit in Tuha Basin can be divided into four kinds of facies regions and features development of three kinds of reservoir beds; namely, volcanic rock, clastic rock, and biolithite. Moreover, the Carboniferous System-Lower Permian series in most areas of this region have undergone long-term weathering and denudation; therefore, weathered crust has become an important type of reservoir body. Carboniferous-Lower Permian source rocks may have developed at the Takequan structural belt, which offers the basic geologic conditions for forming Carboniferous self-generation-self-storage reservoirs. The Takequan area is a profitable region for exploring the hydrocarbon accumulation conditions in Carboniferous-Lower Permian lower assemblages of superimposed basins and for actively preparing new exploration reserves replacement areas. The main target layer of this play may form structural and buried hill reservoirs. For example, 15 traps have been discovered at the Ludong and Dikaner structural belts in Tainan, which is a profitable target region.

In addition, a series of Carboniferous Basins are also developed along the periphery of northern Xinjiang, all of which have some resource potential.

5.4   OTHER POTENTIAL REGIONS

A significant amount of volcanic rocks is distributed in other hydrocarbon-bearing basins across China, including Sichuan, Tarim, and Ordos Basins. The overall degree of exploration and understanding level are very low, however, because no dedicated exploration has been conducted on target layers. Our next steps should be to strengthen and renew our understanding and overall evaluation of volcanic rocks in these basins, and comprehensively search for profitable regions and layer series of volcanic hydrocarbon accumulations.

5.4.1   Sichuan Basin

The tectonic movements in Sichuan Basin were polycycle and multistage. The Hercynian movement had a quite intensive taphrogen, forming the famous Emeishan basalt. As the product of basic magmatic eruptive-effusions in late

Permian continental rift valleys, the Emeishan basalt is composed of a set of basaltic volcanic eruption rocks. Its top is mainly black-gray and dark green-gray basalt, intercalated with dark purple and mauve tuff and mottled volcanic breccia. Its middle and base are green-gray, dark green-gray, and black-gray basalt, intercalated with a small amount of mauve tuff. Its base is an obvious division between dark green-gray basalt and dark gray limestone at Member 4 of the underlying Maokou Formation. Petrographically, it is classified as close to alkalic tholeiite, which is mainly distributed at the western part of the basin and has a maximum thickness of 350 m. It gradually becomes thinner, branched, and pinches out from the perimeter to the interior of the basin. After pinchout, the basalt gradually and laterally transitions into Lopingian coal measures.

The Zhougong 1 well drilled an explorative target layer on the Zhougongshan structure in September 1992, into the Sinian System. While freshwater was being produced in the Sinian and Lower Permian strata, a commercial gas flow of $25.61 \times 10^4 \, m^3 \, d^{-1}$ was obtained in tested Upper Permian basalt. This initiated volcanic exploration in Sichuan Basin and indicated that volcanic rocks can form good reservoir conditions and provide high yields. Currently, there are two wells (Zhougong 1 and Zhougong 2) in western Sichuan that have drilled through volcanic rocks. Drilling has shown that the major types of Emeishan basalt include porphyritic basalt, dense plagiobasalt, diabase, and volcanic breccia, intercalated with thin layers of sedimentary tuff. The types of reservoir spaces in basalt mainly include vesicles, intragranular dissolved pores, matrix dissolved pores, residual pores, and fractures, with surface porosity ranging between 0.1% and 1%.

The Zhouhui 1 well shows the development of microfractures. Based on SEM examination, fractures show a netlike or X-shaped occurrence and overlap each other, proving that this region has been affected by multiple stages of tectonic stress. Cycles 1-5 feature the highest fracture density, are usually fully filled or half-filled with organic matter, and serve as favorable reservoirs. Cycles 6-10 feature dense basalt, low fracture density, and relatively poor reservoir conditions. Based on statistics of the physical properties of cores from the Zhougong 2 well, the average porosity is 0.76%, average permeability is $0.164 \times 10^{-3} \, \mu m^2$, and reservoir conditions are poor. Preliminary research findings indicate that the distribution area of Upper Permian volcanic rocks in western Sichuan is approximately $6 \times 10^4 \, km^2$ and the thickness of volcanic rocks is 500-1000 m. Permian source rocks are developed in the Permian System in Sichuan Basin, which are the major hydrocarbon source beds deep in the basin. The source rocks have a typical thickness of 40-120 m, high percentage of organic carbon, hydrocarbon-generation intensity of between 10 and $30 \times 10^8 \, m^3 \, km^{-2}$, and superior

source rock conditions, and the volcanic rocks have hydrocarbon accumulation conditions. The natural gas resources are preliminarily estimated to be $3000\text{-}5000 \times 10^8 \, m^3$, offering some exploration potential.

5.4.2 Tarim Basin

Tarim Basin features the development of Permian basalt, which is mainly distributed east of the Keping-Pishan line and includes mottled clastic rock and basalt of continental facies. Along its western rim, the Upper Permian series features clastic rock of continental facies, and the Lower Permian series features neritic clastic rock and carbonate rock deposits intercalated with basalt.

During the early Permian epoch, large-scale magmatic eruptions occurred because of fault activity west of the Tarim platform and around the Bachu uplift, including Ketuer, Karayulgun, Aqia-Tumuxiuke, and Mazhatake. At this time, the activities of the Luntai fault and the Qiemo-Luobuzhuang fault east of the platform continued. Affected by this fault activity, the middle and top part of the Permian System show red formations of volcanic rock, volcaniclastic rock, and clastic rock but are basalt primarily. The distribution area of the volcanic rocks exceeds $8 \times 10^4 \, km^2$. The profitable exploration range is mainly concentrated at the northern part of the volcanic rock development region.

Research indicates that the lithology of the Permian volcanic reservoirs in Tarim Basin mainly features dacite and basalt. The maximum porosity is 14.2%, and average porosity is 6.77%; maximum permeability is $14 \times 10^{-3} \, \mu m^2$, and average permeability is $1.57 \times 10^{-3} \, \mu m^2$. The reservoir thickness is typically 50-70 m, and the reservoir conditions are medium to poor. Current exploration results indicate that the SG&O (oil and gas shows) of the oil trace-oil patch level is observed in volcanic massifs, as revealed in multiple wells such as S114-1, S79, T205, and T208. A small amount of crude oil has been produced from the Haitan 1 well. Lower-generation upper-storage conditions are present in the Paleozoic Erathem of the Tarim Basin, because of the development of multiple sets of source rocks and quite good supply-source conditions. Only SG&O is observed and no breakthrough has been made so far because the degree of exploration degree is low due to the sense that the development of Permian volcanic rocks would damage the reservoir. Given their wide distribution, Permian volcanic rocks offer some promise of hydrocarbon resources.

5.4.3 Ordos Basin

As affected by the basin-mountain coupling effect of the Qinling-Qilian orogenic belt and the Ordos Basin, the Indosinian tectonic and magmatic activities along the southwestern part of Ordos Basin were quite intensive, where subvolcanic massifs of certain scales and distribution ranges were formed. For example, the Longmen subvolcanic massif was confirmed by drilling. Preliminary research findings indicate that the distribution area of the profitable exploration region in the volcanic rocks southwest of the basin is up to 1200 km^2. It is inferred that this Indosinian thermal event promoted the maturation and retransformation of the basin's Lower Paleozoic source rocks and also promoted the large-scale migration and accumulation of the basin's regional hydrocarbons from southwest to northeast.

The Long 1 and Long 2 wells drilled at the southwestern part of the basin at the end of the 1970s both encountered magmatic rock. The highest horizon of its emplacement was the bottom of the upper Triassic Yanchang Formation, and it was inferred that it was formed during the Indosinian epoch. The magmatic massif at the Long 1 well features plutonic intrusion characteristics and very thick, massive shapes. Its thickness is 154.6 m, measuring from 1584 m to the bottom of the hole. Its lithology mainly features intrusive bodies of carbonated nepheline syenite. Its top shows a small amount of acidic magmatic rock debris, without porphyritic texture. The Long 2 well features hypabyssal veins and a widespread porphyritic texture. It shows quite significant changes in mineral and chemical components, mainly including nepheline orthophyre, malignite orthophyre, melanite orthophyre, sussexite, hornblende ivernite, and beschtauite. Emergence of the magmatic massif is observed in Taoshaoao, Tongcheng Xiang, and Chongxin counties, about 25 km northwest of Long 1 and Long 2 wells. Field geologic surveys during 2005 and 2006 confirmed that this region has three east-west-trending vein massifs that intruded into the Jixianian algal dolomite. The vein body features alkalic-intermediate quartz orthophyre and alkalic pseudoleucitophyre orthophyre. We believe after analyses and comparison of lithology, intrusive horizons, and rare earth elements that the Tongcheng and Longmen massifs have similar genetic backgrounds and both are products of the Indosinian epoch. In addition, the Longmen region in the southwestern part of the basin shows obvious aeromagnetic and gravity anomalies. The aeromagnetic anomaly has an amplitude as high as 200 nT and is distributed in a nearly east-west direction in an elliptic, netlike shape, indicating that this region may have a large-area distribution of concealed magmatic massifs.

During drilling of the Long 2 well, the 3322- to 3324.64-m well interval encountered diorite-porphyrite, where the drilling rate accelerated significantly and 52 m^3 of mud was lost. An obvious gas logging anomaly was observed from 3430 to 3442 m, where the gas logging peak value was 4.5445% and the base value was 0.0895%.

The Shanxi Formation in the basin features the deposition of effective coal-measure source rocks and locally contacts with volcanic veins directly. Its volcanic rocks have a certain reservoir capacity. Whether or not effective hydrocarbon accumulation is present, however, still awaits further research.

Chapter 6

Prediction and Evaluation Technology for Volcanic Rock

Volcanic rock has special geophysical properties, such as strong magnetism and resistivity, and high density. Therefore, gravity, magnetic, and electrical nonseismic prospecting techniques play an important role in the comprehensive prediction of volcanic rock, which has already been proven to be effective in overseas and domestic experiences of volcanic rock exploration for half a century. In this chapter, we summarize the integrated gravity, magnetic, electrical, and seismic prospecting as a series of four steps, based on the volcanic hydrocarbon exploration in Songliao Basin and Junggar Basin. First, distributed provinces of volcanic rock are predicted based on gravimetric and magnetic data to determine target areas; a comprehensive evaluation is conducted based on seismic, electrical, drilling, logging, and geologic data to further confirm target areas. Second, seismic data are taken as the primary dataset combined with electrical and other data to describe the shape of the volcanic rock and predict the favorable target zones. Third, the volcanic reservoir is predicted using seismic attribute analysis, coherence volume, and poststack seismic inversion on the basis of seismic data combining with logging data. Fourth, fluid detection is conducted based on seismic or logging data to determine drilling targets with lower exploration risk. These four steps are generally mature techniques.

6.1 CURRENT VOLCANIC ROCK EXPLORATION TECHNOLOGIES

6.1.1 Gravity, Magnetic, and Electrical Prospecting

Conventional gravity, magnetic, and electrical prospecting played an important part in the early prospecting of oil and gas. But they were overshadowed with the rapid development of seismic prospecting. Since the 1980s, however, these nonseismic, comprehensive geophysical prospecting methods have developed at a rapid rate. The separation and inversion of gravity and magnetic anomalies are now regarded as the two most important aspects of nonseismic technologies, both at home and abroad.

For the separation of gravity and magnetic anomalies, polynomial fitting, position-variable filtering, circular averaging, and interpolation cut are commonly used in the spatial domain; Wiener filtering, matched filtering, and optimum linear filtering are used in the wave-number domain. But in practical application, polynomial fitting is accessible to human errors. Phase-variable filtering is a kind of nonlinear filtering. It is difficult to choose the point location and point number automatically. The Wiener filtering, matched filtering, and optimum linear filtering have high computation speed but low accuracy, so they pose problems for quantitative computation.

In China, gravity and magnetic anomaly inversion were realized through the regularization method and singular value decomposition (SVD) introduced from the former Soviet Union at an early date. Since the 1980s, some methods that are generally used in the West have been employed in anomaly inversion, such as convolution, Euler deconvolution, analytic signal approximation, and continuous wavelet transform.

Most inversions, especially those developed since the 1960s, have to be realized by solving large or even super-large systems of equations, which makes it difficult for a real inversion process to arrive at a stable solution. So those practically effective inversion approaches have long been major and challenging interests to researchers.

From the perspective of application conditions and objectives, in modern gravity and magnetic data processing and interpretation, great importance has been placed on rapid inversion approaches because of the large data volume and requirements on inversion speed, leading to the extensive application of the Euler deconvolution and analytic signal approximation. On the other hand, the Euler deconvolution is based on the Euler equation, and the Euler structure index must be known or already established. In addition, the potential field as well as its first-order derivative should be known in advance. Even a modified Euler deconvolution would arrive at an approximate solution on the condition that the correlation coefficient between the resultant field and background field reaches the minimum. This is a disadvantage to the Euler deconvolution.

Volcanic Reservoirs in Petroleum Exploration. http://dx.doi.org/10.1016/B978-0-12-397163-0.00006-3

In analytic signal approximation, only the modulus of the analytic signal, instead of its modulus and phase together, is included in the approach, which has limited its application to a great extent.

Another approach to potential field inversion since the 1960s has used the Fourier transform dealing with the scale properties of the potential field. Fourier transform has been applied extensively to potential field data processing, e.g., potential field conversion, filtering, and identification of source features.

Wavelet transform has been used in data processing recently. Compared with Fourier transform, which is only capable of global analysis, wavelet transform is superior in its capability of local field analysis of potential as well as noise depression, which cannot be realized with conventional methods such as local Euler deconvolution. Even though they have high computation speed, Fourier transform and wavelet transform fail to output a unique solution with considerable reliability, which is attributed to the difficulties in applying effective prior information and constraints to inversion. Therefore, we must to combine the outputs with other data for effective interpretation.

The magnetotelluric method and electromagnetic sounding in the frequency domain have been adopted in electrical prospecting in the West as well as in China, while the magnetotelluric method and electromagnetic sounding in the time domain have been adopted in the Countries of Independent States (CIS). At present, there are various types of data processing software introduced or self-developed with large differences in outputs in China. Therefore, special attention should be paid to the comparison of different electrical data processing and interpretation software for a good selection.

In summary, gravimetric, magnetic, and electrical data processing and interpretation could be carried out in both the time-space domain and the wave number-frequency domain, the former of which features high precision and resolution and the latter features high computation speed and is suitable for qualitative evaluation. In exploration activities at home and abroad, these two kinds of techniques can be used successively in data processing. Methods in wave number-frequency domain would be employed first for qualitative evaluation of an anomaly distribution, and then in accordance with a specific anomaly area, methods in the time-space domain would be used to conduct quantitative analysis, which is helpful to improve the accuracy and effectiveness of data processing and interpretation.

6.1.2 Seismic Prospecting

Compared with sedimentary rock, volcanic rock features high seismic wave velocity, magnetic susceptibility, and resistivity, as well as large density variation and absorption of seismic energy, which lay the foundation for volcanic rock characterization with geophysical techniques. At present,

geophysical techniques have been extensively applied to volcanic rock identification, volcanic lithofacies classification, and volcanic reservoir characterization in foreign countries. The overall volcanic rock is identified with the comprehensive use of gravimetric, magnetic, electrical, and seismic data. Based on the identification result, seismic waveform classification and attribute analysis, together with logging facies, are used to classify a volcanic rock facies belt. Furthermore, the reservoir prediction technique is applied to volcanic reservoir characterization and fluid detection.

In volcanic reservoir exploration in China, gravimetric, magnetic, and electrical technologies, along with seismic technologies, have been used to describe the overall configuration and distribution of volcanic rock. These techniques have already been applied to the exploration of Songliao Basin and Junggar Basin. Poststack seismic attribute analysis and inversion, aimed at lithofacies and reservoir prediction, have been applied to exploration in the east. For volcanic prospecting in the west, current data quality does not satisfy the need of reservoir characterization because of complex subsurface geologic conditions and poor seismic imaging of deep formations. Effective volcanic reservoir characterization and fluid detection have mainly dealt with poststack and prestack seismic data and technologies at present.

6.2 PROCEDURES AND APPLICATIONS OF VOLCANIC ROCK CHARACTERIZATION TECHNOLOGY

6.2.1 Distributive Province Prediction

6.2.1.1 Foundation of Technical Applications

Predicting volcanic rock distribution with magnetic data is based on the strong magnetic properties of the volcanic rock, as shown in Table 6.1. The magnetic susceptibility decreases gradually with lithology change from intermediate and basic rock, to volcanic breccia, andesite, dacite, tuff, and then to conglomerate, sandstone and mudstone.

Lithology prediction with gravimetric data is based on the density variation of different volcanic rocks. From Table 6.2, rock density decreases with lithology change from diabase, basalt, mudstone, sandstone, conglomerate, andesite, volcanic breccia, dacite, tuff, and felsite.

The resistivity of volcanic rock is high in general, which makes it possible to identify volcanic rock with electrical data. From Table 6.3, volcanic breccia has the highest resistivity of 2450 Ω m on average.

6.2.1.2 Technologies

Gravity and magnetic exploration detect volume information. Gravity and magnetic anomalies are comprehensive

TABLE 6.1 Magnetic Susceptibility of Volcanic Rock in Junggar Basin (Units: 10^{-5} SI)

Lithology	Core from Luxi-Mobei		Core from Di'nan-Baijiahai		Outcrop in Kelameili		Composite susceptibility
	Sample number	Average susceptibility	Sample number	Average susceptibility	Sample number	Average susceptibility	
Conglomerate	22	150	5	34	50	42	77
Mudstone	14	40	21	39			39
Sandstone	1	17	22	46	45	29	40
Tuff	21	85.3	35	34			85
Volcanic breccia	7	455	54	675	30	75	454
Andesite	84	57	42	366	65	258	205
Dacite	21	175					175
Basalt	28	727	41	600	30	1160	850
Diabase	17	813	30	750	30	730	810
Rhyolite			6	10	60	108	108
Felsite	10	25	29	32	30	12	30

TABLE 6.2 Volcanic Rock Density, Junggar Basin (Units: g cm^{-3})

Lithology	Core from Luxi-Mobei		Core from Di'nan-Baijiahai		Outcrop in Kelameili		Composite density
	Sample number	Average density	Sample number	Average density	Sample number	Average density	
Conglomerate	22	2.4	5	2.555	50	2.663	2.56
Mudstone	14	2.585	21	2.527			2.58
Sandstone	1	2.576	22	2.542	45	2.716	2.57
Tuff	21	2.422	26	2.442			2.44
Volcanic breccia	7	2.535	41	2.545	30	2.672	2.53
Andesite	84	2.383	32	2.629	65	2.66	2.54
Dacite	21	2.43					2.5
Basalt	28	2.695	35	2.701	30	2.75	2.7
Diabase	17	2.708	26	2.747	30	2.66	2.71
Rhyolite			6	2.523	60	2.525	2.52
Felsite	10	2.231	21	2.409	30	2.554	2.41
Granite					30	2.643	2.52
Diorite porphyrite					30	2.76	2.65

additive effects of various field sources from the surface to the subsurface. When gravity and magnetic data are used to recognize geologic targets, the data shall be transformed to highlight and separate the information from different field sources.

Gravity anomalies are complicated because of the density boundaries of sedimentary cap rocks, internal structures, relief of the sedimentary basement, and Mohorovicic discontinuity. Magnetic anomalies may also be complicated by shallow volcanic rocks, intrusive bodies, heterogeneous

TABLE 6.3 Volcanic Rock Resistivity, Junggar Basin (Units: Ω m)

Epoch	Lithology	Sample number	Resistivity			Epoch	Lithology	Sample number	Resistivity		
			Max.	Min.	Average				Max.	Min.	Average
K	Sandstone	2	564	263	414	C	Sandstone	3	1084	296	690
J	Mudstone	1	635	635	635	C	Conglomerate	2	2336	2287	2310
J	Sandstone	37	1355	186	690	C	Felsite	5	1564	376	827
J	Conglomerate	2	646	635	641	C	Tuff	10	3608	275	1450
T	Mudstone	4	730	339	588	C	Andesite	11	3246	822	1800
T	Sandstone	1	527	527	527	C	Volcanic breccia	9	4480	1918	2450
P	Sandstone	8	681	559	620	C	Volcaniclastic rock	3	3161	923	1985
P	Mudstone	7	879	275	552	C	Diabase	3	3402	2436	3200
C	Mudstone	2	846	841	843	C	Basalt	6	4772	1745	3353

magnetic bodies, magnetic sedimentary rocks, magnetic basement relief, and Curie surfaces. Therefore, gravity and magnetic data should be processed and interpreted according to different geologic tasks and targets.

In gravimetric and magnetic data interpretation, gravity anomaly measurements have often been regarded as the composite of regional anomaly and local anomaly, the former referring to that resulting from deep geologic agents with large distribution and the latter referring to that resulting from geologic agents with relatively small distribution (such as geological structure, ore body, or rock mass) and low magnitude. The residual anomaly, derived by subtracting the regional anomaly from the total measurement, is considered the local anomaly. If we need information from a certain source, the first step is to extract the anomaly purely caused by this geologic agent from the total measurements. Without separation, it is impossible to get information about a single target we are interested in, let alone do interpretations and quantitative computations.

Within the total measurements of gravity anomalies, the regional anomaly is caused by a residual mass that is deeply buried and much bulkier than local sources in size. This kind of widely distributed anomaly with a small rate of horizontal change often shows itself as the background. In contrast, a local anomaly, caused by a residual mass that is shallowly buried or small in size, is often distributed in a limited area with an evident rate of horizontal change. The stacking of local anomalies of interest and certain regional anomalies will lead to a variation in local anomaly configuration as well as an offset of the anomaly center. Gravity anomaly separation has usually been achieved using the graphical method, circular method, averaging, trend analysis (polynomial fitting), and prior model building.

Filtering in the frequency domain, analytic continuation, field partition, and derivation have often been used to separate regional anomalies and local anomalies, while the general horizontal gravity gradient method, interpolation cut, matched filtering, and linear gravity inversion have often been used to separate anomalies from various sources with different properties. These seven methods are described as follows.

In general, subdued anomalies caused by deep geologic bodies are characterized by low frequencies, while local anomalies by shallower bodies are characterized by high frequencies. Observational errors and anomalies caused by near-surface heterogeneity feature high frequencies as well. Different anomalies would be provided with different frequencies, which is the prerequisite of anomaly separation. Anomalies with large frequency differences are liable to be separated.

Regional anomalies may be studied by filtering in the frequency domain. Through comparison of frequency spectrums with different frequency thresholds, as well as integration of prior geologic and geophysical information, we may determine the frequency scope of regional anomalies and thereby establish regional field effects.

Conversion of anomalies from observation surfaces to different altitudes may cause variations in magnitude and gradient, but anomalies of geologic bodies with various scales would change with depth at different rates. The anomalies of shallow and small bodies would attenuate more quickly than those of deep and large bodies in the process of upward continuation. Therefore, upward continuation would suppress shallow anomalies and stand out from deep anomalies, while downward continuation is quite the opposite.

The derivatives of weak and widely distributed anomalies are small, while those of strong and locally distributed anomalies are relatively large. So anomalies caused by deep and bulky geologic bodies may be suppressed by the derivative anomaly, while shallow and local geologic bodies may be highlighted. The horizontal derivative is often used to identify a fracture system, and the vertical derivative is used to separate local fields and side superimposed anomalies in the background.

Gravity and magnetic fields contain structure information that cannot be identified with surface geologic surveys, and also help identify shallow and deep fracture systems after some data processing. Gravity and magnetic anomalies arising from fractures on a large scale may be well marked because of differences in petrophysical properties between rocks from both sides of the fractures.

Fractures generally appear as gradient belts extending in certain directions or contour distortions on gravity anomaly maps, but there may be some uncertainty about fault locations derived directly from gravity anomalies. Conversion of a gravity gradient belt to a gravity gradient anomaly extremum would improve the resolution of the gravity anomaly to make fault locations more clear and further highlight secondary faults.

A general horizontal gravity gradient map from an anomaly computation highlights subsurface faults and is superior to a Bouguer anomaly map in its resolution. Peaks of general gradients have often been used to find the location and strike of faults.

Interpolation cut, based on interpolation between the field value of a selected point and circular average of four surrounding points, is considered to extract high-order components of the field around a selected point by constructing a second-order difference with these five points. It is a kind of nonlinear filtering in the space domain, and the magnitude may be adjusted automatically in accordance with potential field features.

If geologic bodies with various burial depths, extensions, and scales differ largely in spectrum features, matched filtering would then be applied to the separation of spectra of different bodies by constructing forward-modeling operators with different spectrum features.

The system of equations for gravity and magnetic anomalies for all observation points is established by forward modeling of all cuboids with different physical properties (density or susceptibility), which are obtained by partitioning subsurface space using vertical hexahedron units. The system of equations is solved to obtain the physical property distribution features of cuboids that simulate subsurface space media. Excessive partitioning may cause problems in solving large linear equations that could not be handled by conventional solutions. Damped least squares (DLSQR), a concept from seismic tomography, could invert subsurface properties with acceptable accuracy and speed. In solving linear equations with 222 lines and 1665 columns, the computation time of DLSQR amounted to 1/30 of that of SVD, and in solving linear equations with 555 lines and 1665 columns, the computation time of DLSQR amounted to 1/47 of that of SVD. DLSQR has been proven to speed up the solving process by several dozens of times.

6.2.1.3 Applications

An aeromagnetic survey is a kind of geophysical prospecting, which makes use of airborne geophysical surveying gauges installed in an aircraft at a certain flight height range to acquire magnetic field strength from the subsurface.

In 50 years of airborne geophysical surveys in China, the aeromagnetic survey, one of the classical methods in airborne geophysical prospecting, has been extensively applied to geotectonic study, geologic mapping, and solid minerals, as well as oil and natural gas prospecting, city stability evaluation, geologic hazard prediction, and other geoscience studies.

Figure 6.1 shows the countrywide aeromagnetic anomaly ΔT (ΔT is the modulus difference between the observed geomagnetic field and normal geomagnetic field with the unit of Tesla) on a scale of 1:5,000,000, which was mapped by China Aero Geophysical Survey & Remote Sensing Center for Land and Resources (AGRS) with reprocessed data in 2004. The map represents a collection of aeromagnetic measurements and achievements from 432 survey areas with different scales and precision. This was accomplished by more than 10 domestic research institutes over more than 40 years and proved to be of great value in the study of crustal structures and depositional basins. It provides information from both different geologic units in the lateral direction and to a great depth inside the crust. Most vestiges of geologic structures, whether in outcrop or buried in the subsurface, would be reflected on an aeromagnetic map.

Figure 6.1 differs from all previous maps in three aspects. The first is onshore aeromagnetic data covering 97% of the land area, especially filling up the vacancy on the Qinghai-Tibet Plateau. The second is the application of all high-precision measurements of digital recording after 1980 and Global Positioning System (GPS) positioning after 1989. The third is overall digitized mapping with data from each survey area except the Taiwan Strait.

Comparing the distribution of major depositional basins with that of aeromagnetic anomalies, we can see that the boundaries of major basins (Songliao, Erlian, Haila'er, Bohai Bay, southern North China, Ordos, Sichuan, Junggar, Tuha, Tarim, Qaidam, and Cuoqin) all exhibit high magnetic anomalies along fault belts. Especially the discordogenic fault belts in the basin manifest strong magnetic anomalies, which indicate the development of igneous rock

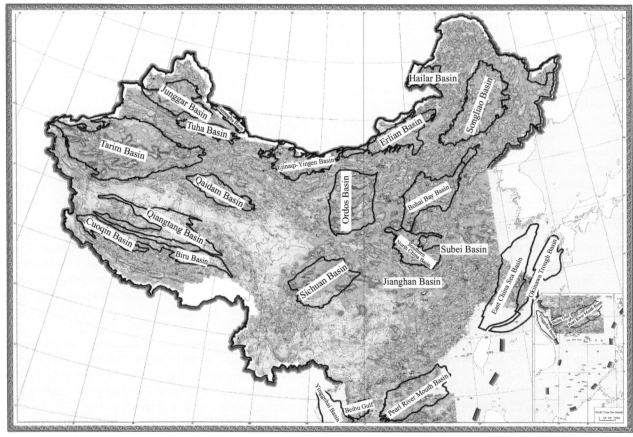

FIGURE 6.1 ΔT aeromagnetic map and major sedimentary basins in China. Taiwan data left vacant for the time being. *After AGRS (2004).*

(volcanic rock) inside the basins. Some high magnetic anomaly belts, including the nearly south-north-trending high magnetic anomaly belt in central Songliao Basin, northwest-trending high magnetic anomaly belt in central Junggar Basin, east-west-trending high magnetic anomaly belt in central Tarim Basin, nearly east-west-trending high magnetic anomaly belt in northern Ordos Basin, and northeast-trending high magnetic anomaly belt in central and southern Ordos Basin, as well as the northeast-trending high magnetic anomaly belt in central Sichuan, have proven to be favorable target zones of volcanic reservoirs. According to exploration experiences in Songliao and Junggar Basins, there would be natural gas fields on a scale of more than 1000×10^8 m^3 (e.g., Qingshen gas field and Kelameili oil and gas field) along high magnetic anomaly belts inside the basins, and oil and gas fields on a scale of more than 100 MMt (e.g., oil and gas field at the northwest margin of Junggar) along anomaly belts at the basin margins.

According to the countrywide distribution of aeromagnetic anomalies, high anomaly belts in eastern basins have been controlled by northeast faulted structures, and belts in western basins by northwest and nearly east-west structures.

As for basins in central China, high anomaly belts have been dominated by faulted structures in both the northwest and northeast directions.

Gravity and magnetic anomalies from surface observations represent the comprehensive effect of various subsurface geologic bodies, such as depositional basins, and basement and crust with different structures and petrophysical properties. In order to isolate the features of interest, anomaly separation should be performed based on the differences among geologic bodies at various depths.

Figure 6.2 demonstrates regional-local aeromagnetic anomaly separation in northern Xinjiang using interpolation cut, which shows regional anomalies of crustal geologic bodies. In detail, high anomalies widely distributed from the hinterland of Junggar Basin to eastern Junggar correspond to the ancient Junggar landmass, anomalies in southern Junggar correspond to an intracrustal anatectic magma belt in the underthrust zone at the southern margin, and anomalies in the southern margin of Tuha correspond to an intracrustal anatectic magma belt in the underthrust zone at the southern margin. On the other hand, local aeromagnetic anomalies highlight features of the fold basement. For example, anomalies at the northwest margin of Junggar

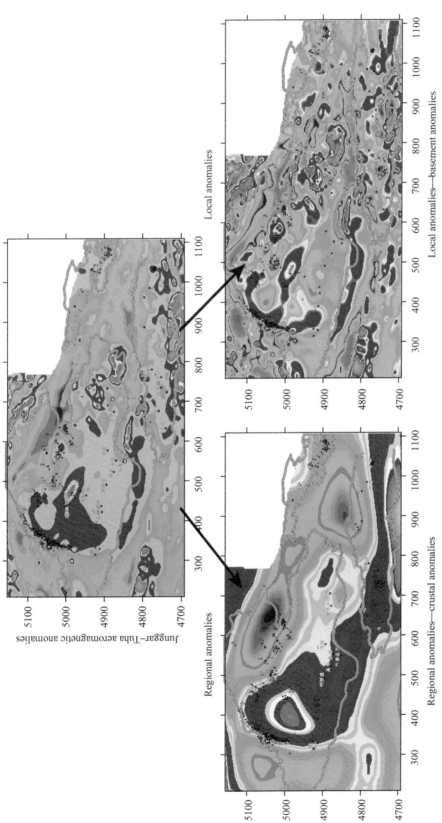

FIGURE 6.2 Separation of aeromagnetic anomalies from shallow-deep sources, northern Xinjiang. Taiwan data left vacant for the time being. *After AGRS (2004).*

reflect volcanic rock distribution in nappe structures, and high anomalies from the hinterland to eastern Junggar correspond to the distribution of island arc volcanic rock. At the southern margin of Junggar and Tuha, magnetic anomalies correspond to volcanic rock distribution in collision belts. At the northern margin of Junggar and southern margin of Santanghu, anomalies correspond to volcanic rock distribution in the belts resulting from collision of the Junggar-Tuha plate and Siberian plate.

Superimposed on the total gravity field of the Earth, the gravity effect of subsurface geologic bodies is generally less than 0.01% of the total field. Only with high-precision gravity, instruments can these weak responses be captured. A gravity anomaly is usually denoted in units of milligal (mGal; $1 \text{ mGal} = 10^{-5} \text{ m s}^{-2}$) in oil and gas exploration. In general, the gravity anomaly of a geologic body is less than 25 mGal, and the observational accuracy of a typical gravimeter is less than 0.5 mGal.

Different from geomagnetic fields with arbitrary direction, the earth's gravity field is always downward. A gravity field is a scalar field, while a magnetic field is a vector field, both magnitude and direction of which are of great importance to the magnetic anomaly interpretation.

Typically, rock (igneous, metamorphic, and sedimentary) density varies between 1.60 and 3.20 g cm^{-3}. The density of sedimentary rock varies from 1.80 to 2.80 g cm^{-3} in general. It is difficult to detect such small density variations among different sedimentary rocks with gravimeters with low precision, the error of which may be up to 5-10%, whereas the susceptibility difference between geologic bodies and their wall rocks may reach several orders of magnitude, which is liable to be detected by magnetic geophysical methods. In this respect, magnetic prospecting is superior to gravity prospecting in its relatively high resolution.

Gravity prospecting would be helpful to locate geologic structures with density variance in a lateral direction. For example, fault zones and structural boundaries may be determined through a sudden change of gravity anomaly gradient or contour distortion because there would often be an evident density difference between geologic bodies from the two sides of a large fracture or tectonic unit boundary.

Figure 6.3 demonstrates how to identify large regional fractures based on the first-order derivative of total gravity in northern Xinjiang. In other basins, such as Junggar, Tuha, and Santanghu, the distribution of the major fault belts that control basin development (Da'erbute fault belt, Karamay-Mosuowan fault belt, Ebinur Lake-Xingxingxia fault belt, Bogda mountain front fault belt, and Kelameili-Moqinwula fault belt) has an apparent relationship to the anomaly of the first-order derivative of total gravity. Some discordogenic fault belts inside basins (Hongshanzui-Chepaizi fault, Karamay-Mosuowan fault, Dishuiquan fault, Baijiahai fault, Buerjia fault, and northeast thrust fault) also manifest

strong abnormal responses in the first-order derivative of total gravity.

Gravity and magnetic anomalies contain responses from different volcanic rocks of different scales and at different burial depths, which are affected by factors such as burial depth, lithology, and others. Therefore, the interpretation of anomalies for volcanic rock identification should be calibrated with all available core and outcrop data.

From our study, intermediate and basic volcanic rock (basalt and diabase) features high magnetic anomalies, while intermediate and acidic rock (andesite, dacite, and rhyolite) features medium to low anomalies. Sedimentary rock often demonstrates low magnetic anomalies or negative anomalies. But it is difficult to distinguish among acidic volcanic rock, tuff, and sedimentary rock effectively without integrating other data.

6.2.2 Target Zone Prediction

6.2.2.1 Technologies

A band pass filter operator of return upward continuation has been introduced to modify the conventional derivation and the improved processing method for gravity and magnetic anomalies at exploration target depth; namely, the vertical derivative target processing technology using gravity and magnetic return upward continuation with high resolution has been developed, which improves the identification of volcanic rock bodies.

The generalized n-order derivative by multiple upward and downward continuations may suppress high-frequency noise and enhance the information of target intervals. Its application to aeromagnetic data processing aimed at deep targets in the Xujiaweizi fault depression, an area with the highest degree of exploration in Songliao Basin, has been checked with drilling data to clarify the features of volcanic rock distribution.

The continuation return filter factor is combined with the generalized vertical n-order derivative filter factor to constitute a band pass filter. Frequency may be modulated by adjusting the n-order of the generalized vertical n-order derivative, and continuation upward return height and times, so as to highlight the information of exploration target intervals. The operator in the wave-number domain is expressed as follows:

$$F = \left\{ \exp\left(-2\pi h\sqrt{u^2+v^2}\right)\left[6 - \exp\left(-2\pi h\sqrt{u^2+v^2}\right) - 2\cos(\pi h u) - 2\cos(2\pi h v)\right] \right\}^p \left(2\pi\sqrt{u^2+v^2}\right)^n$$

where h is continuation height, and u and v are wave numbers in the x and y direction, respectively. The parameter n is the times of derivation and would be a positive rational number with value from 0.1 to 5.0, which breaks through the limitation in conventional derivations

Regional fault interpretation on the total gravity derivative anomaly

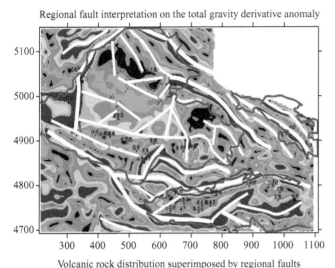

Volcanic rock distribution superimposed by regional faults

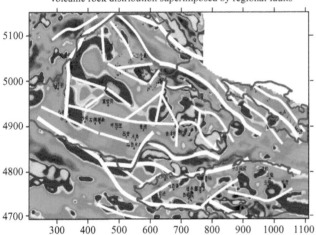

FIGURE 6.3 Total gravity first-order derivative anomaly interpretation and superimposed map of regional fault and volcanic rock distribution in northern Xinjiang.

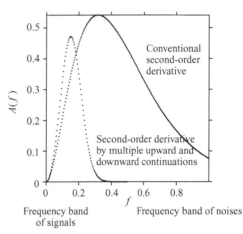

FIGURE 6.4 Comparison of the frequency spectrum between a modified filter and a conventional filter.

where n should be a positive integer. By choosing different n's, anomalies from geologic bodies at different depths would stand out.

Figure 6.4 compares the spectrum of a modified method and that of a conventional filter. The former shows itself as a narrow filter with low dominant frequency ($f_0 = 0.178$ Hz). It has a good band pass effect to suppress noises and amplify signals. Therefore, we would observe evident useful anomalies on the data after processing. The conventional filter constructed from a second-order derivative is a broadband filter with relatively high dominant frequency ($f_0 = 0.318$ Hz), which would amplify both signals and noises. So processed data would still contain much noise and be difficult to interpret. In addition, the larger gradient of the modified filter spectrum in comparison with that of the conventional one, especially at frequencies ranging from 0.08 to 0.18 Hz, implies that signals within this frequency band (for example, signals from a deep buried hill) would be efficiently enhanced.

From the frequency spectrum of a modified filter in Figure 6.5, the dominant frequency would move upward with an increase in parameter n, which implies that it is possible to highlight anomalies from different target intervals by choosing different values of n.

Data processing for field sounding is conducted in the following steps:

1. Prestack processing: check received signals, expressed as $E(t)$, through visualized playback and data analysis. Remove signals with apparent distortions. Perform Fourier analysis to get information about frequency components of signals.
2. Synchronous stacking: apply synchronous stacking and averaging, filtering, and smoothing, and other methods to reduce noise. Calibrate signals and eliminate step response, based on which converts $E(t)$ into a standardized function $F(t)$, which denotes the subsurface propagation of electromagnetic energy.
3. Poststack processing: starting from $F(t)$, work out the geoelectrical parametric curve for each observation

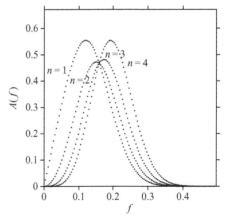

FIGURE 6.5 Frequency spectra of modified filters.

point and finally obtain the time section and depth section of the anomaly field, which characterizes the electrical property of the subsurface strata.

4. Forward modeling and inversion: One-dimensional (1D) forward modeling and inversion is a mature technology and has been used extensively. Two-dimensional (2D) forward modeling and inversion has recently made much progress in theoretical studies but with few practical applications.

Solving the geoelectrical parameter to depth section using the processing mentioned above is an approximation based on certain assumptions. For a real quantitative interpretation, it is necessary to start with the computation of a theoretical sounding formula for a 1D model and make use of an automatic iterative inversion to realize a theoretical model that approaches the actual measurements.

Different from other electrical methods such as magnetotelluric method (MT), the theoretical field sounding formula for a horizontal stratified homogeneous medium is relatively rational, which is constituted by the double infinite integral of the product of the model's kernel function (layer thickness and resistivity) and Bessel's function. Bessel's function therein only has relationship to the geometric scale. Besides, because of its slowly damped oscillation behavior, a conventional numerical integration would be time consuming and would not guarantee the accuracy as well. In the 1980s, Koefoed, a Dutch scholar, put forward a solution through linear digital filtering, which has simplified the theoretical calculation of field sounding for a multilayered medium and laid the foundation for 1D field sounding inversion (Koefoed, 1979). A reasonable selection of filter factor would also improve the accuracy. At present, the quantitative data processing for field sounding has mainly been achieved through 1D inversion. 2D inversion is still restricted to theoretical study.

6.2.2.2 Applications

Local magnetic anomalies after original aeromagnetic anomaly separation still contain complicated signals from various magnetic bodies inside the basement. To derive efficient information from target bodies, a second derivative would be used to detach the top anomaly and side superimposed anomaly from composite anomalies.

Figure 6.6 shows the aeromagnetic anomalies after second derivative processing in Junggar Basin. According to preliminary statistics, the extensive distribution of anomalies has amounted to 6800 km^2. Calibrated with the Carboniferous volcanic rock drilled in more than 50 wells in Ludong-Wucaiwan, the goodness of fit between magnetic anomaly and volcanic rock development has been beyond 85%.

Inside Junggar Basin, magnetic anomalies demonstrate medium to low magnitude with sheet-like distribution, while those at the basin margin demonstrate strong magnitude with bead-like distribution or distributing along fault belts. In eastern Junggar, there are mainly chaotic strong anomalies.

In Junggar Basin, Ludong-Wucaiwan is a highly explored area and is second only to the northwest margin. A density model built for strata above the Permian System has been used in forward gravity modeling. Then, shallow- and medium-gravity anomalies have been subtracted from the original gravity anomalies to eliminate the effect of shallow layers and strengthen the responses of the Carboniferous volcanic rock.

Figure 6.7 shows the case of enhancing the responses of the Carboniferous volcanic rock using gravity and magnetic forward modeling and stripping. Initially, the original subdued gravity anomaly contours reveal little information about the volcanic rock, and it is difficult to separate deep anomalies from shallow and medium anomalies. After processing, geologic bodies with various densities inside the Dishuiquan depression and Di'nan low uplift have emerged. In particular, the low-density belt from the Dixi5 to Dixi8 wells, and then to the Dixi10 well, implies the presence of intermediate and acidic volcanic rock.

Volcanic rock characterization using the anomaly field mentioned above is equivalent to observations on the ground. If we want to further approach the targets, downward continuation would be helpful to pick up gravity and magnetic anomalies from target strata and strengthen the signals of local geologic bodies.

Regularized downward continuation has been used to detect magnetic anomalies along the top boundary of the Carboniferous in Junggar Basin, the output of which is shown in Figure 6.8. The original magnetic anomalies in Ludong-Wucaiwan demonstrate sheet-like distribution because of their distance from the top of Carboniferous. In comparison, data after downward continuation processing contain more detailed information. The anomalies along the top of the Carboniferous show a complex distribution of two belts in a nearly east-west extension with an annular band between them. In the central area, the anomalies are distributed as a northeast and northwest belt crossing each other, instead of the original single belt in a northwest extension.

For a geologic body with various petrophysical properties, e.g., elastic modulus, resistivity, density, and magnetism, any geophysical measurement is only the observation of a single property from a certain perspective. In general, there is no absolute one-to-one correspondence between a geologic body and a petrophysical property because of observational error and complicated geologic conditions, which leads to the nonuniqueness in geophysical data interpretation. Therefore, only through the

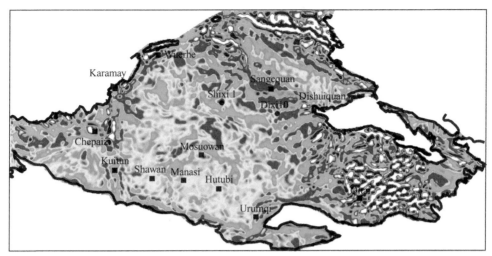

FIGURE 6.6 Second derivative aeromagnetic anomaly and volcanic rock distribution, Junggar Basin.

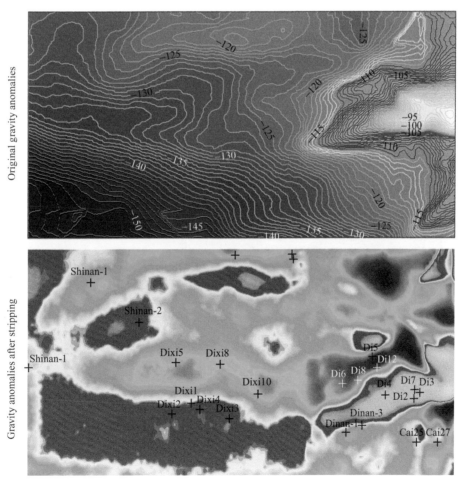

FIGURE 6.7 Gravity anomalies after stripping the responses of sedimentary cover (Q-P), Ludong-Wucaiwan, Junggar Basin.

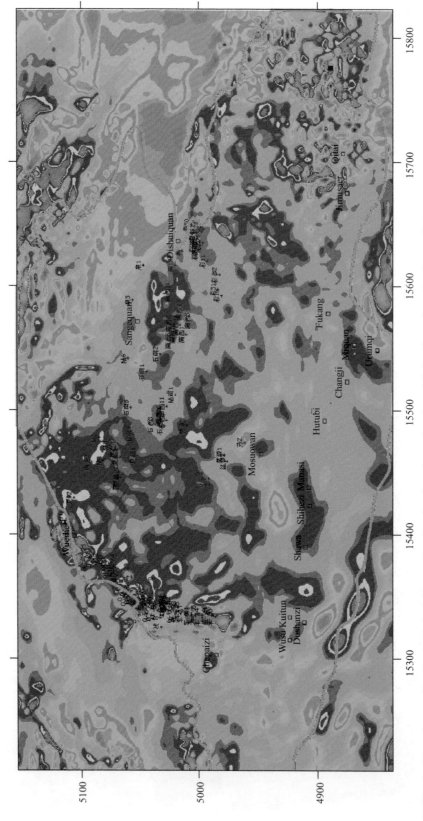

FIGURE 6.8 Magnetic anomalies after regularized downward continuation along the top boundary of Carboniferous formation, Junggar Basin.

integrated interpretation of all kinds of geophysical information can volcanic rock characterization be effective and reliable.

Figure 6.9 shows the case of volcanic rock identification with gravity, magnetic, electrical, and seismic data in Ludong, Junggar Basin. On seismic line L2007-05 passing through the Dishuiquan depression, there is an anomalous seismic reflection with continuous strong energy and low frequency on the central part of seismic profile, the property of which could not be determined solely through seismic interpretation. Therefore, regularized downward continuation was conducted in aeromagnetic anomaly processing and gravity anomaly processing. On the cross section of magnetic anomaly downward continuation, there is a magnetic high at the center, corresponding to low density on the gravity section and high electrical resistivity on the field sounding inversion with the same location of that on seismic profile. The combination of medium to high magnetism, high resistivity, and low density has been proven to be the attributes of intermediate and acidic volcanic rock. With the integrated interpretation of gravity, magnetic, electrical, and seismic data, the anomaly is identified as intermediate and acidic volcanic rock.

According to volcanic oil and gas exploration at home and abroad, the development of volcanic reservoirs are mainly influenced by fault belts, weathered crust, lithology, and lithofacies. Fault belts are the most dominant factors controlling the development of structural fractures. Discordogenic faults might not only act as the passageway between source rock and reservoir rock but also play an important role in changing the petrophysical properties of volcanic reservoirs. Weathering and leaching mainly contribute to the development of secondary pores in volcanic reservoirs, which tend to be well developed in the interval from 300 m above the weathered crust to 300 m below the crust, and serve as major reservoir space for oil and gas. As for lithology and lithofacies, intermediate and acidic volcanic rock of eruptive-effusion facies are the most favorable.

Based on the appraisal criteria of distributing close to the source rock and the weathered crust with a developed fault belt, there are several favorable target zones in Junggar Basin, i.e., Luliang uplift, Da-Mo-Bai, Wulungu, northwest margin, Zhangbei, and Zhundong. Among these, there have been discoveries of large volcanic oil and gas fields in the Luliang uplift and the northwest margin, and hydrocarbon shows in wells in the Da-Mo-Bai, Zhangbei, and Zhundong

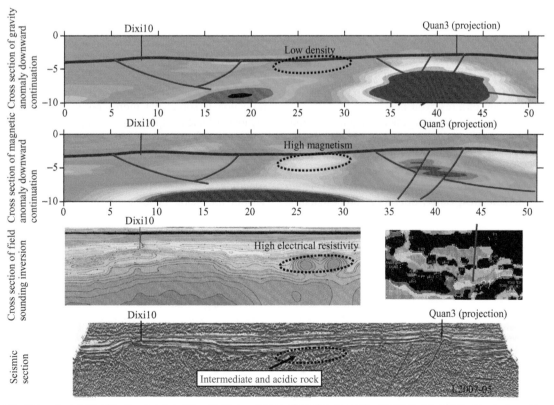

FIGURE 6.9 Volcanic rock identification using an integrated interpretation of gravity, magnetic, electrical, and seismic data, Ludong-Wucaiwan.

areas. More progress will be made by intensifying volcanic reservoir exploration in these areas.

6.2.3 Volcanic Reservoir Prediction

Volcanic reservoir prediction mainly deals with the study of lithofacies and volcanic reservoirs with seismic data. Changes in lithology, petrophysical properties, and fluids in pore space would lead to a consequent variation in seismic velocity, amplitude, phase, frequency, and waveform. On the condition that these variations could be detected, it is possible to delineate properties of the geologic body from seismic reflections. Just like sedimentary and carbonate rocks, the development of volcanic reservoirs is affected by many factors such as tectonic processes, lithofacies, and diagenesis, among which volcanic lithofacies is the internal factor and has great influence on the properties of a volcanic reservoir. On this account, identification of the volcanic lithofacies is the basis of volcanic reservoir characterization.

6.2.3.1 Applications

In general, the rock type and primary pore condition of a reservoir bed are affected by lithofacies, so in many cases the prediction of rock type and its development could be achieved indirectly by prediction of lithofacies. Especially for complex lithologic reservoirs, such as volcanic, carbonate, and fractured reservoirs, it is difficult to gain insight into the properties of a reservoir bed, which could then be fulfilled by lithofacies prediction. Based on the study of seismic facies, sedimentary facies, and sedimentary systems, lithofacies prediction aims at finding out those determinants of a reservoir bed, as well as favorable facies for the development of a reservoir bed.

Different internal and external reflection configurations correspond to different volcanic lithofacies. There are various kinds of volcanic lithofacies because of the different patterns of volcanic rock generation and emplacement. For volcanic rock of the extrusive facies, which has been intensively studied, there are a variety of criteria and designation

in volcanic rock classification. According to the volcanic rock classification scheme (Qiu, 1991) recommended by the International Union of Geological Sciences, volcanic rock has been classified into 5 main facies and 15 subfacies; the five main facies are the explosive facies, effusive facies, volcanic conduit facies, extrusive facies, and volcanic sedimentary facies. The explosive facies has been further divided into hot clastic flow subfacies, hot base surge subfacies, and fallout subfacies. The effusive facies includes an upper subfacies, middle subfacies, and lower subfacies. The volcanic conduit facies includes volcanic neck subfacies, subvolcanic subfacies, and cryptoexplosive breccia subfacies. The extrusive facies includes intrazone subfacies, mesozone subfacies, and outerzone subfacies. The volcanic sedimentary facies includes coal-bearing tuff subfacies, extraclastic-bearing volcanic sedimentary subfacies, and volcanic debris retransportation subfacies. The seismic reflection features of these volcanic facies are shown in Figure 6.10.

An explosive facies centralizing around the crater is usually associated with large faults. Distribution of the volcanic rock is controlled by the crater. The external shape of the explosive facies on a seismic profile assumes mounded reflection, the top of which is a strong reflection boundary between clastic rock and volcanic rock, and the internal of which takes on a chaotic and discontinuous reflection of medium to low frequency. It is easy to identify the explosive facies on a seismic profile.

The effusive facies is subdivided into near-source effusive facies and far-source effusive facies, according to the distance from the facies to the crater. The seismic reflection would change from a mounded shape at the crater to a sheet or wedge shape at the slope or lowland. The top reflection of the effusive facies features good continuity, strong energy, and medium to high frequency. The internal reflection is somewhat more continuous than that of the explosive facies; especially the far-source effusive facies will take on a parallel to subparallel pattern.

The volcanic conduit facies lies at the base of the whole volcanic apparatus and comes into being at the corresponding and later stages of the volcanic cycle. On a seismic profile the facies stands as a nearly vertical column

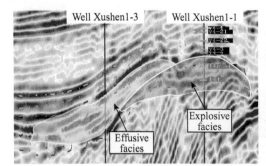
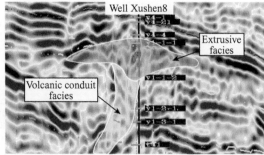

FIGURE 6.10 Seismic reflection features of various volcanic facies in deep Songliao Basin.

at the center of the crater, just below the top of volcanic cone, which is easy to identify. The internal reflection features poor continuity and low frequency.

The extrusive facies lies above the crater and looks like a dome. The external shape takes on an umbrella, mounded, or pillow-like reflection with strong energy and poor continuity. Seismic events may be vertical and diverge outward in a fan shape from the bottom of the facies.

The volcanic sedimentary facies is often associated with volcanic rock and may appear in various stages of volcanic activity. There is a large amount of volcanic debris in the clastic constituents, which is mainly a clastic deposit among volcanic domes. The volcanic sedimentary facies is mainly distributed among different volcanic massifs. The sheet-like reflection features stable waveforms of strong energy, good continuity, and medium to high frequency. The logging responses are similar to those of sedimentary rock.

The study of the seismic reflection features of volcanic rock lays the foundation for volcanic rock identification with seismic data. Discordogenic faults control the generation of volcanic rock. Craters along fractures occur in bead-like shapes and control the distribution of volcanic

massif. The external shape of the crater and volcanic massif distribution further affect the distribution of volcanic rock of different facies.

Volcanic rocks with different colors, textures, structures, and pore spaces demonstrate different features on logging data, e.g., conventional logging, elemental capture spectroscopy (ECS), and formation microimaging (FMI). Conventional logging is usually used to identify the lithology of volcanic rock, and ECS and FMI are helpful in identifying the constitution and structure of volcanic rock, respectively. All these interpretations are combined to diagnose lithology, lithofacies, and logging responses sensitive to volcanic reservoir variation, establishing the basis of seismic reservoir characterization.

In general, the uranium (U), thorium (Th), and gamma ray (GR) logs will show an increase in data values, and the density and resistivity logs will show a decrease with a change from basic to acidic volcanic rock, as shown in Figure 6.11.

Vesicular rhyolite, ignimbrite, and welded breccia can be identified with FMI and sidewall coring data, as shown in Figure 6.12.

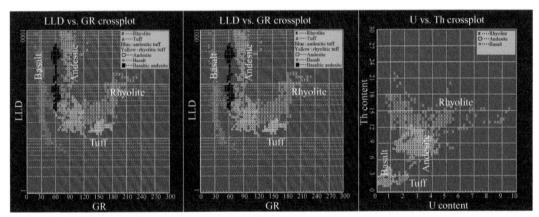

FIGURE 6.11   Volcanic rock logging responses.

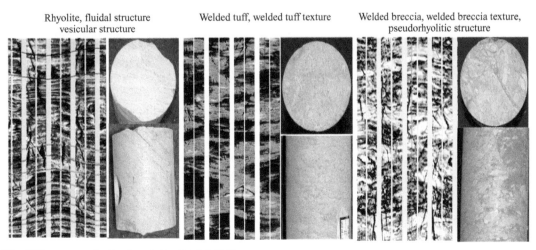

FIGURE 6.12   FMI responses of volcanic rock logging responses.

Generally, the acoustic and density logs are the most sensitive to reservoir pore structure; that is, the lower the density and the higher the interval transit time, the better the pore structure. Acoustic and resistivity logs are the most sensitive to oil in pore space. The higher the interval transit time and absolute resistivity, as well as magnitude difference of triple lateral resistivity, the better the oiliness. As for volcanic rocks, the same kind of rock may differ greatly in resistivity from $n \times 10$ to $n \times 10^4$ Ω m, because of variations in lithofacies and reservoir space. Tight block lava features high electrical resistivity but shows no apparent discrepancies in deep-middle-shallow lateral resistivity logs. Vesicular mandelstone shows low resistivity with small discrepancies in deep-middle-shallow resistivities. Brecciated lava has comparatively high resistivity and some discrepancies in deep-middle-shallow resistivities. Tuff has low discrepancies in deep-middle-shallow resistivities. Density might provide an indication of lithology, provided borehole conditions are acceptable. With an increase in volcanic acidity, density has a tendency to decrease, while acoustic value tends to increase correspondingly.

In China, basalt reservoirs in the petroliferous basins of the Mesozoic and Cenozoic feature low acoustic time, low induction, low natural GR, high resistivity, and high density. The evident steps on electric logs indicate the top and base of the volcanic rock.

Volcanic rocks with similar lithologies at different intervals show different log responses, and volcanic rocks with different lithologies at the same interval also show different log responses. In general:

1. Basalt: low natural GR, relatively high resistivity, medium interval transit time, and low density and neutron.
2. Andesite: medium natural GR, resistivity, and interval transit time; low density and neutron.
3. Rhyolite: relatively high natural GR, relatively low resistivity, relatively high interval transit time, relatively low density, and relatively high neutron.
4. Trachyte: high natural GR and resistivity, low interval transit time and neutron, and high density.
5. Tuff: medium natural GR, low resistivity, high interval transit time and neutron, and low density.
6. Basaltic trachyte: log responses between those of basalt and of trachyte.

According to the above features, a set of criteria can be established to identify different volcanic lithologies in different intervals, followed by logging classification of volcanic lithofacies.

1. Explosive facies: jagged profile with medium to low magnitude on log curves. Hot clastic flow subfacies mainly consist of welded tuff. Hot base surge subfacies consists of crystal tuff. Fallout subfacies consists of agglomerate, breccia, welded breccia, and breccia lava.
2. Effusive facies: blocky and slightly jagged profile on resistivity log with medium to high magnitude. Lower subfacies consists of rhyolite with low porosity. Middle subfacies consists of tight rhyolite. Upper subfacies consists of vesicular rhyolite.
3. Volcanic conduit facies: jagged profile with medium to high magnitude on log curves. The facies mainly consists of welded breccia and breccia lava.
4. Extrusive facies: intrazone subfacies contains perlite with pillow-like and spherical profiles on image logs. Mesozone subfacies contains massive perlite and aplitic rhyolite. Outer-zone subfacies consists of breccia lava with deformed rhyolitic structure.
5. Volcanic sedimentary facies: rhythmic profiles on log curves with large differences in thickness. The lithology is mainly volcaniclastic rock with terrigenous clasts introduced during the volcanic dormancy period.

The electrical and elastic properties of volcanic rock provide a basis for establishing an interpretation model with log and seismic data. This model, after being calibrated with bore core data, would be used to classify facies belts.

The volcanic rock of the Paleogene, discovered by drilling in the Bohai Bay Basin, may be classified into two types—eruptive rock/shallow intrusive rock or subvolcanic rock. Eruptive rock may be further subdivided laterally into crater subfacies and volcanic slope subfacies. The volcanic cone from a volcanic outbreak takes on a small, mounded shape on seismic profiles and annular distribution on horizontal slices, which are separated from the responses of surrounding wall rocks. Seismic responses of the volcanic slope would be weak in reflection energy and be in angular contact with reflections of ambient sedimentary strata. The distribution of slope may be limited. The effusive subfacies from a volcanic eruption may demonstrate continuous events with strong reflection energy, which may be parallel to bed boundaries. Refer to Table 6.4 for the lithofacies, electrical properties, and seismic facies of volcanic rocks in Bohai Bay.

Intrusive rocks usually distribute along strata and unconformably contact the overlying and underlying strata. Shallow volcanic rocks take on a subdued, mounded reflection shape with no evident intravolcanic events. Shallow intrusive rocks are divided into three profiles vertically: i.e., top and bottom corrosively metamorphic section, upper and lower transition section, and central section. There is no evident seismic reflection in the metamorphic section. The transition section may be in contact with the top and bottom metamorphic section and give a strong reflection. The central section will show similar seismic reflection features to the transition section. Laterally, shallow intrusive rocks are classified into central

TABLE 6.4 Lithofacies, Electrical Properties, and Seismic Facies of Volcanic Rocks in Bohai Bay Basin

Rock type	Lithofacies	Assemblage	Electrical properties	Seismic facies features	Example
Shallow intrusive rock	Central subfacies	Alabastrine medium to coarse crystalline structure diabase lithofacies assemblage. Single layer is greater than 50 m	Blocky profile with high resistivity and density, low interval transit time, and natural gamma ray	Strong reflection of low frequency at the top and the base. No evident intravolcanic reflections	Shanghe, Shengli oil field
	Transition subfacies	Medium porphyritic diabase and medium to coarse alabastrine diabase assemblage	Three vertical profiles on electrical logs, middle section similar to those above, top and bottom section similar to those below	Platy strong reflection of low frequency	
	Lip subfacies	Cryptocrystalline to finely textured diabase, vesicular and amygdaloidal structure, lamprophyre and diabase assemblage, most of which belongs to the metamorphic zone of baked corrosion	Low density and resistivity, high natural gamma ray, and interval transit time	No evident reflection at the top and the base	
Volcanic eruptive rock	Crater subfacies	In the center of the volcanic cone. Volcanic breccia and welded brecciform basalt lithofacies assemblage	High density and resistivity, low interval transit time, and natural gamma ray. Large borehole diameter changes may be related to the development of fractures	Dome- and arc-shaped strong reflection	Shanghe, Shengli oil field Oulituozi, Liaohe oil field
	Volcanic slope subfacies	At the slope of the volcanic cone. Volcanic breccia and volcanic tuff assemblage	High resistivity and natural gamma ray. Large variation of interval transit time and density	Weak reflection and limited distribution. In angular contact with reflections of lateral sedimentary strata	
	Effusive subfacies	Basalt of amygdaloidal structure with local diabase	High resistivity, medium interval transit time. Low natural gamma ray with large variation	Continuous strong reflection, parallel to bed boundaries	

After Qiao et al. (2001).

subfacies, transition subfacies, and lip subfacies. The central subfacies features strong seismic reflection and low frequency. The transition subfacies shows platy strong reflection of low frequency. There is no evident reflection at the top and base of the lip subfacies, and reflection horizons appear to decrease outward. There are three kinds of reservoir space in shallow intrusive rock, i.e., fractures, dissolved pores and caverns, and vesicles. In comparison with eruptive rocks, intrusive rocks are inferior in pore space but superior in fractures and dissolved caverns on a relatively large scale.

6.2.3.2   Technologies

Volcanic reservoir prediction includes volcanic lithofacies prediction and reservoir prediction. The former would usually be fulfilled by seismic facies analysis for qualitative evaluation, and waveform and multiattribute classification for semiquantitative evaluation, and the latter would be conducted by seismic inversion. High-angle fractures of eruptive and intrusive rock and alteration zones of intrusive rock may be favorable for hydrocarbon accumulation, and the eruptive facies is superior to the effusive facies in reservoir properties. Therefore, facies belt analysis is very important to reservoir prediction. Appropriate technologies include log analysis, seismic inversion, coherence or variance analysis, and attribute analysis in different frequency bands.

6.2.3.2.1   Seismic Facies Analysis

6.2.3.2.1.1   Seismic Attributes Seismic attribute analysis, based on attribute extraction of amplitude, frequency, phase, and waveform, aims at differentiating volcanic lithofacies from sedimentary lithofacies. Compared with wall rocks, volcanic rocks generally exhibit high density and velocity, leading to large elastic impedance differences

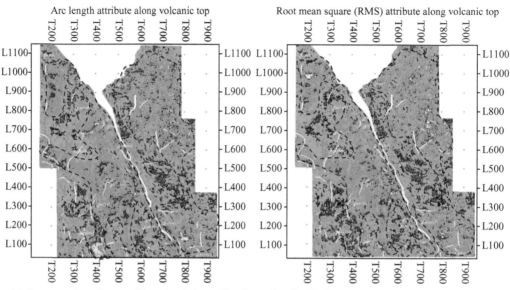

FIGURE 6.13 Attribute map along the top of the volcanic layer, Xingcheng, Songliao Basin.

and subsequent reflection coefficients at the interface between volcanic layers and surrounding beds. Therefore, attributes of reflection amplitude are usually the most effective to qualitatively identify volcanic lithofacies, as shown in Figure 6.13.

6.2.3.2.1.2 Waveform Classification and Clustering
Based on the changes in a seismic trace waveform and its characteristics, real seismic data in a specific reservoir are compared trace by trace to depict the lateral changes in the seismic signal and establish seismic facies patterns. Meanwhile, combining with single-well facies classification results, the relationship between seismic facies and log facies is established to solve the problem of spatial facies sequence assemblages and establish a planar distribution map of seismic facies.

Seismic facies analysis based on waveform classification uses the basic principle of seismic stratigraphy, where the changes in seismic reflection correspond to the changes in depositional facies and lithofacies. Seismic waveforms, instead of some single seismic signal (amplitude, phase, and frequency), would have a much closer relationship with overall seismic reflection variations and be more capable of delineating seismic anomalies. Analysis and classification of seismic waveforms provide information about seismic facies variations based on waveform changes. A neutral network-based classification is superior to a mere visual examination and would be more reliable and objective in identifying complex volcanic lithofacies.

Different strata and rock types have different seismic reflections in strength (amplitude), frequency, and phase. Different sedimentation types would also have different features in seismic reflection because of various sedimentary rhythms and spatial lithofacies changes. Seismic

attributes such as amplitude, frequency, and phase characterize seismic reflections, but in fact, waveform variations in certain periods may be more powerful to delineate changes in seismic reflection, resulting from changes in lithofacies and sedimentation types. The study on seismic waveforms and their classification for horizontal variations of waveforms could in turn be helpful to diagnose the change in rock types and lithologies, provided that there is a certain correspondence between them. Seismic facies may be taken as the image of the sedimentary facies in seismic reflections and refers to the 3D seismic reflection unit with reflection configuration, geometry, amplitude, frequency, continuity, and interval velocity being different from adjacent units. A seismic facies is a seismic reflection of the sedimentary facies and demonstrates the features of lithologic association, bedding, and sedimentation of the deposit generating the reflection. In the process of hydrocarbon exploration and technology, sedimentary facies zonation with all kinds of seismic information would be important to reservoir characterization. Seismic waveform classification is based on the assumption that seismic reflection features (amplitude, frequency, phase, integral energy spectrum, and time-frequency energy) vary with sedimentary facies units; therefore, seismic waveforms indicating the overall change of seismic signal may show waveform anomalies that would be used to map seismic facies and subsequent sedimentary facies. Seismic waveforms may reflect the actual subsurface geologies and would be helpful for geologists to extract useful information from seismic reflections. The technique focuses on seismic facies and attribute analysis aiming at target zones to find out distribution features of seismic facies and extract general geologic information from seismic data, which would be useful for geologists to interpret sedimentary facies. The technique has been proven to reflect the horizontal variation of seismic signals and has been applied extensively to the study of sedimentary facies and reservoir prediction.

There are two approaches to waveform classification in commercial software: One is waveform classification based on seismic data, and the other is multiattribute classification based on various attributes extracted from seismic data. Because of the quick lithofacies variation and strong heterogeneity of volcanic rocks, the single attribute of the waveform may be more practical to yield favorable results with evident regularity. Multiattribute analysis contains information from different formations, along with the complex development of volcanic rocks; therefore, there may be more ambiguities in multiattribute analysis. The single attribute of the waveform has therefore been used in volcanic lithofacies interpretation, which is shown in Figure 6.14.

6.2.3.2.2 Reservoir Prediction

Volcanic reservoir prediction can be achieved through seismic inversion, seismic attribute analysis, and seismic prediction of reservoir geologic parameters, based on the conventional exploration method of seismic reflection P-wave. Prior to seismic reservoir prediction, the criteria of the seismic interpretation are set up through log interpretation and evaluation. Then, seismic inversion and attribute analysis would be combined to identify different rock types and further diagnose reservoir rocks.

Seismic inversion is considered the key and one of the most effective techniques in seismic reservoir prediction, which focuses on reservoir heterogeneity and the lateral

and vertical variation of reservoir beds, and predicts reservoirs and their parameters based on seismic information. Integrated seismic interpretation is based on seismic inversion and information extraction, including elaborate reservoir calibration, seismic attribute analysis or seismic coherence analysis, impedance inversion based on the acoustic log (seismic reservoir inversion), and reservoir parameter inversion based on the density log (reservoir characteristics inversion). Seismic inversion is the process of imaging underground physical structures and physical properties from surface seismic measurements, with geologic rules, drilling, and logging data acting as constraints. Compared with some statistical methods such as pattern recognition, neutral network, and thickness estimation with amplitude-frequency relationships, impedance inversion yields an output with clear physical significance and consequently leads to a deterministic interpretation of lithology and reservoir properties.

Volcanic rock is characterized by great burial depth and complex rock types. The development of volcanic reservoirs is controlled by many factors, such as facies and structure. It is difficult to predict volcanic reservoirs because of their great burial depth and great vertical thickness, rapid lateral variation of rock type and lithofacies, poor stratification, complex distribution, and difficulties in tracking seismic horizons and defining time windows for seismic attribute analysis, and model building for seismic inversion. In order to predict and delineate volcanic reservoirs with acceptable accuracy, we must integrate multiple disciplines in reservoir prediction and employ various techniques to approach true reservoir development based on major controlling factors. Directed toward

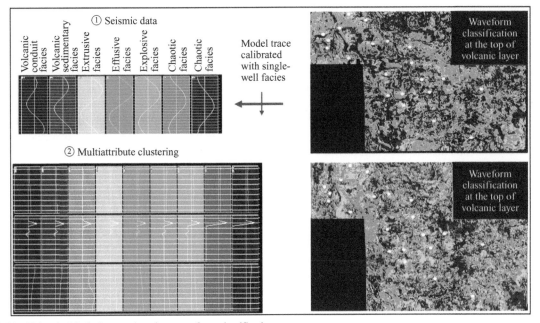

FIGURE 6.14 Volcanic lithofacies map based on waveform classification.

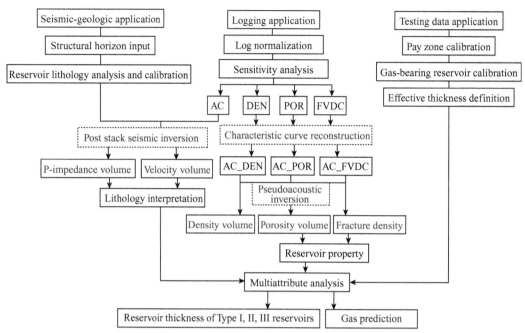

FIGURE 6.15   Workflow of volcanic reservoir characterization.

volcanic reservoir characterization, a workflow using multiple disciplines is shown in Figure 6.15.

6.2.3.2.2.1   *Conventional Seismic Inversion*   In view of quick lateral changes of the volcanic reservoir in Xingcheng, Songliao Basin, more focus should be placed on normalizing the acoustic log, wavelet extraction, reservoir calibration, and initial model in poststack seismic inversion.

6.2.3.2.2.2   *Reservoir Characteristics Inversion*   Conventional P-impedance inversion may fail to discriminate among different volcanic rock types because of their strong heterogeneity and quick facies changes. An alternative parameter of P-impedance, which may be sensitive to rock types (density) or reservoir properties (porosity) based on analysis of reservoir property responses, can be constructed to replace P-impedance in seismic inversion. The alternative reservoir characteristic parameter should first be scaled to the dimension of the acoustic data to yield pseudoacoustic dimension and then be used in reservoir characteristics inversion through conventional seismic inversion or geostatistical inversion, which is conducted in the following steps: (1) scale the alternative parameter to the dimension of the acoustic log, (2) conduct the reservoir calibration and extract the wavelet, (3) build the initial model for inversion, (4) conduct the inversion to yield pseudoimpedance volume, and (5) conduct reservoir characteristics inversion to yield density volume and porosity volume. Next, scale the characteristic parameter to the acoustic, set up the initial geologic model, and then conduct inversion to generate the pseudoimpedance volume, which is closely

related to reservoir characteristics. Waveform coherence or statistical methods are then used to establish a relationship between the pseudoimpedance and the reservoir parameter, and another inversion yields the reservoir parameter volume. A reservoir parameter inversion is conducted using two methods to derive reservoir parameters. Then, well-constrained seismic inversion yields impedance volume. A reservoir characteristics inversion yields density volume and porosity volume. These three data volumes are sensitive to volcanic rock types and properties and may be used in volcanic reservoir characterization and thickness estimation.

6.2.3.2.2.3   *Fracture Development Prediction*   Various structural fractures and diagenetic fractures that developed during the multiple stages of tectonic movement, along with secondary pores, serve as the major reservoir spaces and pathways of oil and gas seepage. The development of fractures is crucial to the properties of a volcanic reservoir.

Fracture length usually ranges from several millimeters to a few meters, fracture width is usually millimeter-sized, and fracture density may vary from rock to rock and from one fracture type to another. A density of several fractures per meter is common in volcanic reservoirs. It is difficult to directly detect fractures of such scale (from millimeter to meter) through seismic data with a wavelength from a meter to dozens of meters; instead, core and logging data may be more effective to give a reliable description of the fractures. On the other hand, there should be a certain relationship between the regional development of fractures and tectonic movements, or fault system or rock type, and a regional development of fractures may affect the properties of strata, which may be reflected in seismic responses. For example, dense fracture zones may demonstrate a stripe-like distribution with low coherence on coherence slices, while those areas with underdeveloped fractures may show good coherence. Therefore, fracture development and distribution may be delineated with seismic attributes such as coherence (Figure 6.16).

According to the seismic scattering theory, seismic wave attenuation would be affected by the spatial variation of the fracture density field. Attenuation of seismic waves propagated along a fracture strike would be slower than those perpendicular to the fracture strike. Vertical fractures, oblique fractures, and net fractures will have a different influence on seismic energy attenuation and distribution. For oil- and gas-bearing reservoir rock, reservoir properties and fluids will intensify the anisotropic attenuation of seismic waves. Consequently, for a fractured reservoir, seismic responses may contain information about fractures, provided that the existence of fractures has a detectable influence on strata properties.

6.2.3.2.2.4 Reservoir Thickness Estimation The estimation of volcanic reservoir thickness and effective reservoir thickness may be carried out with either density data or porosity data as constraints, among which porosity data, derived from density and neutron data, contain more comprehensive information about the variation of volcanic reservoir rocks and would be more powerful than density data.

1. Thickness estimation constrained by density: the output of velocity data volume from poststack seismic inversion convolved by the T0 difference of volcanic strata may be taken as bed thickness, which may then be converted into volcanic reservoir thickness under the constraint of the lower density limit as the threshold.

   This method may be used to determine the gross thickness of total volcanic rock or volcanic rock of different stages. It is difficult to classify volcanic reservoir with a density constraint.

2. Thickness estimation constrained by porosity: the output of velocity data volume from seismic inversion convolved by T0 difference may be taken as bed thickness, which may then be converted into reservoir thickness and effective reservoir thickness under the constraint of the lower porosity limit of the reservoir bed as the threshold.

This method may be used to determine the gross thickness of the total volcanic rock or volcanic rock of different stages

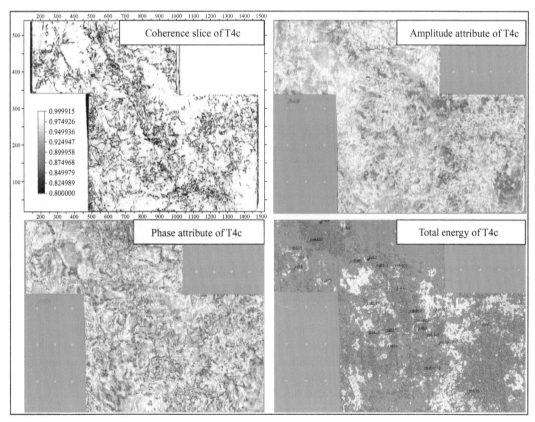

FIGURE 6.16 Qualitative analysis of fracture development with poststack seismic attributes.

and also may be capable of classifying reservoir beds into different grades that could be verified by drilling.

6.2.3.3 Cases Study

6.2.3.3.1 Volcanic Reservoir Prediction of Xushen Gas Field in Deep Songliao Basin

The deep Songliao Basin includes three tectonic units: the eastern fault zone, central paleouplift zone, and western fault zone. Exploration activities in the deep Songliao Basin, with a favorable exploration area of 28,860 km^2, have centered in the Xujiaweizi fault depression, central paleouplift, Shuangcheng fault depression, and Gulong fault depression. Of these, Xujiaweizi has been highly explored, and the other three have not been explored extensively. According to the third resources assessment, the total deep gas resources were estimated to be $11,740 \times 10^8$ m^3. Resources in the Xushen gas field, located in the northern part of the eastern fault zone with an area of 5350 km^2, were estimated to be 6772×10^8 m^3. During the 10th Five-Year Plan period, natural gas exploration made a great breakthrough in the deep Songliao Basin and reported proven reserves of 1258.54×10^8 m^3 in Xushen field, which accounts for 18.5% of resources. Natural gas exploration in Xudong, Anda, and Xingcheng-Fengle in Xujiaweizi field is currently concentrating on preliminary prospecting and evaluation to verify the amount of reserves. Exploration targets have transitioned from structural traps on structural highs to lithologic traps on slopes and low-lying areas with favorable volcanic reservoirs. As for seismic reservoir characterization, poststack technologies have been used extensively in quantitative prediction, and prestack technologies are just starting to be used.

Considering the features of volcanic rock development and distribution, such as great burial depth, multiple stages of eruption, vertical and horizontal overlaying of different facies belts, large changes in rock type, strong anisotropy, and complex reservoir formation conditions, a multidisciplinary approach has been applied to volcanic reservoir characterization, using the following steps:

1. Identify volcanic massif through gravity and magnetic data as well as seismic data interpretation.
2. Interpret fault system, restore paleogeomorphology (Figure 6.17), and identify possible craters and volcanic conduits (Figure 6.18).
3. Establish geophysical criteria for volcanic rock identification and classify volcanic lithofacies through seismic attribute analysis such as waveform classification (Figures 6.19 and 6.20).

Explosive facies: rock type of crystal welded tuff, strong reflections with middle to high frequencies and chaotic mounded reflection shape.

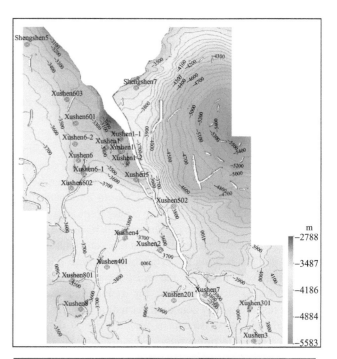

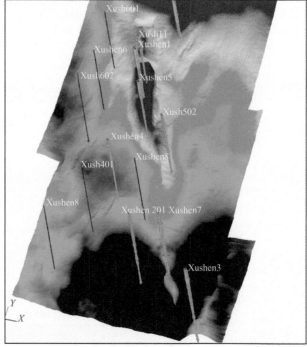

FIGURE 6.17 Fault assemblage at volcanic rock top and crater distribution, Xingcheng gas field, Songliao Basin.

Effusive facies: rock type of rhyolite, strong reflections with middle to high frequencies and parallel to subparallel reflection configuration.

Volcanic sedimentary facies: rock type of volcanic sedimentary rock, weak reflections and parallel to subparallel reflection configuration.

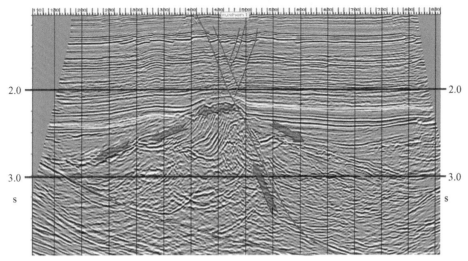

FIGURE 6.18   Seismic profile showing craters along the Xushen1 well, Songliao Basin.

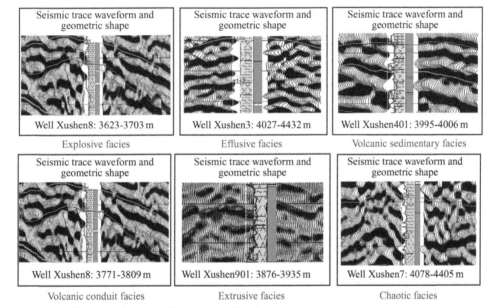

FIGURE 6.19   Waveform classification and volcanic lithofacies identification.

Volcanic conduit facies: rock type of breccia lava, alternating strong and weak reflections with middle to high frequencies and chaotic reflection shape.

Extrusive facies: rock type of rhyolitic crystal welded tuff and breccia lava, strong reflections with middle to high frequencies and spherical reflection shape.

Chaotic facies: rock type of rhyolite and welded tuff, middle to strong reflections with middle to high frequencies and chaotic reflection shape.

4. Predict volcanic reservoir distribution quantitatively through seismic attribute analysis and seismic reservoir characterization.

5. Delineate fracture distribution through multiattribute analysis.

6. Classify effective reservoir by its characteristic parameter (density, porosity) inversion.

Based on the above parameters, we can predict the effective reservoirs in volcanic rocks and evaluate the volcanic reservoir comprehensively from macroscopic to microscopic evaluation, qualitative to quantitative description, nonseismic to seismic evaluation, poststack to prestack processing, and exploration to development.

6.2.3.3.2   Volcanic Reservoir Prediction of Ludong Area in Junggar Basin

Junggar Basin is different from Songliao Basin in its complicated structural evolution as well as complex surface and

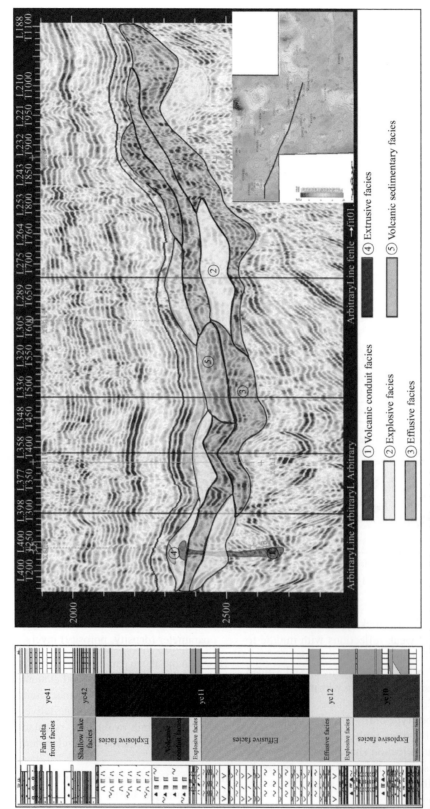

FIGURE 6.20 Volcanic lithofacies classification through well log and seismic facies analysis.

subsurface geologies. Deep seismic reflections are not adequate for interpretation. Therefore, only gravity, magnetic, and electrical prospecting may be effective to identify the volcanic massif and lithofacies. As for volcanic reservoir prediction, more effort should be made to achieve quantitative predictions. At present, qualitative volcanic reservoir prediction is based on lithofacies classification, using the following steps:

1. Interpret fault system and identify volcanic massif through gravimetric, magnetic, electrical, and seismic reflection features (Figure 6.21).

   There are three uplifts and two depressions in the basin. Three nose-like structural belts extend eastward, and volcanic rocks occur along uplift belts.
2. Establish geophysical identification marks for logging and seismic interpretation for volcanic rock (Figure 6.22).

   Core, thin section, and log data are used to establish templates for Carboniferous lithology identification and to determine parameters sensitive to the variations of lithologies and reservoir rocks. For example, the radioactivity of rock may intensify, and rock density and resistivity may decrease when the rock type transitions from basic rock to acidic rock.
3. Classify volcanic lithofacies using seismic attribute analysis such as waveform classification (Figure 6.23).
4. Predict volcanic reservoir distribution quantitatively through seismic attribute analysis and seismic reservoir characterization (Figure 6.24).

   From the impedance inversion through wells Di10, Di102, and Di101, we can see that volcanic rock and sedimentary rock are distinct from each other. Impedances corresponding to the volcanic dormancy period

in the Carboniferous are apparently lower than those related to volcanic rock, which could be used to identify volcanic rock qualitatively.
5. Delineate fracture distribution through multiattribute analysis (Figure 6.25).

   Integration of dip angle, azimuth, and amplitude in multiattribute analysis may delineate fracture distribution qualitatively. The development of fractures is mainly controlled by the fault system and paleostructure, so the fault zone and uplift zone may be favorable areas to explore.
6. Delineate target zones for volcanic rock exploration.

   Based on the analyses of structure, source rock, favorable volcanic lithofacies zone, and hydrocarbon preservation, nine target zones have been determined for volcanic rock exploration in the area.

6.2.4 Fluid Detection

Volcanic trap fluid detection techniques include the poststack and prestack fluid detection techniques.

6.2.4.1 Poststack Fluid Detection

Poststack fluid detection is based on poststack seismic data, in which some poststack seismic attributes, such as amplitude and frequency attenuation attributes, are used for the qualitative interpretation of fluids in reservoir space (Figure 6.26).

With the progress of integrated exploration and development, poststack seismic techniques become more and more incapable of predicting reservoir properties and hydrocarbon-bearing ability. There may be several reasons

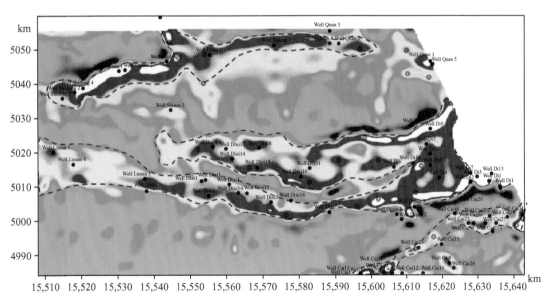

FIGURE 6.21 Volcanic rock distribution at the top of the Carboniferous in the Ludong area, Junggar Basin.

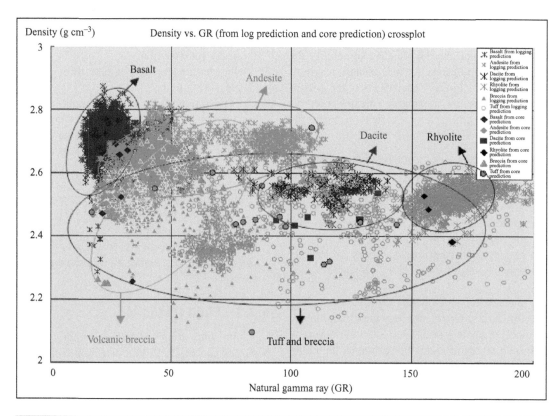

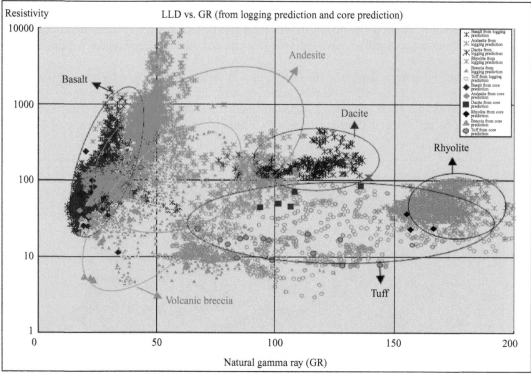

FIGURE 6.22 Template for volcanic rock identification.

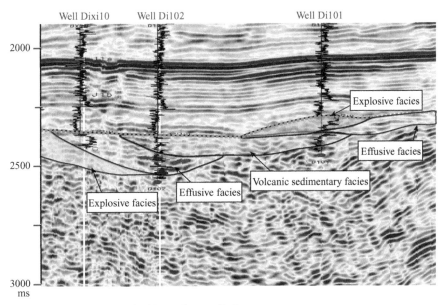

FIGURE 6.23   Seismic facies classification, the Carboniferous, Junggar Basin.

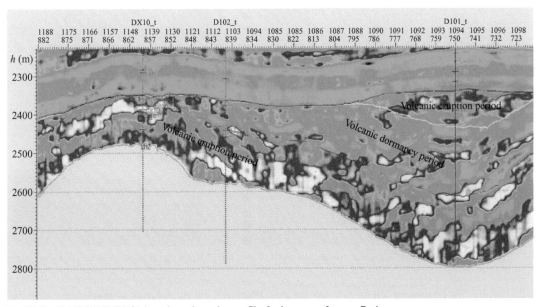

FIGURE 6.24   Wells Dixi10-Di102-Di101 impedance inversion profile, Ludong area, Junggar Basin.

as follows: (1) In most areas, the impedance difference between reservoirs and nonreservoirs is not obvious; the conventional poststack reservoir characterization method does not distinguish them enough. Seismic data stacking results in loss of seismic information and, consequently, contradicts the seismic reservoir prediction. (2) The seismic data acquired, especially prestack seismic data that would avail lithology prediction and fluid detection, have not been fully used in seismic reservoir prediction. (3) Seismic data processing with high signal-to-noise (S/N) ratio, high amplitude preservation, and high resolution has not been

achieved to date, which limits the use of prestack seismic data. Therefore, prestack seismic reservoir characterization based on prestack "three-high" data processing is still in the future.

6.2.4.2   Prestack Fluid Detection

The prestack technique has some advantages for gas reservoir detection. The prestack seismic technique has been used extensively in clastic rock exploration; as for carbonate rock and volcanic rock exploration, the use of the

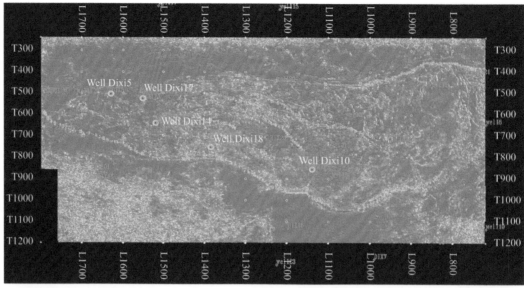

FIGURE 6.25   Fracture detection at the top of the Carboniferous via multiattribute analysis, Di'nan uplift, Ludong area, Junggar Basin.

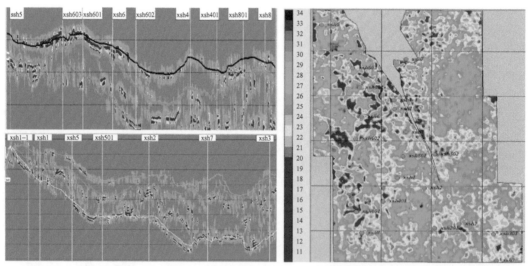

FIGURE 6.26   Seismic reflection amplitude attenuation and absorption coefficient map, Xingcheng gas field in northern Songliao Basin.

prestack technique is just beginning. There are two basic needs in prestack prediction: i.e., amplitude-preserving data processing and petrophysical modeling. As for deeply buried volcanic rocks with rapid rock type and lithofacies variation, it is difficult to detect reservoir fluids with prestack seismic data. However, in the Xushen9 well field, Fengle gas field, Songliao Basin, there is primarily an effusive facies with high reservoir homogeneity. In addition, seismic data quality has been catered to enhance prestack prediction. Therefore, it is possible to conduct prestack fluid detection in this area.

Prestack inversion primarily refers to amplitude versus offset (AVO) inversion. All lithologic parameters may be obtained from AVO inversion, such as rock density, P-wave velocity, S-wave velocity, P-impedance, S-impedance, and

Poisson's ratio. Prestack seismic inversion is based on gather data, while poststack inversion is based on stacking data, which is the essential difference between them. Stacking would improve the data quality and S/N ratio of seismic data. However, stacking is based on the assumption that the seismic reflection characteristics after dynamic correction, such as amplitude and waveform, do not change with variable offsets. In fact, the seismic reflection amplitude from the same reflection point is different at different offsets, and the reflection waveform also changes with variable offsets. Therefore, stacking loses the information of amplitude changing with offsets and actually reduces the ability to solve complicated reservoirs. The changes of seismic reflection amplitude and waveform features are complex with variable offsets, mainly resulting from the fact that seismic waves with different offsets pass through different stratum structures, elastic properties, and lithology assemblages. Stacking destroys true amplitude relationships and loses S-wave information. Prestack inversion could reveal the relationship between lithology and hydrocarbons by analyzing the change in prestack seismic data with variable offsets (AVO), which increases data volume, reduces ambiguity, and improves the reliability of lithology identification, precision of property prediction, and identification ability of pore fluid.

According to the principle of reflection and transmission, when an obliquely incident P-wave strikes an interface between two media with different elastic properties, there would be a reflected P-wave and reflected S-wave emerging in the first medium and a transmitted P-wave and transmitted S-wave in the second medium. Reflection and transmission coefficients bear a complex relationship to incident angle and the elastic properties of the two media, which has been expressed by the Zoeppritz equation and some reduced equations. Elastic parameters, such as P-wave velocity, S-wave velocity, P-velocity to S-velocity ratio, rock density, Poisson's ratio, bulk modulus, and shear modulus, may be extracted from the amplitude variation with incident angle. Prestack seismic inversion based on Zoeppritz and its reduced equations has usually been realized through elastic impedance inversion (single-angle inversion and joint elastic impedance inversion of multiangle), AVO inversion

(two-parameter inversion and three-parameter inversion), or joint P-wave and S-wave inversion in practical application. Prestack inversion based on wave equations has generally not been utilized in practice. Because prestack seismic inversion may yield more elastic properties than poststack inversion, it has been extensively applied to lithologic parameter inversion, fracture detection, formation pressure prediction, reservoir performance detection, oil and gas prediction, and reservoir heterogeneity delineation.

Amplitude-preserved data processing provides the data basis for prestack inversion, and petrophysical analysis offers a judgment of reservoir prediction feasibility, all of which would be crucial to prestack inversion. In the process of inversion, measures should be taken to control both data quality and inversion quality.

In summary, the information of amplitude variation with angle or offset should be properly preserved and reflected in the whole process of prestack inversion. The workflow of inversion is shown in Figure 6.27.

6.2.4.2.1 Field Acquisition of AVO Data and Prestack Amplitude-Preserving Process

AVO analysis aims at lithology and hydrocarbon interpretation according to the reflection amplitude variation with offset inside gathers. In terms of AVO and prestack information, AVO processing aims to recover prestack seismic

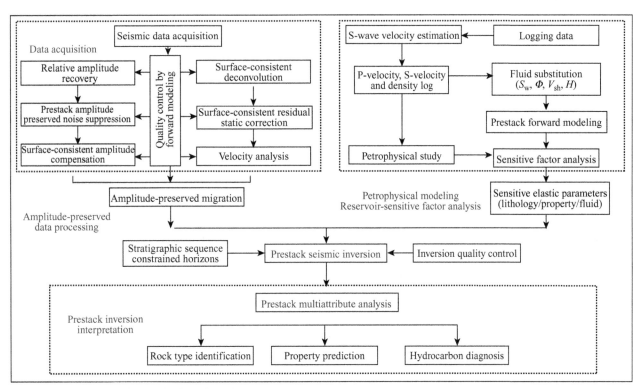

FIGURE 6.27 Workflow of prestack seismic inversion.

information and extract AVO information, which should meet the following demands:

1. preserve relative amplitude relationship;
2. preserve certain S/N ratio;
3. maintain high coverage and large offset as much as possible and also ensure constant offset variation.

The seismic reflection amplitude is affected by many data acquisition conditions and underground geologic conditions. AVO analysis is based on the prestack amplitude recovered to delineate rock type and fluids in pore space, so prestack data processing flow and parameters are crucial to AVO analysis. Seismic reflection energy, attenuated and distorted in the process of propagation, is recovered during data processing. AVO analysis makes use of amplitude variations with offset to describe rock type and oil and gas, and any AVO features irrelevant to underground geologic conditions, including those resulting from nonamplitude-preserving processing, would lead to misinterpretation. Aiming at eliminating amplitude variations caused by non-geologic factors, relative amplitude recovery should focus on offset-related amplitude attenuation, including spherical spreading compensation, absorption and attenuation compensation, surface-consistent amplitude compensation, and residual amplitude compensation.

During amplitude compensation, a single-trace equalization should be avoided in any offset-related amplitude processing, including spherical spreading compensation and absorption and attenuation compensation, so as not to introduce fake AVO features. Surface-consistent processing, such as surface-consistent deconvolution, amplitude compensation, and static correction, should be also taken seriously. Some special processing such as AVO feature extraction, anomaly display, anomaly detection, and data enhancement should be included in AVO processing as well. Coverage compensation for different offsets has been included in data processing and could reduce energy differences arising from coverage variation inside gathers. In the following example, after offset-related coverage compensation and prestack migration, the common reflector point (CRP) gather passing through the well site with AVO features close to type I has been consistent with that from forward modeling (Figure 6.28). Multiwell calibration has shown that offset-related coverage compensation improves the similarity between synthetic data and real data in reflection features and amplitude (Figure 6.29).

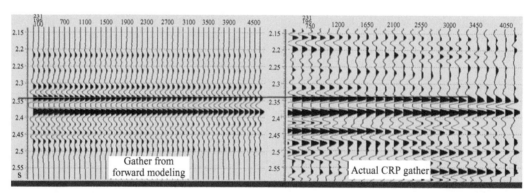

FIGURE 6.28   Pre-compensation gather and AVO features on synthetic seismogram.

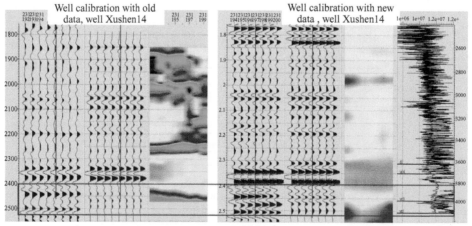

FIGURE 6.29   Compensated gather and AVO features on synthetic seismogram.

6.2.4.2.2 Petrophysical Modeling and Reservoir Sensitive Analysis

Petrophysical studies aim at discovering the relationships between rock properties, such as composition, pore, and pore fluid, and its elastic properties including elastic modulus, bulk density, P- and S-velocity, absorption, and attenuation. Then they set up a connection between seismic responses and rock type, reservoir properties, and fluids so as to characterize the hydrocarbon reservoir with seismic signals quantitatively.

S-wave velocity prediction is the key to petrophysical modeling. S-velocity is used to distinguish between actual and false bright spots, as well as the computation of Poisson's ratio for rock type classification. Both *in situ*, three-component data acquisition and prestack seismic inversion aim to derive S-velocity from seismic data at present. Without S-velocity data, it is impossible to conduct prestack inversion, or convert wave inversion or joint multi-component (simultaneous) inversion. The accuracy of the S-velocity has a controlling effect on inversion. In practice, S-velocity data may be unavailable or of poor quality, which should be estimated according to a P-velocity log using different petrophysical modeling techniques.

S-wave velocity may be derived from empirical equations, theoretical models, or a combination of an empirical equation and theoretical model. Methods with empirical equations make use of linear or nonlinear relationships between S-velocity and P-velocity or other well log parameters with S-velocity data and apply these linear or nonlinear relationships to wells without shear wave data to estimate S-velocity, which has been proven to be effective for some specific rock types or in limited areas. When there is pore fluid, the estimation error would increase. As for the Gassmann model, the Poisson's ratio of dry rock is presumed to be known as a constant. By solving Gassmann equation to get the elastic parameters of the rock frame, one can calculate the S-velocity based on saturation and other information. The approach of the theoretical model has clear petrophysical significance. On the other hand, the Poisson's ratio of dry rock must be given, and a single given constant only yields an S-velocity estimation for single rock type (sandstone in general), which limits the application of the Gassmann model to complicated rock types. An alternative method combines auxiliary parameters from empirical equations with the Gassmann model, which is a good remedy for the Gassmann model and also improves the reliability and accuracy of the auxiliary parameter. The alternative approach takes petrophysical mechanisms into account to make the estimated S-velocity suitable for fluid substitution and is also not confined to a specific petrophysical model. This method has been used in S-velocity prediction in the Xushen9 well field. S-velocity estimations have matched actual measurements (Figure 6.30).

Analyses of factors sensitive to rock types and fluids include analyses of factors sensitive to reservoir properties and factors sensitive to fluids. Detailed reservoir geophysical feature analysis and petrophysical analysis lay the foundation for prestack seismic inversion. Different fields have different reservoir-sensitive factors. A lithology-sensitive factor analysis aims to classify rock type and identify the volcanic reservoir. A property-sensitive factor analysis avails porosity prediction and effective reservoir identification. A fluid-sensitive factor analysis avails saturation prediction and fluid distribution prediction.

In the Xushen9 well field, fluid-sensitive factor analyses have dealt with correlation analysis, crossplot analysis, identification analysis, and regression analysis of more than

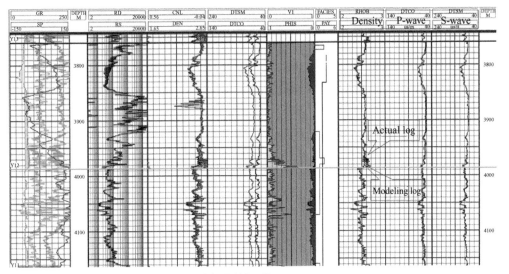

FIGURE 6.30 Petrophysical modeling logs and actual logs of the XS14 well.

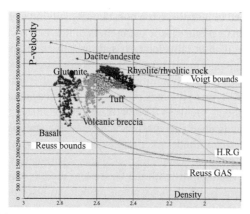

FIGURE 6.31 Volcanic rock type classification by P-velocity vs. density crossplotting.

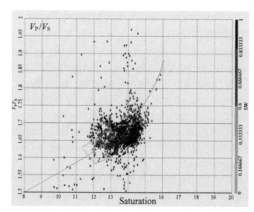

FIGURE 6.32 Correlation of hydrocarbon saturation to by V_P/V_S in volcanic reservoirs.

20 elastic parameters. In conclusion, density has been deemed to be capable of classifying rock types, and a crossplot of P-velocity and density may be used to identify different rock types. Based on rock type classification, density and P-velocity to S-velocity ratio may be sensitive to fluid saturation in reservoir rocks.

6.2.4.2.3 Prestack Inversion and Interpretation

Through petrophysical modeling, fluid-sensitive factor analysis and multi-information analysis, P-velocity, and density have been used together to identify volcanic rocks in the Xushen9 field (Figure 6.31), density has been used to estimate porosity, and the P-velocity (V_P) to S-velocity (V_S) ratio has been used to diagnose pore fluids (Figures 6.32 and 6.33). Seismic reservoir characterization has been proved to be successful in the area.

6.2.5 Technical Applications in Major Volcanic Provinces

In view of the current volcanic reservoir exploration in China, the technologies for volcanic exploration should serve to construct two major oil and gas provinces, four basins, and three areas. The application of volcanic exploration technology should be fulfilled on three levels.

6.2.5.1 Push the Development of Two Major Gas Provinces by Taking Different Technical Measures

1. Songliao Basin: gravimetric, magnetic, and electrical technologies have been proven to be effective in predicting volcanic massif distribution on a large scale. More efforts should be made on volcanic reservoir prediction and prestack fluid detection to find more reserves in volcanic reservoirs. In addition, gravimetric, magnetic, and electrical technologies should be applied to the discovery of new prospects.
2. Junggar Basin: gravimetric, magnetic, and electrical technologies and deep seismic imaging should be used to improve the integrated processing and interpretation of gravimetric, magnetic, electrical, and seismic data in the Ludong-Wucaiwan area. Then, this technique should be applied to volcanic rock prediction in the whole basin.

6.2.5.2 Deepen the Exploration of Four Basins by Actively Popularizing-Related Supporting Techniques for Volcanic Rock Exploration

1. Bohai Bay Basin, Erlian Basin, and Haila'er Basin: gravimetric, magnetic, electrical, and seismic

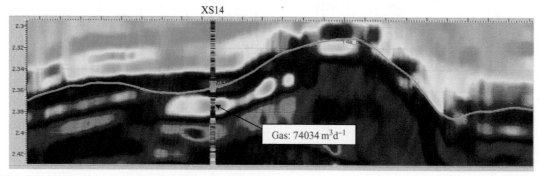

XS14

Gas: 74034 m³d⁻¹

FIGURE 6.33 P-impedance inversion passing through the XS14 well, Songliao Basin.

technologies and poststack seismic prediction technologies should be expanded to find more reserves in volcanic reservoirs.

2. Santanghu Basin: seismic imaging of the Carboniferous strata should be improved to expand the integrated use of gravimetric, magnetic, electrical, and seismic technologies and poststack seismic prediction technologies, so as to explore more volcanic reservoirs and find more reserves in volcanic reservoirs.

6.2.5.3 Strengthen the Deployment and Application of Gravimetric, Magnetic, and Electrical Techniques in Junggar Basin

1. Aeromagnetic prospecting on a scale of 1:50,000 (line spacing of 500 m, point spacing <10 m, precision <1 nT).
2. Gravity prospecting on a scale of 1:50,000 in Junggar Basin, which should be conducted according to the principle of overall design and step-by-step execution (line spacing of 500 m, point spacing of 500 m, precision of 25×10^{-8} ms^{-2}).
3. Exploration with 3D field building method (line spacing of 2 km, point spacing of 250 m).
4. Core susceptibility measurement, observing residual magnetism and induced magnetism.

6.3 FUTURE DEVELOPMENTS

6.3.1 Gravity, Magnetic, and Electrical Prospecting

In the past 20 years, great progress has been made in gravity, magnetic, and electrical prospecting because of the rapid development of instrumentation and techniques and the popularization of the computer. The precision of gravity prospecting has been improved from several microgals to more than 10 mgal; the magnetic precision has improved from several nanoteslas to several millinanoteslas. Electrical measurement instruments have developed from single-channel analog signals to multichannel digital signals, and the application of GPS has improved the precision to the centimeter level. With the development of measurement instruments, techniques, and data interpretation, gravity, magnetic, and electrical technologies will play an important role in volcanic oil and gas exploration.

The gravity, magnetic, and electrical technologies at home and abroad should show future development in the following three aspects: the first is high-precision and elaborate data processing; the second is 3D data processing and interpretation for better reservoir characterization based on 3D merged dataset processing and interpretation on a large scale; and the third is integration of multiple disciplines to take full advantage of gravity, magnetic, and electrical technologies in comprehensive interpretation. Considering the current state of the petroleum industry, exploration activities should be extended into areas with challenging geologic conditions and will encounter more difficulties in such areas. The integrated geophysics of gravity, magnetic, electrical, and seismic methods will be effective in handling complicated underground geologies.

6.3.2 Seismic Prospecting

Poststack seismic technology is the most practical approach to volcanic rock exploration at present, the key of which is a good knowledge of major geologic and geophysical challenges of volcanic reservoir characterization in different basins, and the corresponding requirements on relevant technologies and comprehensive geologic evaluation. Prestack seismic technology is in its infancy, and more attention should be paid to amplitude-preserved data processing with high S/N ratio and fundamental research in petrophysical modeling.

As for the future development of volcanic gas reservoir exploration, in highly explored areas, prestack fluid detection technology should be developed so as to identify volcanic reservoirs and make qualitative and quantitative fluid evaluations. In areas that have not been extensively explored, the integration of gravimetric, magnetic, electrical, and seismic technology should be applied to identify volcanic massifs macroscopically.

The misconception of volcanic rocks as nonreservoir rocks once prevented them from being actively explored. Now things have changed, and volcanic rocks have become an important target in hydrocarbon exploration and development. Great progress has been made in the study of volcanic rock distribution, formation mechanisms, distribution of the volcanic reservoir, oil and gas accumulation in volcanic reservoirs, potential resources, province and target of volcanic rock exploration, and volcanic reservoir characterization and evaluation. Despite these significant research findings, volcanic oil and gas geology still needs further development to guide the activities of volcanic hydrocarbon exploration.

Cai Xianhua, 2002. Pattern of igneous rock distribution and hydrocarbon accumulation in Changling Fault Depression of the south part of Songliao Basin. Geophys. Prospect. Petrol. 37 (3), 291–293.

Cao, Y.C., Jiang Z.X., Qiu L.W., 1999. Study on the type and origin of the reservoir space of igneous oil reservoir in Shang 741 Block, Huimin Depression, Shandong. Acta Petrol. Sin. 15 (1), 129–136.

Chen, H.L., Yang S.F., Dong C.W., Jia C.Z., Wei G.Q., Wang Z.G., 1997. The discovery of Early Permian basic rock belt in Tarim Basin and its tectonic meaning. Geochemica 26 (6), 77–87.

Chen Jianwen, 2002. A rising borderline science—volcanic rock reservoir geology. Mar. Geol. Lett. 18 (4), 19–22.

China Petroleum and Chemical Industry Federation, 2000. Thin section examination of rock, oil and gas industry standard, SY/T 5368-2000, People's Republic of China.

China Aero Geophysical Survey & Remote Sensing Center for Land and Resources (AGRS). Aeromagnetic anomaly (ΔT) map of Continental China and adjoining marine regions. Beijing, Geological publishing House. 2004.

Chuanjin Jiang, Shumin Chen, Erhua Zhang, Deying Zhong, 2004. Methodology and application of seismic prediction of gas-bearing volcanic reservoir. 74th SEG, 10–15 October.

Cui Fenglin, Gou Yongfeng, et al., 2005. The abstraction of seismic attributes and its effectiveness in volcanic prediction. Geophys. Prospect. Petrol. 44 (6), 598–600.

Dai Jinxing, 1988. The carbon isotopic compositions and origin of natural gas from the Liuhuangtang Springs, in Tengchong County of Yunnan Province. Chin. Sci. Bull. 33 (15), 1168–1170.

Dai Jinxing, Qi Houfa, Hao Shisheng, 1989. Introduction to Natural Gas Geology. Petroleum Industry Press, Beijing.

Dai Jinxing, Zeng Guanyun, Chen Xueliang, 1990. Co₂ gas seepages of Zegulong area, Pingyuan County, Guangdong Province. Oil Gas Geol. 11 (2), 205–208.

Dai Jinxing, Pei Xigu, Qi Houfa, 1992. Natural Gas Geology in China (Volume I). Petroleum Industry Press, Beijing, pp. 147–169.

Dai Jinxing, Song Yan, Dai Chunsen, et al., 1995. Natural Gas of Inorganic Origin and Forming Conditions of Gas Reservoirs in East China. The Science Press, Beijing.

Dai Jinxing, Zou Caineng, Zhang Shuichang, et al., 2008. Discrimination between alkane gases of inorganic origin and organic origin. Sci. China Ser. D 38 (11), 1329–1341.

David Chapin A., 1997. Wavelet transforms: a new paradigm for interpreting gravity and magnetic data. 67th SEG GM15, pp. 486–489.

David Williams L., Carol Finn, 1983. Gravity studies in volcanic terranes. USGS, SEG, GM 65, 303–305.

Dun Tiejun, 1995. Situation and development trend of reservoir research. Northwestern Geol. 16 (2), 1–15.

Editorial Committee of "Dictionary of Earth Sciences", 2006. Dictionary of Earth Sciences. Geologic Publishing House, Beijing, pp. 991–1009.

Feng Zhiqiang, Wang Yuhua, et al., 2007. Exploratory techniques and their advancement of deep volcanic gas reservoirs in the Songliao Basin. Nat. Gas Ind. 27 (8), 9–12.

Feng Zihui, Shao Hongmei, Tong Ying, 2008. Controlling factors of volcanic gas reservoir property in Qingshen gas field, Songliao Basin. Acta Geol. Sin. 82 (6), 760–768.

Fisher, R.V., Schmincke H.U., 1984. Pyroclastic Rocks. Springer, Heidelberg, pp. 17–44.

Gao Fuhong, Xu Wenliang, Yang Debin, Pei Fuping, Liu Xiaoming, Hu Zhaochu, 2007. LA-ICP-MS zircon U-Pb dating from granitoids in southern basement of Songliao Basin: Constraints on ages of the basin basement. Science in China(Series D) 50 (7), 995–1004.

Gao Dezhang, 2006. Offshore oil and gas exploration and processing interpretation techniques for gravity and magnetic data. Offshore Oil 26 (3), 1–8.

Gold, T., 1993. The origin of methane in the crust of the earth. In: Cavid, G.H. (Ed.), The Future of Energy Gases. United States Government Printing Office, Washington, pp. 57–80.

Guan Zhining, 2005. Geomagnetic Field and Magnetic Exploration. Geologic Publishing House, Beijing.

Guan Zhining, Zhang Changda, et al., 1993. Theoretical Analysis of Important Issues in Magnetic Prospecting and Their Applications. Geologic Publishing House, Beijing.

Guo Zhanqian, Wang Xianbin, Liu Wenlong, 1997. Reservoir-forming features of abiotic origin gas in Songliao Basin. Science in China, Ser. D 40 (6), 621–626.

He Zhanxiang, 2000. Advances and development tendency of non-seismic exploration techniques. Oil Geophys. Prospect. 35 (3), 354–360.

He Guoqi, Li Maosong, Liu Dequan, 1994. Paleozoic Crustal Evolution and Mineralization in Xinjiang of China. Xinjiang People's Publishing House, Urumqi, pp. 1–437.

He Zhanxiang, et al., 2002. Tendency of non-seismic techniques of hydrocarbon prospecting. Prog. Geophys. 17 (3), 473–479.

He Zhanxiang, Wang Yongtao, et al., 2005. New progress and application of integrative geophysical survey techniques. Oil Geophys. Prospect. 40 (1), 108–112.

Hosgrmez, H., 2007. Origin of the natural gas seep of Cirali (Chimera), Turkey: site of the first Olympic fire. J. Asian Earth Sci. 30 (1), 131–141.

Hou Zunze, Yang Wencai, 1997. Wavelet transform and multi-scale analysis on gravity anomalies of china. Chin. J. Geophys. 40 (1), 85–95.

Huang Yulong, Wang Pujun, Feng Zhiqiang, et al., 2007. Analogy of volcanic edifices between modern volcanoes and ancient remnant volcanoes in Songliao Basin. J. Jilin Univ. (Nat. Sci. Ed.) 37 (1), 65–72.

Huang Liang, Peng Jun, Zhou Kang, et al., 2009. An overview of the study on formation mechanism of volcanic reservoirs. Special Oil Gas Reservoirs 16 (1), 1–12.

185

Hunt, C.W., Collins L.G., Skobelin E.A., 1992. Expanding Geospheres Energy and Mass Transfers from Earth's Interior, Calgary. Polar Publishing, Alberta, p. 421.

International Union of Geological Sciences, 1989. Volcanic Rock TAS Classification Scheme.

Ji Guosheng, Dai Junsheng, Ma Xinben, et al., 2002. Reservoir characteristics for volcanic rocks from Member I, II of Funing Formation in Northern Minbei region in Subei Basin. Oil Gas Geol. 23 (3), 289–292.

Jia Jindou, Kong Fanshu, 2006. Non-seismic exploration techniques for different hydrocarbon exploration stages. Oil Geophys. Prospect. 36 (4), 69–72.

Jin Zhijun, Hu Wenxuan, Zhang Liuping, et al., 2007. Deep Fluid Activities and Their Effectiveness on Hydrocarbon Generation and Accumulation. The Science Press, Beijing.

Jing-kui Mi, Shui-chang Zhang, Shi-zhen Tao, Ting Liu, Xia Luo, 2008. Genesis and Accumulation Period of the CO_2 in Changling Fault Depression of Songliao Basin, Northeastern China. Natural gas geoscience 19 (4), 452–456.

Jokat, W., Uenzelmann-Neben G., Kristoffersen Y., $Rasmussen T., 1992. Lomonosv Ridge – a double sided continental margin. Geology 20, 887–890.

Koefoed, O., 1979. Geosounding Principles. Elsevier Scientific Publishing Comp, Amsterdam p. 276.

Kuang Lichun, Xue Xinke, Zou Caineng, et al., 2007. Reservoiring conditions and enrichment rule of oil reservoirs in formations of volcanic facies—using carboniferous formation at upper wall of Ke-Bai Fault Zone in Junggar Basin as an example. Petrol. Explor. Dev. 34 (3), 285–290.

Li Rong, 2001. Characteristics and classification of Permian reservoirs in Northwestern Margin of Junggar Basin. Oil Gas Geol. 22 (1), 78–87.

Li Chun, Kang Renhua, 1999. Genetic types of reservoir spaces of igneous rocks in the Luo 151 Block, Jiyang Sag. Geol. Rev. 45 (Suppl.), 599–604.

Li Shaojie, Li Shucai, 1992. The division of paleovolcanic apparatus of the Madaer river volcanic rock area. Heilongjiang Geol. 3 (3), 9–25.

Li Xiusheng, Wang Jianguo, 2008. The research actuality and future progress direction of the volcanic reservoir. J. Haikou College Econ. 7 (1), 1–5.

Li Ming, Zou Caineng, et al., 2002a. Identifying and predicting technology of deep volcanic gas reservoir in north of Songliao Basin. Oil Geophys. Prospect. 37 (5), 477–478.

Li Yanli, Wang Guojun, Cao Guoyin, et al., 2002b. Seismic object processing aimed to predict volcanic rocks and new technology for volcanic reservoir prediction. Oil Geophys. Prospect. 37 (5), 541–546.

Li Jinyi, He Guoqi, Xu Xin, 2006. Crustal tectonic framework of Northern Xinjiang and adjacent regions and its formation. Acta Geol. Sin. 80 (1), 148–168.

Lin Rujin, Xu Keding, 1995. The discussion of hydrocarbon potential in Mesozoic volcanic rocks distributed zone in the eastern parts of Zhejiang, Fujian and Guangdong Provinces. Acta Petrolei Sin. 16 (4), 23–30.

Liu Minggao, 1986. Study on Carboniferous volcanic reservoirs in 1st area, Karamay oil-field. Xinjiang Petrol. Geol. 79 (2), 78–90.

Liu Shiwen, 2001. Characteristics and favorable reservoir forming conditions of igneous reservoir in Liaohe Basin. Special Oil Gas Reservoirs 8 (3), 6–10.

Liu Xuejun, Li Dechun, et al., 2001. Predicting deep special target geologic bodies using non-seismic integrated exploration techniques. Oil Geophys. Prospect. 36 (Suppl.), 52–59.

Liu Yunxiang, He Yi, et al., 2005a. Identification of igneous reservoir facies belts using high-accuracy gravity-magnetic-electric exploration data. Oil Geophys. Prospect. 40 (Suppl.), 99–101.

Liu Yunxiang, Sun Weibin, Si Hualu, 2005b. Comprehensive geophysical prospecting technique and its effect for special geological bodies in deep layers. In: SEG Houston 2005 Annual Meeting, GM 16, pp. 707–709.

Liu Yunxiang, He Zhanxiang, et al., 2006. Integrated geophysical techniques for identification of igneous rocks. Prog. Explor. Geophys. 29 (2), 115–118.

Luo Jinglan, Zhai Xiaoxian, Pu Renhai, et al., 2006. Horizon, petrology and lithofacies of the volcanic rocks in the tahe oilfield, northern tarim basin, Chinese Journal of Geology (Scientia Geologica Sinica) 41 (3), 378–391.

Luo Yao, 2007. Characteristic description of volcanic reservoirs at Malang Depression. Tuha Oil Gas 12 (1), 16–19.

Luo Jinglan, Qu Zhihao, Sun Wei, et al., 1996. The relations between lithofacies, reservoir lithology and oil & gas of volcanic rocks in Fenghuadian area. Acta Petrolei Sin. 17 (1), 32–39.

Luo Qun, Liu Weifu, et al., 2001. Distribution peculiarity of deep-buried volcanic reservoirs. Xinjiang Petrol. Geol. 22 (3), 196–198.

Luo Jinglan, et al., 2003. Summary of research methods and exploration technologies for volcanic reservoirs. Acta Petrolei Sin. 24 (1), 31–38.

Lv Bingquan, Zhang Yanjun, Wang Honggang, et al., 2003. Present and prospect of Cenozoic/Mesozoic volcanic rock oil and gas accumulations in East of China. Offshore Oil 23 (4), 9–13.

Ma Xingyuan, 1982. On Extensional tectonics. Earth Science: 18 (3), 15–22.

Ma Qian, Junjie E., Li Wenhua, et al., 2000. Reservoir evaluation of deepseated igneous rocks in Beipu Region, Huanghua Depression. Oil Gas Geol. 21 (4), 340–344.

Meng Qi'an, Men Guangtian, et al., 2001. Prediction method and its application of deep volcanic rock body and facies in Songliao Basin. Petroleum Geol. Oilfield Dev. Daqing 20 (3), 21–22.

Miguel Bosch, Ronny Meza, Rosa Jiménez, Carlos Honig, 2005. Geostatistical inversion of gravity and magnetic data in 3D. Universidad Central de Venezuela, SEG Houston 2005 Annual Meeting, GM 18, pp. 655–658.

Mu Deliang, Li Cheng, et al., 2005. Mesozoic igneous reservoir forecast and exploration achievements in Niuxintuo Area, Liaohe Depression. Petrol. Explor. Dev. 32 (3), 67–68.

Niu Shanzheng, Pang Jiali, 1994. Evaluating the Permian Basalt reservoir of Zhou-1 well. Nat. Gas Ind. 14 (5), 20–23.

Niu Jiayu, Zhang Yinghong, Yuan Xuanjun, et al., 2003. Petroleum geology study, oil and gas exploration future and problems faced in Mesozoic and Cenozoic igneous rocks in Eastern China. Special Oil Gas Reservoirs 10 (1), 7–12.

Peng Caizhen, Guo Ping, et al., 2006. A summary of the development present situation for the volcanic-rock gas reservoir. J. Southwest Petrol. Inst. 28 (5), 69–72.

Qian Zheng, 1999. A conceptual model of Luo 151 igneous reservoir in Jiyang Depression. Petrol. Explor. Dev. 26 (6), 72–74, 94.

Qiao Hansheng, Luo Zhibin, Li Xianqi. 2001. Collection of papers on deep-seated petroleum exploration of Eastern China. Beijing, Petroleum Industry Press, p. 279.

Qiu Jiaxiang, 1985. Magmatic Petrology, Geologic Publishing House, Beijing.

Qiu Jiaxiang, 1991. Applied Magmatic Petrology, China University of Geosciences Press, Wuhan, pp. 1–4.

Qiu Jiaxiang, Tao Huiyuan, Zhao Junlei, et al., 1996. Volcanic Rocks, Geologic Publishing House, Beijing, pp. 10–22.

Qiu Longwei, Jiang Zaixing, Xi Qingfu, 2000. Diagenesis and pore evolution of Lower Es3 volcanic rocks in Oulituozi Region. Oil Gas Geol. 21 (2), 139–143.

Qu Yanming, Shu Ping, Ji Xueyan, et al., 2007. Micro-fabrics of reservoir volcanic rocks in the Qingshen gas field of the Songliao Basin. J. Jilin Univ. 37 (4), 721–725.

Ram Babu, H.V., Prasanthi Lakshmi, M., 2005. A technique for estimating the depth and thickness of a volcanic layer from magnetic data. National Geophysical Research Institute, Hyderabad 500007, India, SEG Houston 2005 Annual Meeting, GM P18, pp. 716–719.

Ringwood, A.E., 1969. Composition and evolution of the upper mantle. Am Geophysics Union Mon. 13, 1–17.

Shao Zhengkui, Meng Xianlu, et al., 1999. Seismic reflection features and distribution law of volcanic rocks in the Songliao Basin. J. Changchun Univ. Sci. Technol. 29 (1), 33–34.

Song Qingxiang, 1991. Tertiary and volcanic hydrocarbon reservoirs in Japan. Discip. Dev. Res. 3, 137–141.

Sruoga P., Nora Rubinstein, 2007. Processes controlling porosity and permeability in volcanic reservoirs from the Austral and Neuquén basins, Argentina. AAPG Bull. 91 (1), 115–129.

Sruoga, P., Rubinstein N., Hinterwimmer G., 2004. Porosity and permeability in volcanic rocks: a case study on the Serie Tobífera, south Patagonia, Argentina. J. Volcanol. Geotherm. Res. 132 (1), 31–43.

Sun Nai, 1985. Igneous Rocks. Geologic Publishing House, Beijing, pp. 10–22.

Surdam, R.C., Boese S.W., Crossey L.J., 1984. The chemistry of secondary porosity [A]. In: McDonald, D.A., Surdam, R.C., Clastic diagenesis [C]. Am Assoc Pet. Geol. Mem. 37, 127–149.

Tan Lijuan, Tian Shicheng, 2001. Faulting and volcanism in Cenozoic of Nampu Depression. J. Univ. Petrol. China (Ed. Nat. Sci.) 25 (4), 1–4.

Tang Jianren, Liu Jinping, Xie Chunlai, et al., 2001. Volcanic rock distribution and gas abundance regularity in Xujiaweizi Faulted Depression, North of Songliao Basin. Oil Geophys. Prospect. 36 (3), 345–351.

Tao Kuiyuan, Yang Zhuliang, Wang Libo, et al., 1998. Oil reservoir geological model of basalt in Minqiao Northern Jiangsu Province. Earth Sci. J. China Univ. Geosci. 23 (3), 272–276.

Tao Kuiyuan, Mao Jianren, Xing Guangfu, et al., 1999. Strong Yanshanian volcanic-magmatic explosion in East China. Mineral Deposits 20 (4), 27–35.

The Group on the Systematics of Volcanic Rocks of the Commission on Petrology of Geological Society of China, 1984. Lava and Volcaniclastic Rock Classification and Naming Scheme.

US Geological Survey, 1987. The importance of total magnetization in aeromagnetic interpretation of volcanic areas: An illustration from the San Juan Mountains, Colorado, K J S Grauch, SEG, GM 17, pp. 109–110.

Virgil Raymond Drexel Kirkham, 1935. Natural gas in Washington, Idaho, eastern Oregon, and northern Utah. Pan-American Geologist, 221–244 June.

Vernik Lev, 1990. A new type reservoir rock in volcaniclastic sequences. AAPG Bull. 74 (6), 830–836.

Wan Minghao, Shunting Qin, et al., 1999. Petrophysical Properties and Their Application in Petroleum Exploration. Geologic Publishing House, Beijing.

Wang Jialin, 2006. Views on the domestic situation and progress of gravity and magnetic petroleum exploration. Prog. Explor. Geophys. 29 (2), 82–86.

Wang Zhixin, Zhao Chenglin, Liu Menghui, 1991. Volcanic lithofacies and their petrophysical properties in Abei oilfield. J. Univ. Petrol. China (Ed. Nat. Sci.) 15 (3), 15–21.

Wang Jinyou, Zhang Shiqi, Zhao Junqing, et al., 2003a. Characteristics of volcanic rock reservoirs in the Linshang area of the Huimin Sag in Bohai Bay Basin. Petrol. Geol. Exp. 25 (3), 264–268.

Wang Pujun, Chen Shumin, Liu Wanzhu, et al., 2003b. Relationship between volcanic facies and volcanic reservoirs in Songliao Basin. Oil Gas Geol. 23 (1), 18–26.

Wang, Pujun, Chi Yuanlin, Liu Wanzhu, et al., 2003c. Volcanic facies of the Songliao Basin: classification, characteristics and reservoir significance. J. Jilin Univ. (Earth Sci. Ed.) 33 (4), 449–458.

Wang Huimin, Jin Tao, Yang Hongxia, 2005a. Characteristics of igneous rock lithofacies and its recognition marks in Chagan Sag, Yin'gen Basin. Xinjiang Petrol. Geol. 26 (3), 249–254.

Wang Xishuang, Wen Baihong, et al., 2005b. PetroChina's latest application examples of non-seismic exploration and prospects. China Petrol. Explor. 5, 36–37.

Wang Pujun, Zheng Changqing, Shu Ping, et al., 2007. Classification of deep volcanic rocks in Songliao Basin. Petrol. Geol. Oilfield Dev. Daqing 26 (4), 17–26.

Wang Pujun, Zhiqiang Feng, et al., 2008. Basinal Volcanic Rocks. The Science Press, Beijing, vol. 1, pp. 251–254.

Wen Baihong, Yang Hui, et al., 2005. Non-seismic techniques for oil and gas exploration in China. Petrol. Explor. Dev. 32 (2), 68–71.

Wen Baihong, Yang Hui, Zhang Yan, 2006. Geophysical features of typical volcanic oil-gas reservoirs and predication of favorable exploration zones in China. China Petrol. Explor. 4, 67–73.

Wilson, M., 1989. Igneous Petrogenesis. London: Unwin Hyman.

Xia Zhao, Jia Chengzao, Zhang Guangya, 2008. Geochemistry and tectonic settings of Carboniferous Intermediate-Basic Volcanic Rocks in Ludong-Wucaiwan, Junggar Basin. Earth Sci. Front. 15 (2), 272–279.

Xiao Xuchang, Tang Yaoqing, Feng Yimin, 1992. Tectonic Evolution of Northern Xinjiang and Its Adjacent Regions. Geologic Publishing House, Beijing, pp. 1–169.

Xiao Long, 2004. Plume Tectonics and Mantle Dynamics: On Their Records in Chinese Continent. Bulletin of Mineralogy. Petrology and Geochemistry 24 (01), 239–245.

Xiao Wenjiao, Han Chunming, Yuan Chao, 2006. Unique Carboniferous-Permian Tectonic-Metallogenic framework of Northern Xinjiang (Nw China): constraints for the tectonics of the Southern Paleoasian domain. Acta Petrol. Sin. 22 (5), 1062–1076.

Xie Qingbin, Han Dexin, Zhu Xiaomin, et al., 2002. Reservoir space feature and evolution of the volcanic rocks in The Santanghu Basin. Petrol. Explor. Dev. 29 (1), 84–86.

Xiong Qihua, Wu Shenghe, Wei Xinshan, 1998. Characteristics of igneous reservoir and controlling factors of reservoir development in the Permian of the Santanghu Basin. Exp. Petrol. Geol. 20 (2), 129–134.

Xu Yigang., Zhong Sunlin, 2001. The Emeishan Large Igneous Province: Evidence for mantle plume activity and melting conditions. Geochimica 30 (1), 1–9.

Xu Liying, Chen Zhenyan, 2003. Application of seismic technique in exploration of volcanic reservoirs in Huangshatuo area, Liaohe oilfield. Special Oil Gas Reservoirs 10 (1), 36–39.

Yan Chunde, Yu Huilong, Yu Fangquan, et al., 1996. Pores development and reservoir properties of volcanic rocks in Jianghan Basin. J. Jianghan Petrol. Inst. 18 (2), 1–6.

Yan Yaomin, Wang Yingmin, Zhu Yanhe, et al., 2007. Model of sedimentary facies in Jiamuhe Formation, Northwestern Margin of Junggar Basin. Nat. Gas Geosci. 18 (3), 386–388.

Yang Maoxin, 2002. Study of tectonic background and reservoir-forming condition of volcanic rock in Songliao Basin. Petrol. Geol. Oilfield Dev. Daqing 21 (5), 15–23.

Yang Xinming, 2005. Forming mechanism and development characteristics of volcanic reservoirs in Minqiao oilfield. Small Hydrocarbon Reservoirs, 4, 1–4.

Yang Jinlong, Luo Jinglan, He Faqi, et al., 2004. Characteristics of Permian volcanic reservoirs in the Tahe Region. Petrol. Explor. Dev. 31 (4), 44–47.

Yang Hui, Zhang Yan, Zou Caineng, et al., 2006. Exploration scheme of gas in deep-seated volcanic rocks in Songliao Basin. Petrol. Explor. Dev. 33 (3), 274–281.

Yang Hui, Song Jijie, Wen Baihong, et al., 2007. Macroscopic prediction of volcanic rock lithology: a case from Xujiaweizi Faulted Depression, Northern Songliao Basin. Petrol. Explor. Dev. 34 (2), 150–155.

Yang Chun, Liu Quanyou, Zhou Qinghua, et al., 2009. Genetic identification of natural gases in Qingshen gas field, Songliao Basin. Earth Sci. J. China Univ. Geosci. 34 (5), 792–798.

Yao Xinyu, He Zhenghuai, et al., 1994. Seismic facies and distribution characteristics of volcanic rocks in Junggar Basin. Xinjiang Petrol. Geol. 20 (3), 214–215.

Yi Peirong, Peng Feng, et al., 1998. Characteristics and exploration techniques of volcanic hydrocarbon reservoirs in foreign countries. Special Oil Gas Reservoirs 5 (2), 65–70.

Yu Yingtai, 1988. The prospect of volcanic rock oil reservoirs in Erlian Basin. Petrol. Explor. Dev. 15 (4), 9–19.

Yu Chunmei, Zheng Jianping, Tang Yong, et al., 2004. Reservoir properties and effect factors on volcanic rocks of basement beneath Wucaiwan Depression Junggar Basin. Earth Sci. J. China Univ. Geosci. 29 (3), 303–306.

Yuan Mingsheng, Zhang Yinghong, Han Baofu, 2002. The geochemical features of volcanic rocks and architectonic environment in Neopaleozoic Era, Santanghu Basin. Petrol. Explor. Dev. 29 (6), 32–34.

Zaixing Jiang, Shangbin Xiao, 1999. Distribution Pattern of Teriary Igneous Rocks in the Bohai Bay Basin. Geological Review. 45 (S1), 618–626.

Zeng Hualin, et al., 1991. Gravity-Magnetic Exploration and Inversion Problems. Petroleum Industry Press, Beijing.

Zeng Hualin, Li Xiaomeng, et al., 1999. Gravity Method for Detecting Hydrocarbon Reservoirs and Its Application. The Geologic Publishing House, Beijing, vol. 7, pp. 32–33.

Zhang Zhaochong, Wang Fusheng, Hao Yanli, 2005. Picrites from the Emeishan Large Igneous Province: Evidence for the Mantle Plume Activity. Bulletin of Mineralogy, Petrology and Geochemistry 23 (03), 17–22.

Zhang Xiaodong, Huo Yan, et al., 2000. The characteristics and distribution rule of the volcanic rock in the northern part of Songliao Basin. Petrol. Geol. Oilfield Dev. Daqing 19 (4), 10–12.

Zhang Shihui, Liu Tianyou, et al., 2003. Identification of volcanic rocks using artificial neural network. Oil Geophys. Prospect. 38 (Suppl.), 85–86.

Zhang Quansheng, Xin Houqin, et al., 2004. New geophysical exploration technique and its application. Miner. Resour. Geol. 18 (4), 371–372.

Zhang Guangya, Zou Caineng, Zhu Rukai, et al., 2010. Petroleum geology and exploration for volcanic reservoirs in the sedimentary basins of China. Eng. Sci. 12 (5), 30–38.

Zhanjun Yang, Yan Wei, 2005. The gravity & seismic data jointed formation separation technique for deep structure study. In: BGP, CNPC, SEG Houston 2005 Annual Meeting, GM 13, pp. 635–638.

Zhao Guolian, Zhang Yueqiao, 2002. Seismic reflection character of volcanic reservoir of Daqing and the comprehensive prediction technology. Petrol. Explor. Dev. 29 (5), 44–46.

Zhao Chenglin, Meng Weigong, Jin Chunshuang, et al., 1999. Volcanic Rocks and Hydrocarbon in Liaohe Basin. Petroleum Industry Press, Beijing, p. L101.

Zhao Hailing, Liu Zhenwen, Li Jian, et al., 2004. Petrologic characteristics of igneous rock reservoirs and their research orientation. Oil Gas Geol. 25 (6), 609–615.

Zhao Zehui, Guo Zhaojie, Han Baofu, et al., 2006a. The geochemical characteristics and tectonic-magmatic implications of the Latest-Paleozoic volcanic rocks from Santanghu Basin, Eastern Xinjiang, Northwest China. Acta Petrol. Sin. 22 (1), 199–214.

Zhao Zhenhua, Wang Qiang, Xiong Xiaolin, 2006b. Two types of Adakites in North Xinjiang, China. Acta Petrol. Sin. 22 (5), 1249–1265.

Zheng Changqing, Wang Pujun, Liu Jie, et al., 2007. Types and distinguishing characteristics of Cretaceous volcanic rocks in Songliao Basin. Petrol. Geol. Oilfield Dev. Daqing 26 (4), 9–18.

Zhou Fangxi, 2003. Application of nonlinear deformation to volcanics fissure prediction in Min 7 Block of Jinhu Sag. Petrol. Geol. Recovery Efficiency 10 (1), 12–13.

Zhou Dingwu, Liu Yiqun, Xing Xiujuan, 2006. Restoration of forming paleotectonic environments and tracing of regional tectonic settings of Permian Basalt in Tuha and Santanghu Basins, Xinjiang. Sci. China D 36 (2), 143–153.

Zhu Jingyi, 2003. Overview of integrated geophysics of oil & gas exploration. Prog. Geophys. 18 (1), 19–23.

Zou Caineng, Qu Wenzhi, Jia Chengzao, et al., 2008. Formation and distribution of volcanic hydrocarbon reservoirs in sedimentary basins of China. Petrol. Explor. Dev. 35 (3), 257–271.

Багдасарова, М.В., 2000. Современные гидротермальные и их связьсформированием месторождении нефтн и газа Фундаментальный базис новых технологий нефтяной и газовой. Москва, Наука, pp. 100–105.

Бека, К, Высоцкий, Й., 1976. Геология нефтн и газа, НЕДРА, Москва.

Вержбицкий, Е.В., 2002. Геотермический режим тектоника дна и температурные условия генерации УВ восточной части Баренцева моря. Геотектоника, vol. 1, pp. 86–96.

Гептнер, А.Р., 2002. Полнцнклнческие арматнческие УВ в покровных отложениях и тефре Ислаандии (состави особенности распредения). Литогия и полезные ископаемые, vol. 2, pp. 172–181.

Зорвкий, Л.М, Старобинец, И.С.И., Стаднин, Е.В., 1984. Геохимия природных газов нефтегазоносных бассейнов Недра.

Лесовой, Ю.И., 2001a. О генезисе УВ и механизме формирования ихзалежей в кристаллических пород Геол, геофиз н разрб нефт мест-ний, vol. 12, pp. 8–12.

Лесовой, Ю.И., 2001b. Генезис нефтегазовы УВ как результат эволюции нестабильных элементов Геол, геофиз н разрб нефт мест-ний, vol. 8, pp. 28–33.

Index

Note: Page numbers followed by *f* indicate figures and *t* indicate tables.

Printed and bound by CPI Group (UK) Ltd, Croydon, CR0 4YY

08/05/2025

01864936-0001